Lecture Notes in Networks and Systems

Volume 555

The series “Lecture Notes in Networks and Systems” publishes the latest developments in Networks and Systems—quickly, informally and with high quality. Original research reported in proceedings and post-proceedings represents the core of LNNS.

Volumes published in LNNS embrace all aspects and subfields of, as well as new challenges in, Networks and Systems.

The series contains proceedings and edited volumes in systems and networks, spanning the areas of Cyber-Physical Systems, Autonomous Systems, Sensor Networks, Control Systems, Energy Systems, Automotive Systems, Biological Systems, Vehicular Networking and Connected Vehicles, Aerospace Systems, Automation, Manufacturing, Smart Grids, Nonlinear Systems, Power Systems, Robotics, Social Systems, Economic Systems and other. Of particular value to both the contributors and the readership are the short publication timeframe and the world-wide distribution and exposure which enable both a wide and rapid dissemination of research output.

The series covers the theory, applications, and perspectives on the state of the art and future developments relevant to systems and networks, decision making, control, complex processes and related areas, as embedded in the fields of interdisciplinary and applied sciences, engineering, computer science, physics, economics, social, and life sciences, as well as the paradigms and methodologies behind them.

Indexed by SCOPUS, INSPEC, WTI Frankfurt eG, zbMATH, SCImago.

All books published in the series are submitted for consideration in Web of Science.

For proposals from Asia please contact Aninda Bose (aninda.bose@springer.com).

Rituparna Chaki · Agostino Cortesi · Khalid Saeed ·
Nabendu Chaki
Editors

Applied Computing for Software and Smart Systems

Proceedings of ACSS 2022

Springer

Editors
Rituparna Chaki
School of Information Technology
University of Calcutta
Kolkata, India

Agostino Cortesi
Department of Environmental Sciences,
Informatics and Statistics
Ca' Foscari University
Venice, Italy

Khalid Saeed
Faculty of Computer Science
Bialystok University of Technology
Bialystok, Poland

Nabendu Chaki
Department of Computer Science
and Engineering
University of Calcutta
Kolkata, India

ISSN 2367-3370 ISSN 2367-3389 (electronic)
Lecture Notes in Networks and Systems
ISBN 978-981-19-6790-0 ISBN 978-981-19-6791-7 (eBook)
https://doi.org/10.1007/978-981-19-6791-7

This Springer imprint is published by the registered company Springer Nature Singapore Pte Ltd.
The registered company address is: 152 Beach Road, #21-01/04 Gateway East, Singapore 189721, Singapore

Preface

In this edition of the proceedings of the International Symposium on Applied Computing for Software and Smart systems (ACSS), we present the readers with a collection of papers in five different domains, namely Algorithms, Machine Learning, Smart Healthcare, Software and Systems Engineering and Smart systems and Networks. These papers have been accepted for presentation in the 9th version of ACSS, which has evolved from a doctoral symposium over the years to a full-fledged symposium this year. The success of ACSS in bringing together students and scholars interested in Computing has been to motivate many young graduates to strive towards the challenging career of a researcher. This year, looking at the changing paradigm of the research in Computing, we renamed the symposium as Applied Computing for Software and Smart systems. The symposium was held in Kolkata during September 09-10, 2022. The Symposium was organized by the University of Calcutta in collaboration with Ca Foscari University of Venice, Italy, and Bialystok University of Technology, Poland. ACSS is aimed specially to facilitate Computing students to showcase their unique ideas. Each contributed paper was subjected to a double-blind review by experts in respective domains. The accepted papers were presented during the symposium which enabled further discussion of the innovative ideas among the research community peers.

Over the years, the overall quality of the papers submitted to ACSS has been improving and their subjects reflect and somehow anticipate the emerging research trends in the area of applied computation for software and smart systems. The symposium had included the topics such as the call for papers listed the following topics of interest related to Applied Computation: Smart Computing, Requirements and formal specifications, Artificial Intelligence, Data Science, Biometrics and Algorithms.

The editors are greatly indebted to the members of the international program committee for sharing their expertise in reviewing the papers within time. Their reviews have allowed the authors not only to improve their articles but also to get new hints towards the completion of their research work.

The dissemination initiatives from Springer have drawn a large number of high-quality submissions from scholars primarily but not exclusively from India. The reviewers mainly considered the technical quality and the originality of each paper, specifically considering the clarity of the presentation. The entire process of paper submission, review and acceptance process was done online. After carefully considering the reviews, the Program Committee selected only 17 papers for publication out of a total of 58 submissions.

We take this opportunity to express our sincere gratitude towards the members of the Program Committee and Organizing Committee, whose sincere efforts before and during the symposium have resulted in a strong technical program and in effective discussions. We thank Springer Nature for sponsoring the best paper award. In particular, we appreciate the initiative from Mr. Aninda Bose and his colleagues in Springer Nature for their strong support towards publishing this post-symposium book in the series **"Lecture Notes in Networks and Systems"**. We would also like to thank ACM for the continuous support towards the success of the symposium. Last, but not the least, we thank all the authors without whom the Symposium would not have reached up to this standard.

On behalf of the editorial team of ACSS 2022, we sincerely hope this volume will be beneficial to all its readers and motivate them towards better research work.

Rituparna Chaki
Agostino Cortesi
Nabendu Chaki
Khalid Saeed

Contents

Editors and Contributors

About the Editors

Rituparna Chaki a full professor in the A. K. Choudhury School of IT, University of Calcutta, India, since 2013. She is actively involved in the research related to the domain of wireless networking for over last 12 years. Her field of research encompasses optical network topology to adhoc network routing, wds for WSN. She is also actively involved in the promotion of Internet of Things and has recently authored a book on IoT security by CRC press. She is an active member of ACM India and is currently chairing the ACM-Kolkata professional Chapter. Besides wireless networking and IoT, she is also involved in research related to Systems Software, Software security and testing. She has well over 150 international publications to her credit. Professor Chaki has delivered invited lecture in different Universities, and Conferences in India and abroad. She is also a Visiting Professor in the AGH University of Science and Technology, Poland since October 2013.

Agostino Cortesi, Ph.D., is a full professor of computer science at Ca' Foscari University, Venice, Italy. He has extensive experience in the area of software engineering, static analysis and verification techniques. His main research interests concern programming languages theory, software engineering, and static analysis techniques, with particular emphasis on security applications. He has been the adviser of several doctoral and postdoctoral students from Italy and abroad (India, Cuba), and has published more than 150 papers in high-level international journals and proceedings of international conferences.

Khalid Saeed is a full Professor of Computer Science at Bialystok University of Technology and a half-time full professor at Universidad de La Costa, Barranquilla, Colombia. He was with Warsaw University of Technology in 2014–2019 and with AGH Krakow in 2008–2014. He received his B.Sc. Degree from Baghdad University in 1976, M.Sc. and Ph.D. (distinguished) Degrees from Wroclaw University of Technology in Poland in 1978 and 1981, respectively. He received his D.Sc. Degree

(Habilitation) in Computer Science from the Polish Academy of Sciences in Warsaw in 2007. He was nominated by the President of Poland for the title of Professor in 2014. He has published more than 250 publications including about 120 journal papers and book chapters, about 100 peer reviewed conference papers, edited 50 books, journals and Conference Proceedings, written 13 text and reference books (h-index 17 in WoS base and 14 in SCOPUS base). He supervised more than 15 Ph.D. and 150 M.Sc. theses. He was selected as IEEE Distinguished Speaker for 2011–2016. Khalid Saeed is the Editor-in-Chief of International Journal of Biometrics with Inderscience Publishers (since 2008).

Nabendu Chaki is a Professor in the Department of Computer Science and Engineering, University of Calcutta, Kolkata, India. He is sharing the responsibility of the Series Editor for Springer Nature book series on Services and Business Process Reengineering jointly with Prof. Agostino Cortesi of Venice, Italy. Besides editing close to 50 conference proceedings with Springer, Dr. Chaki has authored eight text and research books with CRC Press, Springer Nature, etc. He has published more than 250 Scopus Indexed research articles in Journals and International conferences books (h-index is 16 in SCOPUS base). Professor Chaki has served as a Visiting Professor in different places including US Naval Postgraduate School, California, and in different Universities in Poland and in Italy. He has been the founder Chair of ACM Professional Chapter in Kolkata and served in that capacity for three years during 2014–17. He has been active during 2009–2015 towards developing several international standards in Software Engineering and Service Science as a Global (GD) member for ISO-IEC.

Contributors

Sawan Aich Ramakrishna Mission Vivekananda Educational and Research Institute, Dist Howrah, West Bengal, India

Shubham Kant Ajay National Institute of Technology Patna, Patna, Bihar, India

B. K. S. P. Kumar Raju Alluri VIT-AP University-Amaravati, Amaravati, AP, India

Aditya Bagchi Indian Statistical Institute, Kolkata, India

Ujjwal Bhattacharya CVPR Unit, Indian Statistical Institute, Kolkata, India

Barun Biswas AKCSIT, University of Calcutta, Kolkata, India

Rahma Boucetta Department of Physics, Faculty of Sciences, University of Sfax, Sfax, Tunisia

Bidyut B Chaudhuri TECHNO INDIA UNIVERSITY, Kolkata, India

Sankhayan Choudhury University of Calcutta, Kolkata, India

Anil Bikash Chowdhury Department of Computer Applications, Techno India University, Kolkata, WB, India

Vivek Das Novo Nordisk A/S, Maløv, Denmark

Ayan Kumar Das Birla Institute of Technology, Mesra, Patna, India

Kishore Babu Dasari Department of CSE, Acharya Nagarjuna University, Andhra Pradesh, India

Debapratim Das Dawn Kolkata, India

Rajat K. De Machine Intelligence Unit, Indian Statistical Institute, Kolkata, India

Satyajit De Department of Computer Science, Maheshtala College, Maheshtala, WB, India

Nagaraju Devarakonda School of Computer Science and Engineering, VIT-AP University, Amaravati, India

H. Mamata Devi Department of Computer Science, Manipur University, Imphal, India

B. Gowtham VIT-AP University-Amaravati, Amaravati, AP, India

Amarnath Gupta University of California San Diego, La Jolla CA, USA

Raju Halder Indian Institute of Technology, Patna, India

Saksham Jha Indian Institute of Technology, Patna, India

Ananya Kanjilal BPPIMT, Kolkata, India

Abhinandan Khan Kolkata, India

Yanglem Loijing Khomba Khuman Department of Computer Science, Manipur University, Imphal, India

Hari Kishan Kondaveeti School of Computer Science Engineering, VIT-AP University, Amaravati, Andhra Pradesh, India

Pallavi Malavath School of Computer Science and Engineering, VIT-AP University, Amaravathi, India

Amit Kr Mandal Department of Computer Science and Engineering, SRM University AP, Amaravati, India

Saikat Pahari Omdayal College of Engineering and Architecture, Howrah, India

Anita Pal National Institute of Technology, Durgapur, India

Rajat Kumar Pal Department of Computer Science and Engineering, University of Calcutta, Acharya Prafulla Chandra Roy Siksha Prangan, Saltlake City, Kolkata, India

A. S. Venkata Praneel Department of Computer Science and Engineering, GITAM (Deemed to Be University), Visakhapatnam, Andhra Pradesh, India

Chouhan Kumar Rath Department of Computer Science and Engineering, National Institute of Technology Durgapur, Durgapur, India

Rishikesh National Institute of Technology Patna, Patna, Bihar, India

Paweł Romaniuk Faculty of Computer Science, Bialystok University of Technology, Bialystok, Poland

Pratik Roy Department of Computer Engineering and Applications, GLA University, Mathura, UP, India

Soumen Roy Department of Computer Science and Engineering, University of Calcutta, Acharya Prafulla Chandra Roy Siksha Prangan, Saltlake City, Kolkata, India

Sanjib Roy Birla Institute of Technology, Mesra, Patna, India

Utpal Roy Department of Computer System Sciences, VisvaBharati, Santiniketan, India

Khalid Saeed Faculty of Computer Science, Bialystok University of Technology, Bialystok, Poland

Swagatika Sahoo Indian Institute of Technology, Patna, India

Anirban Sarkar Department of Computer Science and Engineering, National Institute of Technology Durgapur, Durgapur, India

Somenath Sarkar Indian Institute of Technology, Patna, India

Santosh Kumar Satapathy Department of Information and Communication Technology, Pandit Deendayal Energy University (PDEU), Gandhinagar, Gujarat, India

Dibyendu Bikash Seal A. K. Choudhury School of Information Technology, University of Calcutta, Kolkata, India

Taniya Seal Kolkata, India

Sanjit Kumar Setua Kolkata, India

O. Imocha Singh Department of Computer Science, Manipur University, Imphal, India

T. Romen Singh Department of Computer Science, Manipur University, Imphal, India

Devadatta Sinha Department of Computer Science and Engineering, University of Calcutta, Acharya Prafulla Chandra Roy Siksha Prangan, Saltlake City, Kolkata, India

Ditipriya Sinha National Institute of Technology Patna, Patna, Bihar, India

H. Subramani VIT-AP University-Amaravati, Amaravati, AP, India

D. Sumathi VIT-AP University-Amaravati, Amaravati, AP, India

Soumi Tokdar BPPIMT, Kolkata, India

Algorithms

Predicting Useful Information From Typing Patterns Using a Bootstrapped-Based Homogeneous Ensemble Approach

Soumen Roy, Utpal Roy, Devadatta Sinha, and Rajat Kumar Pal

Abstract Nowadays, the way a user connects with computing devices is being analysed with the goal of extracting useful information for some interesting applications beyond user authentication. It enables the development of next-generation intelligent human-computer interaction, auto-profiling, and soft biometrics. In this paper, a bootstrapped-based homogeneous ensemble model has been proposed without wasting rare samples in each bootstrapped training set to overcome the uneven distribution of classes (common in keystroke dynamics) for predicting users' traits, fine motor skills, and cognitive deficiency automatically using the user's daily typing habit. This model is lightweight, faster, and could be implemented in low-configured devices like smartphones. The proposed model has been verified with a more realistic evaluation and several shared and authentic keystroke dynamics (KD) datasets and achieved 93.21% of accuracy in predicting age group, 65.35% in identifying gender, 87.14% for handedness, 77.14% in the case of hand(s) used determination, 91.25% in predicting qualification, 74.44% in recognising typing skill, 58.45% in observing lie, 84.12% in the determination of Parkinson's disease (PD), and 99.14% in predicting emotional stress (ES). The proposed model might be used for a wide range of more interesting applications, including automatic user profiling in social networking, age-restricted access control to protect kids from Internet threats, age- and gender-specific product recommendations in e-commerce, medical diagnostics at-home environment for better treatment and therapy management, soft biometric traits in improving biometric models, learner's cognitive deficiency in effective online teaching and learning, cognitive deficiency marker in a competitive examination, unbiased online feedback collection, and online inquiry correctness measurement, to mention a few.

S. Roy (✉) · D. Sinha · R. K. Pal
Department of Computer Science and Engineering, University of Calcutta, Acharya Prafulla Chandra Roy Siksha Prangan, JD-2, Sector - III, Saltlake City, Kolkata 700106, India
e-mail: soumen.roy_2007@yahaoo.co.in

U. Roy
Department of Computer System Sciences, VisvaBharati, Santiniketan 731235, India

R. Chaki et al. (eds.), *Applied Computing for Software and Smart Systems*, Lecture Notes in Networks and Systems 555,
https://doi.org/10.1007/978-981-19-6791-7_1

Keywords Bootstrapping · Ensemble classification · Keystroke dynamics · Traits recognition

1 Introduction

Some of the few personal traits (e.g., age, gender, education level, etc.) have a major influence on social networking. However, among other factors, handedness, hand(s) used, and typing competence are important in user profiling and digital identification. These details, as added features, improve the performance of KD user authentication [38]. On the other hand, deficiency of cognitive capability and fine motor skills are critical neural disorders that must be diagnosed early for effective treatment [17, 41]. Each of these pieces of information might be extracted from the user's typing rhythm [12, 41]. According to earlier research [3], how users interact with a desktop keyboard or smartphone provides potential features in the form of time, pressure, and rotational factors that might be used to exploit user-specific qualities. Every day, smartphone users create millions of keystroke and touch events, leading to the creation of a myriad of huge features (i.e., temporal, spatial, and orientation). On the other hand, smartphones with superior attached sensors create forces (degree of vibration), angular velocity, rotational, and orientation information while typing a text, browsing through social networking postings, or swiping photographs. This way of building passive digital footprints invisibly makes computing devices smarter, which motivates this study.

1.1 Challenging Issues

The number of individuals utilising social networking sites to engage with friends and family is steadily growing. Each user's digital identification necessitates the examination of fake or duplicate identities. Furthermore, digital identification is critical in forensic or surveillance systems [37]. Similarly, a multi-user device's passive digital traces in the form of cookies cannot forecast the specific user, necessitating automated detection. Furthermore, as with soft biometric features, this information improves user authentication performance [32]. On the other hand, rapid improvements in mobile, wearable Internet of Things (IoT) devices, wireless technology, and applications, provide the opportunity to incorporate automatic recognition information into existing services [34].

In Biometric science, researchers collect samples from a user regularly to monitor intra-class variability. Similarly, the patterns in several sessions of each subject are collected to develop adaptation methods to address ageing. Therefore, researchers collect data in various postures and settings to measure the external influences. If we utilise the k-fold (5-fold or 10-fold) cross-validation evaluation method, samples from a subject may be distributed in the training and testing sets, resulting in unrealistic findings [37]. When a user checks his or her gender/age group/handedness/disease/stress, the data from that user should not be included in the training set. It requires careful consideration in the machine learning (ML) model evaluation. Studies [40] applied Leave-One-User-Out Cross-Validation (LOUOCV) to address this problem. However, the non-uniformity of KD raises the additional difficulty of class imbalance in LOUOCV.

On the other hand, in LOUOCV, the evaluation is conducted n times (subject size) and in each iteration, $(n-1)$ subjects are selected for the training set [37]. If we use bootstrapping in ensemble learning, the samples from $(n-1)$ subjects are distributed in several training sets and there is a high chance of uneven distribution in each bootstrapped sample. However, there are many promising machine learning techniques, like support vector machines, that are influenced by imbalance issues in the training set. In addition, the rare samples are wasted in the distributed samples in bootstrapping. It needs to be addressed for a reliable and robust ensemble model using bootstrapping.

1.2 Major Goals and Contributions

The main goal of this study is to extract useful information from regular typing patterns for the best user experience and soft biometrics. To tackle the uneven distribution of classes in bootstrapped samples, we suggest a unique approach and avoid over-fitting. To enhance the reliability and accuracy, we suggest a homogeneous ensemble setting. The primary purpose of this research is three-fold as follows:

- Propose a novel ML setting in a bootstrapped-based homogeneous ensemble.
- Extraction of a large number of useful information from typing rhythm.
- The performances of five classifiers have been presented and analysed in the same ML setting.

Other contributions of this work include (a) the implementation of a unique approach for a reliable and accurate framework for binary classification and (b) showing appropriate statistical evidence in selecting a suitable classifier that was previously unclear. To our knowledge, this is the first study that outlined an effective way of predicting a large number of useful items from typing clues. Furthermore, the study validated the proposed approach with publicly available datasets and compared different ML models with the same homogeneous ensemble framework. This will aid in the development of next-generation intelligent human-computer interfaces, including m/e-Teaching-learning, m/e-Commerce, m/e-Banking, m/e-Social-networking, and m/e-Health.

2 Literature Review

KD is a fool-proof biometric trait for user authentication [9]. However, there is information that could be extracted from these KD attributes [14]. In this section, we summarised the latest studies in predicting traits [32], lies [26], user's mental stress [41], and PD [12] using these attributes to understand the latest trends, approaches, evaluation options, results, and challenges towards achieving a modest version of KD-based predictive models. Table 1 shows the latest improvement of predicted models using KD characteristics.

The information in the table shows that there is a huge amount of information that can be extracted from typing tendencies, including age, gender, culture, qualification, etc. There were huge efforts to extract age and gender, but much of this information has not been analysed more. To extract this information, a large number of datasets were developed, because no single dataset was labelled with all this important information. Therefore, different researchers used unique datasets resulting in different combinations of features and used different machine learning techniques. A very less number of techniques have been adopted to extract useful information.

On the other hand, k-fold cross-validation (CV) is common in system evaluation. Where overlapping of the samples from the same subject in training and testing sets may be distributed can produce impractical results. Some researchers have taken care of these problems even in cross-validation, which was not applied for extracting all this information. On the other hand, in cross-validation, the main two issues are (a) rare samples are wasted and (b) imbalanced fold in class distribution.

In the KD domain, the predicted model can be divided into two main categories—(a) Prediction in Static Mode (PiSM) and (b) Prediction in Dynamic Mode (PiDM). PiSM is developed using the patterns developed for pre-defined inputs, where the user needs to type the same input at the time of validation. Whereas, PiDM is developed using patterns for the inputs that are not restricted. Therefore, the usability of PiDM is much higher than that of PiSM. However, due to the unstructured data stream, PiDM is less accurate than PiSM within the same typing duration. In the previous efforts, both were not taken care of equally. This is a study bias that needs to be explored more.

2.1 KD as Traits Determination Technology

- KD-based age group determination—Nowadays, a significant number of computer and smartphone users are children due to online classes, e-assessment, social networking, etc. Without proper treatment, they have been forced to use these computing devices during social distancing regulations. A study [35] proposed a method to protect children from Internet threats by recognising child users by their typing patterns. A recent study [28] achieved 73.3% of accuracy using the typing patterns for free text typing on a touch screen. A similar study [46] discovered an 89.2%

Table 1 Latest studies in predicting useful information from typing patterns using machine learning techniques

Study	Year	Model	Device	Features (see Table 2)	Subject	Information	Method	Aclc. (%)	Evaluation
Davarci et al. [7]	2017	PiDM	S	Accelerometer	200	Age	KNN	88.33	5-Fold CV
Tsimperidis et al. [44]	2017	PiDM	D	DD, UD, DU, UU	153	Age	ANN	73.07	10-Fold CV
Roy et al. [39]	2018	PiSM	D	KH, DD, UD	65	Age	FRNN	88.35	10-Fold CV
Roy et al. [32]	2019	PiSM	D	KH, DD, UD	65	Age	FRNN	94.93	10-Fold CV
Roy et al. [32]	2019	PiSM	S	Temporal	51	Age	FRNN	87.91	10-Fold CV
Roy et al. [33]	2020	PiSM	S	KH, DD, UU, UD, DU	92	Age	Fusion	98.74	T/T
Udandarao et al. [47]	2020	PiDM	D	DU, UU, DD, UD	117	Age	CNN	80.00	5-Fold CV
Udandarao et al. [47]	2020	PiDM	S	DU, UU, DD, UD	117	Age	CNN	82.92	5-Fold CV
Udandarao et al. [47]	2020	PiSM	D	DU, UU, DD, UD	117	Age	CNN	85.37	5-Fold CV
Udandarao et al. [47]	2020	PiSM	S	DU, UU, DD, UD	117	Age	CNN	78.04	5-Fold CV
Abinaya and Sowmiya [1]	2021	PiSM	D	UD, DD, DU, UU	118	Age	PSO-NN	94.44	T/T
Oyebola and Adesina [28]	2021	PiDM	S	Temporal, Spatial, P	50	Age	RF	73.30	10-Fold CV
Tsimperidis et al. [46]	2021	PiDM	D	KH, DD, Digraph	385	Age	RBFN	92.00	10-Fold CV
Tsimperidis et al. [46]	2021	PiDM	D	KH, DD, Digraph	387	Age	RBFN	89.20	10-Fold CV
Roy et al. [37]	2022	PiSM	S	KH, DD, UU, UD, DU	92	Age	Ensemble	91.49	LOUOCV
Yaacob et al. [48]	2021	PiSM	D	Temporal	148	Culture	SVM	77.49	10-Fold CV
Yaacob et al. [48]	2021	PiSM	D	Temporal	148	Education level	SVM	83.31	10-Fold CV
[11]	2019	PiDM	S	Timing, Statistical	22	Emotion	RF	78.00	10-Fold CV
Ghosh et al. [32]	2019	PiSM	D	KH, DD, UD	65	Gender	FRNN	94.97	10-Fold CV

(continued)

Table 1 (continued)

Study	Year	Model	Device	Features (see Table 2)	Subject	Information	Method	Aclc. (%)	Evaluation
Roy et al. [32]	2019	PiSM	S	Temporal	51	Gender	FRNN	83.87	10-Fold CV
Roy et al. [33]	2020	PiSM	S	KH, DD, UU, UD, DU	92	Gender	Fusion	88.08	T/T
Udandarao et al. [47]	2020	PiDM	D	DU, UU, DD, UD	117	Gender	CNN	86.04	5-Fold CV
Udandarao et al. [47]	2020	PiDM	S	DU, UU, DD, UD	117	Gender	CNN	83.72	5-Fold CV
Udandarao et al. [47]	2020	PiSM	D	DU, UU, DD, UD	117	Gender	CNN	86.04	5-Fold CV
Udandarao et al. [47]	2020	PiSM	S	DU, UU, DD, UD	117	Gender	CNN	88.37	5-Fold CV
Abinaya and Sowmiya [1]	2021	PiSM	D	UD, DD, DU, UU	118	Gender	PSO-NN	92.00	T/T
Oyebola and Adesina [28]	2021	PiDM	S	Temporal, Spatial, P	50	Gender	RF	71.30	10-Fold CV
Tsimperidis et al. [45]	2021	PiDM	D	KH, DD, Digraph	387	Gender	RBFN	89.70	10-Fold CV
Roy et al. [37]	2022	PiSM	S	KH, DD, UU, UD, DU	92	Gender	Ensemble	62.07	LOUOCV
Roy et al. [32]	2019	PiSM	D	KH, DD, UD	110	Hand(s)used	FRNN	98.90	10-Fold CV
Roy et al. [32]	2019	PiSM	D	KH, DD, UD	65	Handedness	FRNN	97.77	10-Fold CV
Monaro et al. [25]	2018	PiDM	D	Temporal	40	Lie	RF	95.00	10-Fold CV
Monaro et al. [26]	2021	PiDM	D	Temporal	50	Lie	RF	90.00	30 subsects
Giancardo et al. [12]	2016	PiDM	D	KH	85	Parkinson's	Support vector	AUC 81.0	T/T
Iakovakis et al. [20]	2018	PiDM	S	KH, UD	33	Parkinson's	LR	AUC 92.0	LOSO
Iakovakis et al. [19]	2018	PiDM	S	KH, UD	33	Parkinson's	Regression	78.0	LOSO
Milne et al. [24]	2018	PiDM	D	KH	85	Parkinson's	LR	AUC 85.0	CV
Pham [30]	2018	PiDM	D	KH	85	Parkinson's	Fuzzy	AUC 98.0	2-Fold CV
Iakovakis et al. [18]	2019	PiDM	S	KH, UD	33	Parkinson's	CNN	AUC 89.0	LOSO
Hooman Oroojeni et al. [15]	2019	PiDM	D	KH	85	Parkinson's	Tensor Train	AUC 88.0	CV

(continued)

Table 1 (continued)

Study	Year	Model	Device	Features (see Table 2)	Subject	Information	Method	Aclc. (%)	Evaluation
Dhir et al. [8]	2020	PiDM	D	Temporal	41	Parkinson's	LSTM	AUC 91.0	2-Fold CV
Lim et al. [23]	2020	PiDM	D	DD	190	Stress	FFBP	86.02	T/T
Sağbaş et al. [41]	2020	PiDM	S	Sensory, Statistical	46	Stress	KNN	87.56	10-Fold CV
Dacunhasilva et al. [5]	2021	PiDM	D	KH, DD, UD, DU, P	25	Stress	Near. neig.	74.50	T/T
Udandarao et al. [47]	2020	PiDM	D	DU, UU, DD, UD	117	Style	CNN	93.18	5-Fold CV
Udandarao et al. [47]	2020	PiDM	S	DU, UU, DD, UD	117	Style	CNN	85.37	5-Fold CV
Udandarao et al. [47]	2020	PiSM	D	DU, UU, DD, UD	117	Style	CNN	90.38	5-Fold CV
Udandarao et al. [47]	2020	PiSM	S	DU, UU, DD, UD	117	Style	CNN	87.80	5-Fold CV
Roy et al. [32]	2019	PiSM	D	KH, DD, UD	65	Typing skill	FRNN	95.19	10-Fold CV

D->Desktop/laptop, S->Smartphone, KNN->K-Nearest Neighbour, FRNN->Fuzzy Rough Nearest Neighbour,
ANN->Artificial Neural Network, LSTM->long short-term memory,
FFBP->Feed-forward Back Propagation, RBFN->Radial Basis Function Network
T->Timing, R-> Rotational, NT->No temporal, P->Pressure, S->Statistical, MD->Mouse dynamics
LOSO->Least-One-Subject-Out, LOOCV-> Least-One-Out CV,CV->Cross-Validation,
T/T->Training versus Testing

accuracy for the pattern developed using a traditional keyboard. With the use of recent smartphone sensor data, a more advanced study achieved 91.49% accuracy [37].

- KD-based gender determination—A study [13] introduced that gender is predictable information through the way of typing on a traditional keyboard. They achieved 91% of accuracy with the help of a Support Vector Machine (SVM). In the case of considering patterns for free text, it is 86.1% [46]. A study [34] achieved 62.63% accuracy with only timing features for touch tapping behaviour. The accuracy rate is reduced to 58.26% in the practical scenario [40]. By incorporating sensory features, it could be 62.07% [37]. They also confirmed that it could be 64.73% with the selection of the best 15 features. A recent study [28] used the best five features from timing, pressure, rotational, and statistical features and found 71.3% of accuracy with 10-fold cross-validation.
- KD-based handedness determination—Dominating the hand (left-handed/right-handed) of a user is one of the vital attributes to enhance the performance of KD-based authentication design [36]. A recent study [40] achieved 60.59% of accuracy in recognising handedness using smartphone temporal features. Another study [29] proposed a tree-based ML approach and achieved 99.5% of accuracy.
- KD-based hand(s) used determination—The number of hand(s) used while typing is one of the soft biometric features that requires greater attention to gain the authentication performance [34]. A study [32] achieved 98–99% accuracy in the determination of this trait. A level of accuracy of 78.62% lower than the previous has been achieved through the use of smartphone [40].
- KD-based typing skill determination—Typing skills can be separated into two types—"Touch" and "Others". An accuracy of 88–95% has been achieved in extracting this feature [32]. A study [36] used this as an extra feature and observed a positive impact on KD authentication. It could also apply to smartphone authentication [34].
- KD-based qualification determination—A recent study [43] proposed an ML model to predict a user's education level with typing data developed through a traditional keyboard and observed an accuracy of 84.5%. They have considered five unique education levels for the purpose.
- KD-based lie determination—Providing fake information is the major unsolved issue in the recent age of social networking. There is no practical solution in today's world. A study [25] observed 95% of accuracy in the determination of a lie using the keystrokes developed on a traditional keyboard. They reported 97.5% of accuracy with the Random Forest in their recent study [26].

2.2 *KD-Based Disease Determination*

- KD-based ES determination—A recent study [23] presented a feed-forward back-propagation (FFBP) neural network to identify university students' learner stress. Another study [11] proposed (*EmoKey*)—a form of Android-based keyboard for

detecting four mental disorders. In contrast, a research [41] used smartphone sensors in the prediction of ES and found 87.56% accuracy.
- KD-based PD determination—A research [12] created *neuroQWERTY*—a score based on the ML approach in the diagnosis of PD by developing a series of key hold times (KH) for PD patients and control desktop users. Another model, *Tappy*, was proposed by a study [2] in research, and they made use of a variety of timing elements, including KH and UD. According to a research [15], *FRESH-TT* based on KH and *Tensor Train (TT)* decomposition produced excellent results. Fuzzy recurrence plots (FRP) were used in a research [30] to get the best accuracy. Another research [24] used the same dataset and discovered an AUC of 0.85 by extracting and combining statistical characteristics. According to a study [18], a study employed Convolutional Neural Networks (CNN) as a classifier and KH and UD as feature groupings to detect early indicators of PD using a smartphone. A research [19], on the other hand, exploited just time features via typing on a touch screen.

3 Proposed Method

The suggested model has been separated into six primary subparts, as shown in Fig. 1, and each portion is critical to trustworthy and accurate prediction performance. The model's uniqueness is a new approach to bootstrapping without losing any samples from the minority class for the model's dependability, as well as homogeneous ensemble learning for more accurate performance. Each section has been explored in detail below.

3.1 Pre-processing Towards ML Suitable Patterns

The goal of this phase is to convert the raw keystroke data into an ML-ready format. Because of the non-uniform data acquisition device (e.g., shape, size, type, etc.), inputs of various types (e.g., numeric, simple, complex, logically complex, etc.) and lengths (e.g., short, long, sentence, paragraph, etc.), unique environments (e.g., desktop, laptop, smartphone, tablet, etc.), sensors attached in various numbers, and so on, there are multiple settings to develop the raw data. Table 2 displays the potential features for users' typing habits. Some of the few characteristics are action-specific, while others are produced at a preset sampling rate that is appropriate for that device. There are a lot of outliers in the patterns due to unavoidable factors.

Sliding windowing is used for generating repeated samples from raw samples with a fixed window length. These samples are labelled with class names. The classes and the procedure for labelling them are presented in Table 3. During the learning model, samples are labelled using self-reported information or by establishing artificial stress and lie conditions, or by classes suggested by expert doctors with

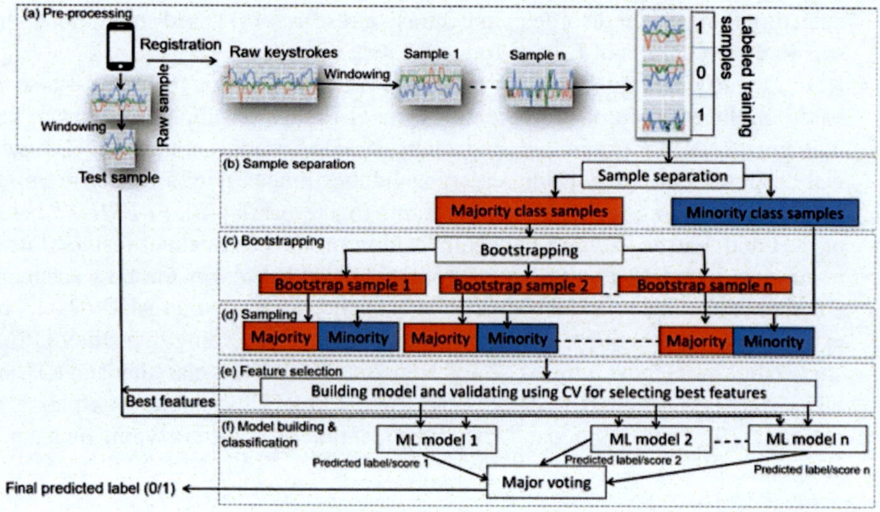

Fig. 1 The proposed bootstrapped-based homogeneous ensemble classification architecture and the predictive contrivance of KD-based predictive model

cross-validated diagnoses. The samples with labelled classes are now ready to be used to create a supervised ML model. Similarly, the sample is prepared for testing without class identification. In the case of continuous prediction, these samples are generated again at some regular interval. So that it could be used as an extra feature in continuous/implicit/adaptive authentication to minimise session hijacking.

3.2 *Sample Separation*

KD datasets are imbalanced in different ways. It needs to be balanced to build a strong learner. Random sampling with replacement may lose the rare samples. Therefore, the samples are divided into two groups—the majority class organises all samples from easily available samples, and the minority class keeps all the rare samples. This method of separation can select random samples from the majority classes based on the size of the minority class.

3.3 *Bootstrapping and Sampling*

In our proposed ML setup, an ensemble method is used. The foundation model is a standard binary classifier, and numerous training sets are investigated. To create these training sets, we employed bootstrapping, which creates a large number of

Table 2 Events (press/touch/release) and corresponding temporal and sensory raw data. Features—T, P, and A are developed for each action, whereas x, y, z are captured at reasonable sampling rate continuously

Action	Event	Timestamp	Pressure (P)	Area (A)	Rotational (R)			Timing (T)
Key press	e_1	t_1	p_1	a_1	x_1	y_1	z_1	KH=$\{t_{i+1} - t_i : \forall i = 1, 2, 3, \ldots, n-1\}$
Key release	e_2	t_2	p_2	a_2	x_2	y_2	z_2	DD=$\{t_{i+2} - t_i : \forall i = 1, 2, 3, \ldots, n-1\}$
Key press	e_3	t_3	p_3	a_3	x_3	y_3	z_3	UD=$\{t_{i+2} - t_{i+1} : \forall i = 1, 2, 3, \ldots, n-1\}$
Key release	e_4	t_4	p_4	a_4	x_4	y_4	z_4	UU=$\{t_{i+3} - t_{i+1} : \forall i = 1, 2, 3, \ldots, n-1\}$
Key press	e_5	t_5	p_5	a_5	x_5	y_5	z_5	Digraph=$\{t_{i+3} - t_i : \forall i = 1, 2, 3, \ldots, n-1\}$
:	:	:	:	:	:	:	:	:
Key release	e_n	t_n	p_n	a_n	x_p	y_p	z_p	Four-graph$\{t_{i+5} - t_i : \forall i = 1, 2, 3, \ldots, n-1\}$

KH->Key Hold duration, DD->Down-Down key latency, UD->Up-Down key latency, UU->Up-Up key latency

samples by randomly selecting samples with replacements from the core training dataset. These bootstrap training sets may be imbalanced. To solve this issue, a one-of-a-kind sampling approach is used. When the same number of samples of the majority class is randomly selected, the samples of the minority classes are the same in all bootstrap training sets using the under-sampling approach. This bootstrapping approach provides a large number of balanced training sets while avoiding the loss of rare samples.

3.4 Feature Selection

The ranks of each feature depend on the $GainRatio$ that is calculated for each training set that measures penalised attributes by incorporating *Spilt Information* as per Eqs. 1 and 2. Once the $Gain$ Ratios are calculated, the attributes with the highest $GainRatio$ have been selected. In our model, the top ten most gained attributes have been used for each dataset.

$$Split\ Information\,(S, A) = -\sum i = 1^c \frac{\|S_i\|}{\|S\|} \log_2 \frac{\|S_i\|}{\|S\|} \tag{1}$$

$$Gain\ Ratio\,(A) = \frac{Gain(A)}{Split\ Information\,(S, A)} \tag{2}$$

Table 3 Information and corresponding classes with labels and class labelling methods. The bold faced text indicates the high probability of minority classes

Knowledge	Class and assigned labels	Sample labelling method
Age group	Child = 0, Adult = 1	Self reported, age below 18, has been considered a "Child".
Gender	Female = 0, Male = 1	Self reported, third gender has not been considered
Handedness	Left = 0, Right = 1	Self reported, users from "Ambidextrous" has not been considered
Hand(s) used	One Hand = 0, Both Hands=1	Self reported, typing and holding the phone in one hand or holding by one hand and typing by another considered as "One Hand"
Typing skill	Touch = 0, Other = 1	Self reported, searching for next key without releasing previous key considered as "Touch"
Education level	Madhymic = 0, Post Madhymic = 1	Self reported, those who had just completed Madhymic considered as "Madhymic"
Lie	Yes = 0, No=1	Users were bound to false inputs for "Yes" and Right inputs for "No"
Parkinson's	Parkinson's = 0, Controlled = 1	Experienced doctors and cross-validation diagnosis
Stress	Stress = 0, Calm = 1	Stressful jobs for "Stress" and non-stressful jobs for "Calm"

3.5 *Classification and Decision Level Score Fusion*

In this part, many ML models are trained using balanced bootstrap training sets. A homogeneous ensemble approach is proposed as defined in Eq. 3. When the same technique ($h(x_i)$) is used on different training sets (x_i). Here, $h(x_i)$ is also a strong learner. Since the same technique is used, a_i is set to 1. To increase the confidence level of the results, $f(x)$ is used. In this case, the claim sample is evaluated with multiple learners to get a variety of decisions. The ultimate score is derived using these decision-level ratings. The predicted score P_S is computed using Eq. 4.

$$\text{Strong-learner } f(x) = \sum_{i=1}^{n} a_i h(x_i) \tag{3}$$

Table 4 Publicly available datasets and data acquisition protocols were used in our study

Dataset	Year	Study	#Subject	Sensing device	Environment	Features	Information
Traits	2022	Roy et al. [37]	87	Attached sensors	Smartphone	Sensory	Age, gender, handedness, hand(s) used, education
CMU	2012	Killourhy [21]	63	Conventional keyboard	Desktop	Timing	Typing skills
Lie	2018	Monaro et al. [25]	40	Conventional keyboard	Desktop	Timing, Non temporal	Lie
Stress	2020	Sağbaş et al. [41]	190	Attached sensors	Smartphones	Sensory	Emotional stress
Parkinson's	2016	Giancardo et al. [12]	85	Conventional keyboard	Laptop	Hold time	Parkinson's disease

Table 5 Windowing and sampling technique in our consideration depending on the data acquisition protocols

Dataset	Window length	Sampling rate	Windowing technique
Traits	3.5 s	20/s	Time-based sliding window
Stress	5 s (100 data points)	20/s	Time-based sliding window
Parkinson's	100 keystrokes	–	Count-based sliding window

$$\text{Predicted score } P_S = \frac{1}{n}\sum_{1}^{n} C_i \tag{4}$$

$$\text{Predicted label } P_L = RoundOff(P_S, 0) \tag{5}$$

Where C_i is the predicted score for the ith classifier based on the ith bootstrap training set. By rounding off with the zero decimal point, the predicted label P_L is created. The value of n should be at least three for a trade-off between the stability of the result and the speed of learning. In this way, the other ML techniques (SVM, RF, KNN, and NB) have been evaluated for a fair comparison.

4 Datasets, Implementation, and Evaluation

In this section, detailed information about the suggested approach's implementation, experimental settings, and the outcome assessment technique has been provided. The following are the steps we took in our research.

4.1 Datasets Preparation

Table 4 presents the authentic and shared datasets used in our study. We used multiple datasets for this purpose because of the unavailability of all the useful information in a single dataset.

4.2 Windowing and Sampling

Huge raw data in a continuous form is not suitable for designing an ML model. Therefore, the sliding window method has been used, in which the samples are built within a reasonable window length. For ES, data points developed in 5 s. have been taken into account. In the case of traits, it is 3.5 s. The reason for this variation is the availability of the patterns in the shared datasets. To recognise PD, a count-based sliding window of 100 keystrokes has been used. Table 5 describes the nuances of windowing strategies. Depending on the datasets, different windowing approaches have been used. The Stress dataset was collected at a rate of 20 inspections per second. As a result, if we take the constitutive 100 samples, we should estimate 5 seconds of composing action. Parkinson's dataset, on the other hand, was acquired without respect to the examination rate. We tallied the constitutive 100 keystrokes to generate the 100 samples.

4.3 Outlier Detection and Removal

Outliers are unusual data points that often occur in behavioural patterns. It should be clean. We controlled for the deviations (esteem not in the middle of the 1st quartile and 3rd quartile or the worth not in the middle of the IQR) as recommended by the study [31] by the median of the data points.

4.4 Statistical Features Extraction

We have considered the following statistical features—$Minmim$ is utilised for the smallest data value in the information, $Maximum$ is used for the large data value in the information, to recognise the focal inclination we used $Average$, Sd is applied for the variety in the information, 50% of information is estimated by $Median$, Q_1 and Q_3 measure the 25 and 75% of information, where $Kurtosis$ and $Skewness$ measure the peakedness and the lopsidedness in the information stream, and a $histogram$ has been utilised to get the thickness of various information esteem inside a reach.

4.5 Normalisation

Then, we used *min-max* normalisation method for faster computing process as defined in Eq. 6. Here, x_i is normalised.

$$x_i = \frac{(x_i - \min(x))}{(\max(x) - \min(x))} \tag{6}$$

4.6 Training Dataset Preparation

Patients' samples are generally infrequent, and we don't want to lose these rare samples through traditional k-fold cross-validation or training vs testing data spitting in model creation and assessment. We utilised LOUOCV because both public datasets were tiny. Where samples from one subject are put together to form a test set, while the remaining samples from the other subjects are used as the training set. This spitting procedure is repeated for each participant. In this manner, data spitting offers various advantages, including the construction of a huge training dataset and the observation of the model's real performance. As a result, samples of $(n - 1)$ participants were chosen for the training set in each iteration.

Another issue is the training set's unequal distribution of classes. A balanced dataset is essential for building a high-performance ML model, and we must take care of it. We might employ a variety of sampling approaches. However, in our work, we adopted a novel sampling approach. We constructed bootstrap training samples by retaining all rare samples (minority class samples) and randomly selecting the equivalent number of uncommon samples from the majority class. This enables us to build a large number of balanced training sets while avoiding the loss of unusual data.

4.7 Feature Selection

We trained XGBoost with the balanced bootstrapped training samples and found the gain of each feature. In the boosted tree model, we considered each of the features and each gain. The average gain has been recorded. The highest gained features (top 10) have been selected to train the model. This process has been evaluated for each bootstrapped training sample.

4.8 Classifier Selection and Arrangements

In our investigation, we considered the most recent and popular five classifiers that are appropriate for low-configuration devices such as smartphones or tablets. Each classifier operates on a different principle. The classifiers that we used to build the prediction model are mentioned as follows:

- *Extreme Gradient Boosting (XGBoost)*—It is an ensemble classification model that is built on boosted decision trees and an objective function that is linked to the gradient of a loss function. It is a scalable, lightweight, open-source, sparsity-aware, missing data handler that is faster than a bagging classifier, and can run on a low-configuration device, making it suitable for use on smartphones as standalone apps. For this classifier, the "XGBoost" package from R [4] was used. We set the operational parameters eta to 0.3 to regulate the learning rate, $lambda$ to 0.1 to use regularisation terms as weights, $gamma$ to 0 to use the smallest loss reduction necessary to construct a further partition, and iteration to 11.
- *K nearest neighbour (KNN)*—It is a similarity metric and a non-parametric supervised classifier that classifies coordinates and identifies which group the test data points nearest to it belong to. To identify the nearest sites, distance measuring methodologies (here, Euclidean distance) and a majority vote are applied. In this method, the nearest k (here k = 4) neighbours are examined. This classifier was built with the R package "class".
- *Naive Bayesian (NB)*—It is a probabilistic classifier based on Bayes' theorem, with the assumption of predictor variable independence and metric predictors having a Gaussian distribution. This classifier was built with the well-known machine learning package "e1071."
- *Random Forest (RF)*—It is a classifier based on trees that work as part of an ensemble. It builds a large number of individual decision trees from random training samples, and the average scores are used to make a common decision. As a result, while constructing a model takes time, over-fitting is reduced. In terms of predictability, it outperforms the descriptive method. We used R's "randomForest" package to train and test our datasets.
- *Support Vector Machine (SVM)*—It is founded on a strong mathematical foundation. It determines the optimal choice boundary (hyperplane) between two different classes. The R package "e1071" was used to evaluate this classifier. We utilised the default parameter settings for the execution.

4.9 Bootstrapping

In our assessment, we used the under-sampling bootstrapping approach. Bootstrap training sets are constructed by preserving the same number of tests from the rare or minority classes and arbitrarily selecting tests from the majority class, while keeping in mind that samples from both the minority and majority classes are the same in each training set.

4.10 Scores Fusion

Three assessment scores were utilised for three bootstrapped samples using the same classifier. The average was then computed, and in order to get the anticipated level (0/1), we employed a significant digit of 0.

4.11 Model Evaluation

The LOUOCV model evaluation test option was used in this case. The dataset is divided into n folds (the number of participants is n). In each assessment, samples from one subject are used as the test set, while the remaining samples from other $(n - 1)$ individuals are prepared for the training set. As a result, the model has been assessed n times and the average performance has been calculated.

4.12 Tools and Techniques

For the implementation of the proposed setting and results analysis, the R (version 3.6.3) statistical programming language has been used. All the approaches in the form of applications have been uploaded to https://rstudio.cloud/project/3763364 in order to produce outcomes with different parameters and potential for future implementation.

5 Experimental Results

In this section, the novel ensemble technique's implementation in determining age, gender, handedness, hand(s) used, typing skill, education level, lie, ES, and PD has been evaluated and reported.

5.1 Performance Assessment Metrics

Because of the high likelihood of unequal distribution of the classes due to uncommon samples gathered from unusual individuals, accuracy alone is not an appropriate metric in multi-class classification, particularly in medical research to assess the prediction model in illness determination where rare samples are common. Sensitivity, specificity, area under curve (AUC), and Receiver Operating Curve (ROC) were other important metrics in our study. Sensitivity shows how well positive classes are recognised, whereas specificity shows how well negative classes are recognised. The ROC curve is a line graph that depicts how the apparent true positive rate changes as the false positive rate changes. An AUC is a measure of the area under the ROC curve that is typically used to assess the model's overall performance.

5.2 *Performance of Ensemble Approach*

The results in Table 6 are presented with a Confidence Interval (CI) at a 95% significant level for each classifier and predictor separately. CI indicates the confidence level of the obtained results in 100 repeated evaluations by considering randomly taken samples. In order to understand the performance difference with the modest approach XGBoost, the performance of modern binary classifiers in ensemble strategy was compared with the ANOVA indicated by 's'. No symbol indicates the performance is similar to XGBoost. No single classifier is effective for predicting information. XGBoost is effective in determining handedness and PD, KNN is effective in determining gender and ES, NB is effective in determining hand(s) used and skill prediction, and SVM is effective in determining age group, lie, and educational level.

In recognising the age group, it has been observed that 74.66% of accuracy could be achieved while using XGBoost as a classifier, where SVM achieved 82.12% of accuracy, outperforming XGBoost significantly. On the other hand, KNN is the top performer for gender prediction. However, there is no significant difference between the performance of XGBoost and KNN, XGBoost and RF, and XGBoost and SVM. In the case of handedness recognition, XGBoost achieved the highest performance with an accuracy of 81.78%, which is significantly similar to SVM. It has been discovered that typing patterns for smartphones for free text are unsuitable, even when advanced sensory features are included; liar detection is also difficult in practical evaluation in the case of hand(s) used determination. The performance of XGBoost in recognising PD is impressive. In the case of ES determination, XGBoost, RF, and SVM could be used. However, SVM achieved 79.17% of accuracy in identifying education levels.

The ROCs comparison is depicted in Fig. 2. It demonstrates how the true positive rate varies when the false positive rate changes. It shows that gender, hand(s) used, and lie prediction are all more difficult than others. Stress, on the other hand, may be predicted with great accuracy. Not a single classifier is sufficient enough for the prediction of all sorts of information. The average results of XGBoost and SVM, on the other hand, are outstanding.

Table 6 The performance of the ensemble technique of contemporary ML models is not yet known. The symbol 's' suggests a considerably different technique than XGBoost, whereas no symbol implies performance that is significantly similar to XGBoost. The text in bold indicates the highest measurement

Information	Method	Accuracy ± CI (%)	Specificity ± CI (%)	Sensitivity ± CI (%)	AUC ± CI (%)
Age group	XGBoost	74.66 ± 1.07	62.33 ± 1.54	84.25 ± 0.70	73.29 ± 1.11
	KNN	60.91 ± 0.85s	45.16 ± 1.12s	72.49 ± 1.23s	58.83± 0.72s
	NB	74.61 ± 0.88	60.70 ± 0.75	86.24 ± 0.74s	73.47 ± 0.75
	RF	76.42 ± 0.77s	63.74 ± 0.82	86.50 ± 0.78s	75.12 ± 0.72s
	SVM	**82.12 ± 0.49**s	**69.31 ± 0.69**s	**92.36 ± 0.18**s	**80.83 ± 0.44**s
Gender	XGBoost	57.02 ± 1.97	64.98 ± 1.64	46.46 ± 3.39	55.72 ± 2.23
	KNN	**58.50 ± 1.05**	**65.81 ± 0.72**	**48.49 ± 1.74**	**57.15 ± 1.19**
	NB	51.22 ± 1.53s	63.88 ± 0.98	43.33 ± 2.00	53.61 ± 1.02
	RF	55.62 ± 2.67	62.01 ± 2.30s	44.98 ± 3.28	53.49 ± 2.54
	SVM	54.06 ± 0.90	62.43 ± 1.50	44.04 ± 1.41	53.24 ± 0.75
Handedness	XGBoost	**81.78 ± 0.14**	92.91 ± 0.10	**42.75 ± 0.22**	**67.83 ± 0.06**
	KNN	66.03 ± 1.55s	89.26 ± 0.95s	18.74 ± 1.51s	54.00 ± 0.65s
	NB	66.17 ± 0.68s	**96.24 ± 0.47**s	25.92 ± 1.57s	61.08 ± 0.75s
	RF	79.03 ± 1.40s	92.43 ± 0.84	36.61 ± 2.28s	64.52 ± 1.47s
	SVM	80.42 ± 1.89	93.98 ± 0.50	36.12 ± 3.76	65.05 ± 2.08s
Hand(s) used	XGBoost	51.73 ± 1.03	58.45 ± 0.92	43.66 ± 1.09	51.05 ± 1.01
	KNN	46.95 ± 1.82s	54.30 ± 2.47s	38.52 ± 2.35s	46.41 ± 1.88s
	NB	**55.27 ± 0.95**s	59.65 ± 1.55	**49.41 ± 0.71**s	54.53 ± 0.87s
	RF	49.00 ± 2.13s	56.24 ± 2.21	39.92 ± 2.65	48.08 ± 2.02s
	SVM	54.14 ± 1.34	**60.71 ± 1.70**	46.51 ± 2.87	**53.61 ± 1.55**
Typing skill	XGBoost	66.06 ± 1.91	70.58 ± 2.52	60.35 ± 2.14	65.47 ± 1.83
	KNN	68.69 ± 0.19s	72.07 ± 0.18	63.50 ± 0.85s	67.78 ± 0.34s
	NB	**74.44 ± 1.08**s	**74.17 ± 1.52**s	**74.92 ± 0.29**s	**74.55 ± 0.90**s
	RF	69.19 ± 1.35s	72.51 ± 1.51	64.43 ± 2.72s	68.47 ± 1.50s
	SVM	71.67 ± 0.99s	73.51 ± 1.65	68.82 ± 0.65s	71.17 ± 0.84s
Lie	XGBoost	57.08 ± 0.10	56.61 ± 0.09	57.62 ± 0.12	57.12 ± 0.11
	KNN	49.96 ± 0.13s	49.96 ± 0.13s	49.96 ± 0.12s	49.96 ± 0.13s
	NB	55.18 ± 0.28s	54.48 ± 0.25s	56.12 ± 0.32s	55.30 ± 0.28s
	RF	58.19 ± 0.17s	**57.69 ± 0.17**s	58.53 ± 0.22s	58.13 ± 0.20s
	SVM	**58.45 ± 0.08**s	57.21 ± 0.08s	**60.21 ± 0.07**s	**58.71 ± 0.07**s
Parkinson's disease	XGBoost	**84.12 ± 1.98**	**88.05 ± 2.58**	80.76 ± **1.45**	**84.40 ± 2.01**
	KNN	59.62 ± 4.44s	60.51 ± 4.58s	58.84 ± 5.17s	59.67 ± 4.45s
	NB	75.62 ± 2.77s	83.04 ± 3.95	71.18 ± 3.64s	77.11 ± 2.61s
	RF	79.88 ± 2.13	85.27 ± 3.67	75.56 ± 3.13	80.41 ± 2.21
	SVM	76.75 ± 2.39s	79.79 ± 3.24s	74.28 ± 3.86	77.04 ± 2.41s
Emotional stress	XGBoost	98.12 ± 0.45	98.17 ± 0.56	98.08 ± 0.45	98.13 ± 0.45
	KNN	**99.14 ± 0.33**	**99.31 ± 0.30**	**98.98 ± 0.41**	**99.14 ± 0.32**
	NB	67.66 ± 1.99s	67.88 ± 5.91s	76.58 ± 10.45s	72.23 ± 3.36s
	RF	99.51 ± 0.26	99.51 ± 0.35	99.51 ± 0.23	99.51 ± 0.26
	SVM	97.53 ± 0.91	98.31 ± 1.49	96.88 ± 1.22	97.59 ± 0.86
Education level	XGBoost	72.58 ± 1.00	80.97 ± 0.97	61.63 ± 1.43	71.30 ± 0.85
	KNN	58.14 ± 2.56s	68.67 ± 2.08s	44.55 ± 3.48s	56.61 ± 2.70s
	NB	75.02 ± 0.32s	84.73 ± 0.45s	63.61 ± 0.28	74.17 ± 0.36s
	RF	75.83 ± 1.31s	85.04 ± 1.00s	64.54 ± 2.79	74.79 ± 1.58s
	SVM	**79.17 ± 0.17**s	**90.99 ± 0.20**s	**67.13 ± 0.19**s	**79.13 ± 0.23**s

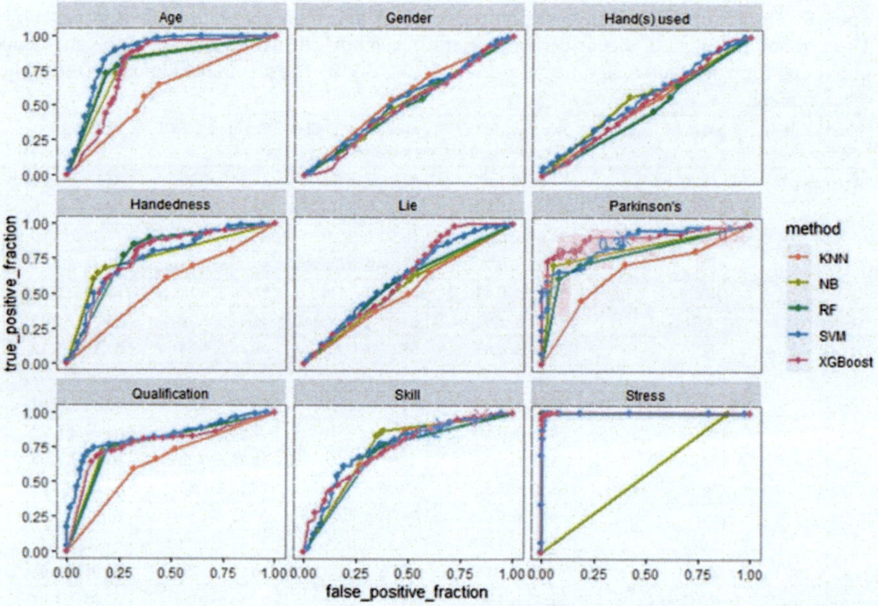

Fig. 2 Comparing ROCs for the classifiers in the proposed approach

6 Discussion

6.1 Performance Analysis and Comparison

The aggregate performance of the classifiers is shown in Fig. 3 to identify the best technique for predicting useful information using KD attributes. SVM, XGBoost, and RF have been found to be the top performers in the given context. In each case, sensitivity is low, indicating a highly trained minority class that demands more attention in other areas.

6.2 Time Complexity Analysis

Table 7 shows the time spent constructing and testing the model. Based on average training and testing times, KNN is quicker in training and XGBoost is faster in testing. However, the testing time for all classifiers is really short and reasonable.

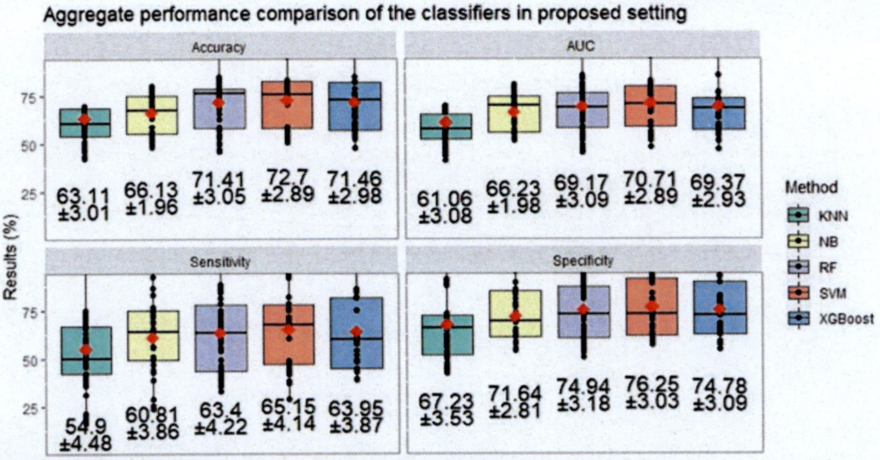

Fig. 3 Comparing the overall performance of classifiers in the proposed setting. Each box represents the classifier's performance for predicting information. Text in each box shows the average and CI

6.3 Comparison of Proposed Approaches with the Latest Literature

It is impossible to compare our approach with that of others because our approach is unique in different ways—feature arrangements, subject selection, input selection, environment, etc. In addition, the evaluation strategy of our approach is more realistic. Table 8 shows how our study differs from the literature.

In the literature, it has been identified that 91.49% of accuracy is observed in a practical evaluation setting in age group recognition. Where our approach for input independent mode is 82.12% bit lower. In the case of input restricted mode, the highest accuracy has been observed. It indicates an age group could be predicted for any input, but the model has been trained accurately for patterns developed for pre-defined inputs.

Similarly, gender recognition models have been compared. In realistic evaluation, the accuracy rate of recognising gender is low. However, our approach is that both modes are unique and interesting. Similarly, the other models have been compared.

Changes in emotions (i.e., from happy to sad, from happy to joyous, etc.) are typical, but feeling hopeless all the time may be a sign of depression. It is a significant mental illness that may have been discovered sooner. ES is a form of mental illness that has emerged in humans as a sickness that is difficult to comprehend. Many people have lost their jobs as a result of the COVID-19 epidemic; many performers have lost their public image, many contracts have been cancelled, and the financial crisis may have contributed to sadness and self-destructive behaviour. Here, our proposed scheme outperforms all previously proposed methods.

Table 7 Time complexity analysis of proposed approach for both building and testing models

	XGBoost		RF		NB		SVM		KNN	
Traits	Train (Se.)	Test (Se.)	Train (Se.)	Test (Se.)	Train (Se.)	Test (Se.)	Train (Se.)	Test (Se.)	Train (Se.)	Test (Se.)
Age group	0.0489	0.0010	0.4358	0.0020	0.0090	0.0080	0.0628	0.0010	0.0010	0.0010
Gender	0.0678	0.0010	0.6383	0.0030	0.0110	0.0080	0.1077	0.0010	0.0010	0.0010
Handedness	0.0359	0.0000	0.1509	0.0010	0.0060	0.0050	0.0090	0.0010	0.0000	0.0000
Hand(s) used	0.0718	0.0000	0.7699	0.0030	0.0100	0.0070	0.1117	0.0010	0.0020	0.0020
Typing skill	0.0469	0.0000	0.1716	0.0020	0.0050	0.0030	0.0199	0.0000	0.0000	0.0000
Qualification	0.0469	0.0000	0.4713	0.0020	0.0090	0.0070	0.0608	0.0010	0.0010	0.0010
Lie	0.1426	0.0000	1.3444	0.0050	0.0110	0.0120	0.3022	0.0020	0.0020	0.0020
Parkinson's	0.0209	0.0000	0.0140	0.0000	0.0070	0.0020	0.0020	0.0010	0.0000	0.0000
Stress	0.9357	0.0010	15.3487	0.0101	0.0459	0.1824	9.4612	0.0259	0.0519	0.0519
Average	0.1575	**0.0003**	2.1494	0.0031	0.0126	0.0260	1.1264	0.0038	**0.0065**	0.0065

Texts in bold face are fastest detectors in building models and verification

Table 8 Comparing approaches, evaluation schemes, and results with existing. The bold face indicates the features were collected through smartphones

Year	Study	#Subject	Features	Method	Evaluation	Accuracy (in%)
Age determination approaches and results						
2021	Tsimperidis et al. [46]	43	T, NT	RBFN	10-fold CV	89.20 (input independent)
2021	Oyebola and Adesina [28]	50	**T, P, R, S**	RF	10-fold CV	73.30 (input independent)
2021	Tsimperidis et al. [45]	43	T, NT	RBFN	10-fold CV	89.20 (input independent)
2022	Roy et al. [37]	92	**T**	RF	LOUOCV	91.49 (input restricted)
2022	Present	87	**R**	SVM	LOUOCV	82.12 (input independent)
2022	Present	87	**T, R**	RF	LOUOCV	93.21 (input restricted)
Gender determination approaches and results						
2018	Roy et al. [40]	92	**T**	Fusion	LOUOCV	58.26 (input restricted)
2021	Tsimperidis et al. [46]	43	T, NT	RBFN	10-fold CV	92.00 (input independent)
2021	Oyebola and Adesina [28]	50	**T, P, R, S**	RF	10-fold CV	71.30 (input independent)
2022	Roy et al. [37]	92	**T**	RF	LOUOCV	62.07 (input restricted)
2022	Present	87	**R**	KNN	LOUOCV	58.50 (input independent)
2022	Present	87	**T, R**	RF	LOUOCV	65.35 (input restricted)
Handedness determination approaches and results						
2017	Shute et al. [42]	100	T	RF	10-fold CV	94.50 (input independent)
2018	Pentel [29]	504	T	RF	10-fold CV	99.50 (input independent)
2018	Roy et al. [40]	92	**T**	Fusion	LOUOCV	60.59 (input restricted)
2019	Roy et al. [32]	Multiple datasets	T	FRNN	10-fold CV	97-99 (input restricted)
2022	Present	87	**R**	XGBoost	LOUOCV	81.78 (input independent)
2022	Present	87	**T, R**	XGBoost	LOUOCV	87.14 (input restricted)
Hand(s) used determination approaches and results						
2018	Roy et al. [40]	92	**T**	Fusion	LOUOCV	78.62 (input restricted)
2019	Roy et al. [32]	Multiple datasets	T	FRNN	10-fold CV	98-99 (input restricted)
2022	Present	87	**R**	NB	LOUOCV	55.27 (input independent)
2022	Present	87	**R**	NB	LOUOCV	77.14 (input restricted)
Typing skill determination approaches and results						
2019	Roy et al. [32]	Multiple datasets	T	FRNN	10-fold CV	88-95 (input restricted)
2022	Present	63	T	NB	LOUOCV	74.44 (input restricted)

(continued)

Table 8 (continued)

Year	Study	#Subject	Features	Method	Evaluation	Accuracy (in%)
Qualification determination approaches and results						
2020	Tsimperidis et al. [43]	43	T	RBFN	10-fold CV	84.50 (input independent)
2022	Present	87	**R**	SVM	LOUOCV	79.17 (input independent)
2022	Present	87	**T, R**	SVM	LOUOCV	91.25 (input restricted)
Lie determination approaches and results						
2018	Monaro et al. [25]	20+20	T	RF	10-fold CV	95.0 (input independent)
2021	Monaro et al. [26]	20+20	T	RF	30 subsects	90.0 (input independent)
2022	Present	60	T	SVM	LOUOCV	58.45 (input independent)
ES determination approaches and results						
2019	Ghosh et al. [11]	22	**T**	RF	10-fold CV	78.00 (input independent)
2020	Lim et al. [23]	190	T, MD	FFBP	T/T	82.88 (input independent)
2020	Sağbaş et al. [41]	46	**R**	k-nn	10-fold CV	87.56 (input independent)
2021	Dacunhasilva et al. [6]	188 samples	T	SVM	T/T	73.30 (input independent)
2022	Present	190	**R**	KNN	LOUOCV	99.14 (input independent)
PD determination approaches and results						
2018	Milne et al. [24]	42+43	T	LR	CV	AUC 85.0 (input independent)
2018	Pham [30]	42+43	T	Fuzzy	2-fold CV	AUC 98.0 to 100.0 (input independent)
2018	Iakovakis et al. [20]	18+15	**T, P**	LR	LOSO	AUC 92.0, Sen. 82.0, Spe. 81.0 (input independent)
2018	Iakovakis et al. [19]	18+15	**T, P**	Regression	LOSO	Accuracy 78 and 70 (input independent)
2019	Iakovakis et al. [18]	18+15	**T**	CNN	LOSO	AUC 89.0, Sen. 79.0, Spe. 79.0 (input independent)
2019	Hooman Oroojeni et al. [15]	42+43	T	Tensor Train	CV	AUC 88.0 (input independent)
2022	Present	42+43	T	XGBoost	LOUOCV	AUC 84.0, Spe. 88.0, Sen. 80.0 (input independent)

T->Timing, R-> Rotational, NT->No temporal, P->Pressure, S->Statistical, MD->Mouse dynamics
LOSO->Least-One-Subject-Out, LOOCV-> Least-One-Out CV,CV->Cross-Validation, T/T->Training versus Testing

Mild symptoms of PD include slow growth, quakes, postural shakiness, comprehension, muscular unbending nature, impaired discourse, loss of stability, and brokenness of muscle coordination and appendages. Because of these factors, fine motor abilities alter, which affects typing proclivity. Our proposed strategy is amazing, but it is not final. It is unusual in that it may be used for remote medical condition monitoring, which is useful in a variety of telemedicine applications. We compare our approach with conventional and usable PD measurement techniques. Here, the performance of our proposed method is slightly better than some of the few recent approaches. However, the proposed method is more time-efficient and could be run with low-configured devices.

6.4 Areas of Application in the Next-Generation Computing

Many prospective and demanded application areas involving the prediction of given information have been found—(a) Human-computer interaction, (b) Forensics, (c) Surveillance, (d) Age-specific access control, (e) Age and user-specific content or advertising, (f) Soft biometric, (g) Psychological pressure recognition, (h) Depression and craziness identification, (i) Mental wellbeing checking to utilise talk meeting, (j) Intelligent game controlling, (k) Measuring ES level of developers for trouble level and length of projects, (l) Recognising student's feeling and commitment in internet learning, (m) Mild cognitive impairment, [27], (n) Clinical incapacity in multiple sclerosis [22], (o) Continuous observing of intellectual status, (p) Quantification of horrible mind injury [16], (q) Identifying Spastic diplegia under cerebral paralysis [10], etc.

7 Conclusion

The study underlined the importance of KD attributes that are formed with every movement while using a desktop or smartphone, have the potential to make a lot of sense, and may be stored and analysed for the enhancement of identifying personal traits, medical diagnosis, and so on to provide the greatest user experience and soft biometrics. In this work, a substantial amount of meaningful information was retrieved from regular typing patterns using a novel technique for practical use. Since the KD dataset is imbalanced in multiple ways, the proposed ML strategy is suitable for some classifiers that are effective on the balanced dataset. It has also been shown that the score of recognising traits is not evenly weighted. As a result, the observed scores in soft biometric implementation must be multiplied by unique weights. Some classifiers are not equally impressive. However, the indication of utilising the proposed technique towards predicting useful information is promising for future research directions.

Traditional ML algorithms in an ensemble framework are excellent and may be used for a longer period in distance-based neural disease diagnosis, health monitoring, cognitive-behavioural treatment, and so on. KD may also be used to diagnose other illnesses that are similar to this technique. We may keep track of the typing habit for future generations of m/e-Health technology because this approach to detecting illness has several advantages. Nowadays, children have been forced to continue their studies using mobiles/computers, students have been forced to study online without understanding the students' cognitive abilities, and patients with neural disorders have been forced to continue their treatment without face-to-face doctor consultation and diagnosis. According to the findings of this study, these are the frequent difficulties for which KD and the proposed approach might be viable solutions.

Typing patterns are heavily correlated with the user's physiological (e.g., finger length, tap area, pressure), behavioural (e.g., personality, confidence), neurological (e.g., discomfort, weakness, paralysis), psychological (e.g., stress, trauma, experience), neurobiological (e.g., nerve system disorder), and neurophysiological (e.g., cognitive load) internal factors. A combination of these factors influences the typing patterns, and we cannot deny the reality. On the other hand, one factor is dependent on others, such as behavioural factors, which are dependent on psychological factors. Similarly, ES and lying both have a positive influence on modifying typing habits because of the user's high cognitive load, which reduces cognitive capability and slows typing speed. Therefore, it is difficult to identify the stress or lie only on cognitive load, and other factors need to be understood for future tuning of the proposed model.

Conflict of Interest The authors declare that they have no conflict of interest.

References

1. Abinaya R Sowmiya R (2021) Soft biometric based keystroke classification using PSO optimized neural network. Mater Today: Proc 1–4. https://doi.org/10.1016/j.matpr.2021.01.733
2. Adams WR (2017) High-accuracy detection of early Parkinson's Disease using multiple characteristics of finger movement while typing. PLoS ONE 12(11):1–20. https://doi.org/10.1371/journal.pone.0188226
3. Bernardi ML, Cimitile M, Martinelli F, Mercaldo F (2019) Keystroke analysis for user identification using deep neural networks. In: Proceedings of the international joint conference on neural networks. https://doi.org/10.1109/IJCNN.2019.8852068
4. Chen T, He T, Benesty M, Khotilovich V, Tang Y, Cho H, Chen K, Mitchell R, Cano I, Zhou T, Li M, Xie J, Lin M, Geng Y, Li Y (2020) Package "xgboost" CRAN 10.1145/2939672.2939785, 'github.com/dmlc/xgboost/issues'
5. Dacunhasilva DR, Wang Z, Gutierrez-Osuna R (2021) Towards participant-independent stress detection using instrumented peripherals. IEEE Trans Affect Comput 1–18. https://doi.org/10.1109/TAFFC.2021.3061417
6. Dacunhasilva DR, Wang Z, Gutierrez-Osuna R (2021) Towards participant-independent stress detection using instrumented peripherals. IEEE Trans Affect Comput 1–18. https://doi.org/10.1109/TAFFC.2021.3061417
7. Davarci E, Soysal B, Erguler I, Aydin SO, Dincer O, Anarim E (2017) Age group detection using smartphone motion sensors. In: 2017 25th European Signal Processing Conference (EUSIPCO), pp 2265–2269 (2017)
8. Dhir N, Edman M, Sanchez Ferro Á, Stafford T, Bannard C (2020) Identifying robust markers of Parkinson's disease in typing behaviour using a CNN-LSTM network. In: Proceedings of the 24th conference on computational natural language learning, pp 578–595. https://doi.org/10.18653/v1/2020.conll-1.47

9. Forsen G, Nelson M, Staron RJ (1977) Personal attributes authentication techniques. Technical report, Rome Air Development Center
10. Gao F, Mei X, Chen AC (2015) Delayed finger tapping and cognitive responses in preterm-born male teenagers with mild spastic diplegia. Pediatric Neurol 52(2):206–213. https://doi.org/10.1016/j.pediatrneurol.2014.04.012
11. Ghosh S, Sahu S, Ganguly N, Mitra B, De P (2019) EmoKey: an emotion-aware smartphone keyboard for mental health monitoring. In: 2019 11th international conference on communication systems and networks, COMSNETS 2019. https://doi.org/10.1109/COMSNETS.2019.8711078
12. Giancardo L, Sánchez-Ferro A, Arroyo-Gallego T, Butterworth I, Mendoza CS, Montero P, Matarazzo M, Obeso JA, Gray ML, Estépar RSJ (2016) Computer keyboard interaction as an indicator of early Parkinson's disease. Sci Rep 6:1–10. https://doi.org/10.1038/srep34468
13. Giot R, El-Abed M, Rosenberger C (2012) Web-based benchmark for keystroke dynamics biometric systems: a statistical analysis. In: Intelligent information hiding andmultimedia signal processing (IIH-MSP), pp 11–15. http://arxiv.org/abs/1207.0784
14. Giot R, Rosenberger C (2012) A new soft biometric approach for keystroke dynamics based on gender recognition. Int J Inf Technol Manag (IJITM) Special Issue Adv Trends Biometrics **11**(August): 1–16. https://doi.org/10.1504/IJITM.2012.044062
15. Hooman Oroojeni MJ, Oldfield J, Nicolaou MA (2019) Detecting early Parkinson's disease from keystroke dynamics using the tensor-train decomposition. In: European signal processing conference, vol 2019-Septe. https://doi.org/10.23919/EUSIPCO.2019.8902562
16. Hubel KA, Yund EW, Herron TJ, Woods DL (2013) Computerized measures of finger tapping: Reliability, malingering and traumatic brain injury. J Clin Exp Neuropsychol 35(7):745–758. https://doi.org/10.1080/13803395.2013.824070
17. Iakovakis D, Hadjidimitriou S, Charisis V, Bostanjopoulou S, Katsarou Z, Klingelhoefer L, Mayer S, Reichmann H, DIas SB, DIniz JA, Trivedi D, Chaudhuri RK, Hadjileontiadis LJ (2019) Early Parkinson's disease detection via touchscreen typing analysis using convolutional neural networks. In: Proceedings of the annual international conference of the IEEE engineering in medicine and biology society, EMBS, pp 3535–3538. https://doi.org/10.1109/EMBC.2019.8857211
18. Iakovakis D, Hadjidimitriou S, Charisis V, Bostanjopoulou S, Katsarou Z, Klingelhoefer L, Mayer S, Reichmann H, DIas SB, DIniz JA, Trivedi D, Chaudhuri RK, Hadjileontiadis LJ (2019) Early Parkinson's disease detection via touchscreen typing analysis using convolutional neural networks. In: Proceedings of the annual international conference of the IEEE engineering in medicine and biology society, EMBS. https://doi.org/10.1109/EMBC.2019.8857211
19. Iakovakis D, Hadjidimitriou S, Charisis V, Bostantjopoulou S, Katsarou Z, Klingelhoefer L, Reichmann H, Dias SB, Diniz JA, Trivedi D, Chaudhuri KR, Hadjileontiadis LJ (2018) Motor impairment estimates via touchscreen typing dynamics toward Parkinson's Disease detection from data harvested in-the-wild. Frontiers ICT 5. https://doi.org/10.3389/fict.2018.00028
20. Iakovakis D, Hadjidimitriou S, Charisis V, Bostantzopoulou S, Katsarou Z, Hadjileontiadis LJ (2018) Touchscreen typing-pattern analysis for detecting fine motor skills decline in early-stage Parkinson's disease. Sci Rep 8(1):1–13. https://doi.org/10.1038/s41598-018-25999-0
21. Killourhy KS (2012) A Scientific Understanding of Keystroke Dynamics. PhD thesis
22. Lam K, Meijer K, Loonstra F, Coerver E, Twose J, Redeman E, Moraal B, Barkhof F, de Groot V, Uitdehaag B, Killestein J (2020) Real-world keystroke dynamics are a potentially valid biomarker for clinical disability in multiple sclerosis. Mult Scler J 135245852096879. https://doi.org/10.1177/1352458520968797
23. Lim YM, Ayesh A, Stacey M (2020) Continuous stress monitoring under varied demands using unobtrusive devices. Int J Hum-Comput Interact 36(4). https://doi.org/10.1080/10447318.2019.1642617
24. Milne A, Farrahi K, Nicolaou MA (2018) Less is more: univariate modelling to detect early Parkinson's Disease from keystroke dynamics. In: Lecture notes in computer science (including subseries Lecture Notes in Artificial Intelligence and Lecture Notes in Bioinformatics), vol 11198 LNAI. https://doi.org/10.1007/978-3-030-01771-2_28

25. Monaro M, Galante C, Spolaor R, Li QQ, Gamberini L, Conti M, Sartori G (2018) Covert lie detection using keyboard dynamics. Sci Rep. https://doi.org/10.1038/s41598-018-20462-6
26. Monaro M, Zampieri I, Sartori G, Pietrini P, Orrù G (2021) The detection of faked identity using unexpected questions and choice reaction times. Psychol Res 85(6):2474–2482. https://doi.org/10.1007/s00426-020-01410-4
27. Ntracha A, Iakovakis D, Hadjidimitriou S, Charisis VS, Tsolaki M, Hadjileontiadis LJ (2020) Detection of mild cognitive impairment through natural language and touchscreen typing processing. Front Digit Health 2(October):1–13. https://doi.org/10.3389/fdgth.2020.567158
28. Oyebola O, Adesina AO (2021) Predicting age group and gender of smartphone users using keystroke biometrics. Malays J Sci Adv Technol 1(4):124–128
29. Pentel A (2017) High precision handedness detection based on short input keystroke dynamics. In: 8th international conference on information, intelligence, systems and applications, IISA 2017. https://doi.org/10.1109/IISA.2017.8316380
30. Pham TD (2018) Pattern analysis of computer keystroke time series in healthy control and early-stage Parkinson's disease subjects using fuzzy recurrence and scalable recurrence network features. J Neurosci Methods 307. https://doi.org/10.1016/j.jneumeth.2018.05.019
31. Rocha R, Carneiro D, Costa R, Analide C (2020) Continuous authentication in mobile devices using behavioral biometrics. Adv Intell Syst Comput. https://doi.org/10.1007/978-3-030-24097-4_23
32. Roy S, Roy U, Sinha D (2019) Analysis of typing pattern in identifying soft biometric information and its impact in user recognition. In: Advances in intelligent systems and computing, vol 699. Springer, Singapore. https://doi.org/10.1007/978-981-10-7590-2_5
33. Roy S, Roy U, Sinha DD (2020) Deep learning approach in predicting personal traits based on the way user type on touchscreen. In: Advances in intelligent systems and computing, vol 999. https://doi.org/10.1007/978-981-13-9042-5_27
34. Roy S, Roy U, Sinha D (2018) Identifying soft biometric traits through typing pattern on touchscreen phone. In: Social transformation—digital way, pp 546–561. Springer, Singapore. https://doi.org/10.1007/978-981-13-1343-1_46
35. Roy S, Roy U, Sinha D (2018) Protection of kids from internet threats : a machine learning approach for classification of age group based on typing pattern. In: Proceedings of the international multiconference of engineers and computer scientists, vol I
36. Roy S, Roy U, Sinha D (2019) Analysis of typing pattern in identifying soft biometric information and its impact in user recognition. In: Information technology and applied mathematics, advances in intelligent systems and computing, pp. 69–83. Springer, Singapore. https://doi.org/10.1007/978-981-10-7590-2_5
37. Roy S, Roy U, Sinha D (2022) Identifying age group and gender based on activities on touchscreen. Int J Biom 14(1):61. https://doi.org/10.1504/ijbm.2022.10042835
38. Roy S, Roy U, Sinha D (2017) User authentication: keystroke dynamics with soft biometric features. In: Group, CPT & F (ed) Internet of Things (IOT) technologies, applications, challenges and solutions, chap. 6, pp 105–124. CRC Press, Boca Raton, FL 33487–2742, 1 edn.
39. Roy S, Roy U, Sinha D (2018) Protection of kids from internet threats: a machine learning approach for classification of age-group based on typing pattern. In: Lecture notes in engineering and computer science
40. Roy S, Roy U, Sinha D (2018) The probability of predicting personality traits by the way user types on touch screen. In: Innovations in systems and software engineering, pp 1–8. https://doi.org/10.1007/s11334-018-0317-6
41. Sağbaş EA, Korukoglu S, Balli S (2020) Stress detection via keyboard typing behaviors by using smartphone sensors and machine learning techniques. J Med Syst 44(4). https://doi.org/10.1007/s10916-020-1530-z
42. Shute S, Ko RK, Chaisiri S (2017) Attribution using keyboard row based behavioural biometrics for handedness recognition. In: Proceedings—16th IEEE international conference on trust, security and privacy in computing and communications, 11th IEEE international conference on big data science and engineering and 14th IEEE international conference on embedded software and systems. https://doi.org/10.1109/Trustcom/BigDataSE/ICESS.2017.363

43. Tsimperidis I, Arampatzis A (2020) The keyboard knows about you. Int J Technoethics 11(2). https://doi.org/10.4018/ijt.2020070103
44. Tsimperidis I, Rostami S, Katos V (2017) Age detection through keystroke dynamics from user authentication failures. Int J Digit Crime Forensics. https://doi.org/10.4018/IJDCF.2017010101
45. Tsimperidis I, Rostami S, Wilson K, Katos V (2021) User attribution through keystroke dynamics-based author age estimation. In: Lecture notes in networks and systems, pp 47–61. Springer, Cham. https://doi.org/10.1007/978-3-030-64758-2_4, https://link.springer.com/chapter/10.1007/978-3-030-64758-2_4
46. Tsimperidis I, Yucel C, Katos V (2021) Age and gender as cyber attribution features in keystroke dynamic-based user classification processes. Electronics (Switzerland) 10(7):1–14. https://doi.org/10.3390/electronics10070835
47. Udandarao V, Agrawal M, Kumar R, Shah RR (2020) On the inference of soft biometrics from typing patterns collected in a multi-device environment. In: Proceedings—2020 IEEE 6th international conference on multimedia big data, BigMM 2020, pp 76–85. Institute of Electrical and Electronics Engineers Inc. https://doi.org/10.1109/BigMM50055.2020.00021
48. Yaacob MN, Syed Idrus SZ, Wan Mustafa WA, Jamlos MA, Abd Wahab MH (2021) Identification of the exclusivity of individual's typing style using soft biometric elements. Ann Emerg Technol Comput 5(Special issue 5):10–26. https://doi.org/10.33166/aetic.2021.05.002

Path Dependencies in Bilateral Relationship-Based Access Control

Amarnath Gupta and Aditya Bagchi

Abstract The Relationship-based Access Control Model (ReBAC) generalizes Role-based Access Control (RBAC) by considering both hierarchical and non-hierarchical relationships between users to specify access control of a set of target resources (objects). This paper extends the ReBAC model by considering relationships between objects as well as between subjects and objects. This generalized model is expressed through the language of dependencies borrowed from data management. We develop a language for bilateral path dependencies which states that a chain of binary relationships over subjects and objects logically implies another chain of binary relationships. We show that this formalism is adequate to capture access control rules with no conflicts. In future work, this formalism will be extended to include conflict detection and resolution.

Keywords Access control · ReBAC · Bilateral relationship

1 Introduction

Access control policies primarily specify how a user (usually referred as subject) can access a resource (usually referred as object) with a set of access rights (read, write, execute, etc.). An atomic access control policy is typically expressed as a 4-tuple structure (s, o, a, v) using a set of $Subject(S)$, $Object(O)$, $AccessRights(A)$, and $Sign(V)$ where $s \in S, o \in O, a \in A$, and $v \in V$. The sign in an access control policy can either be +ve or –ve, where a +ve sign indicates permission for the concerned subject to access the concerned object using the access right specified and a –ve sign indicates denial of such access. Logically, a policy set for a user/subject is expressed by many such atomic policies for different resources/objects and also by logical

A. Gupta (✉)
University of California San Diego, La Jolla CA 92093, USA
e-mail: a1gupta@ucsd.edu

A. Bagchi
Indian Statistical Institute, Kolkata 700108, India

R. Chaki et al. (eds.), *Applied Computing for Software and Smart Systems*,
Lecture Notes in Networks and Systems 555,
https://doi.org/10.1007/978-981-19-6791-7_2

combinations of such atomic policies using Boolean operators [9]. Depending upon the context and policy, *Subject* can be users, group of users, or roles, and *Object* can be any resource within an enterprise that can be accessed by a Subject. Moreover, at times, an Object can also be a Subject for an access policy. For example, when a user u tries to execute a program p, u is the Subject and p is the Object, but when the same program p accesses a file f during its execution, p is the Subject and f is the Object.

In practice, not all authorization rules are explicitly specified. Instead, inference rules are used to derive authorizations using the rules of the access control model used. Specifically, user-group hierarchy ensures that members of a subgroup inherit authorizations from its super-group. Similarly, role hierarchy permits a higher order role to inherit authorizations from the roles below. More recently, authorization models like RPPM [12] and ReBAC [11] have extended authorization inference beyond hierarchies and inheritance and created graph-based authorization schemes.

1.1 Motivation for the Present Work

Recently, multi-graph models are increasingly being used in studying both enterprise security as well as vulnerability analysis [6] because multi-graph models represent multiple types of non-hierarchical relationships that security rules can exploit. However, we find that many models, including ReBAC and RPPM, focus primarily on subject-subject relationships, less on subject-object relationships, and rarely on object-object relationships. However, in real life, all three types of relationships should factor into a security specification. For example, if a user u has read access to a DBMS D, and D resides on server S, u must have access to S to exercise the read access to D. However, *there is no access control formalism today that expresses such non-hierarchical implications*. In this paper, we study such a formalism based on the language of dependencies borrowed from database theory [1].

We try to establish that an extension of ReBAC model is necessary where object/resource side and user/subject-side relationships and hierarchies need to be incorporated. We call this model the Bilateral Relationship-Based Access Control model (BiReBAC). We have introduced a new formalism based on dependency constraints that generalize the standard ReBAC model to give rise to our Bilateral Relationship-Based Access Control model (BiReBAC). However, this being the first proposal for the composite model, we have adopted an assumption to ensure completeness and soundness of our authorization specification. Our proposed model is based on $Closed\ Policy$ and any access is permitted only against explicit positive authorizations or any other authorizations inferred from them. So against any query with any subject-object combination only a positive authorization can be inferred. So for any access request if no positive authorization can be inferred, the access request will be denied. So in a $Closed\ Policy$ no explicit negative authorization is specified. This assumption avoids possible conflicts between positive and negative authorizations and the well-known decidability problem as specified in the HRU model [8]. We will address this issue again in the last section to indicate about our future work.

We now present a running example, to explain our proposed model.

A Running Example Let's consider a modern day *executable electronic textbook* on Data Science that is available over the web. The book has a number of chapters, broken down into a hierarchy of sections and subsections. Some of the textual content are sample problems whose worked-out solutions are provided as executable Jupyter Notebooks that the readers can run. We call these Problem-text-to-Solution links as *forward links* (flinks). Each section also has a set of exercise problems and a reading list. The exercise problems are written in text; however, they are also connected to data sets and solutions (Jupyter Notebooks) provided by the authors. A Notebook may link back to the paragraphs of the book relevant to the exercise problem being solved. We call these Notebook-cell-to-text links as *reverse links* (rlinks). Let us also consider that the authors of the book have created a set of distinct tracks (e.g., "Beginners", "Advanced") which are pathways through the book for different audiences. A "track" is a tree-like structure through the chapters, sections, and subsections of the book.

Clearly, our executable textbook is a *distributed object*, i.e., different parts of the book (e.g., different sections) are on different physical web servers. The book has a set of "principal authors", but sample problems may be written by the graduate students of a principal author. Solutions to the exercise problems may be created by undergraduate students who work in a principal author's lab or have taken the Data Science class offered by one of the principal authors. The universe of these books can be represented as a graph shown in Fig. 1. We can write several access control policies based on this example. We can say that "every author has read, write, and execute access on the Jupyter Notebooks he writes". We can also make relationship-based access policy statements like "a graduate student S who works under the supervision of a professor P and researches on subject X has read access to any textbook written by P on X". This paper introduces a formalism different from [5, 12] to cover access control policies where the objects and their fragments (e.g., a subsection of a book) which are objects themselves are connected through a relationship network.

2 The Language of Dependencies

We formalize graph-based access rules in terms of dependency statements. The intuition behind dependency statements is to identify the critical factors on which the access of a user to a resource depends. Suppose we want to make the statement that user u has access to some portion p of a book b, if and only if he has access to the web server (WS) w that hosts p. We can write this as

$$\forall u : user, p, b : book, m\ hasAccess(u, p, m) \Rightarrow \exists w : WS \mid partOf^{+}(p, b), hosts(w, p), hasAccess(u, w, m) \quad (1)$$

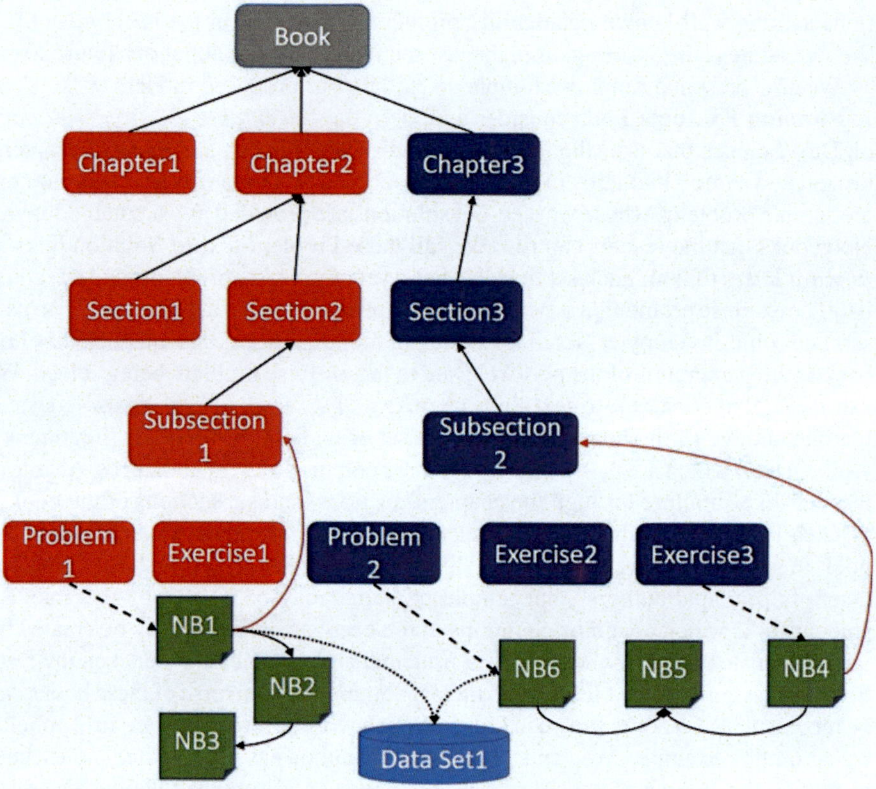

Fig. 1 A Graph-centric View of our object domain. Solid straight black arrows = $partOf$, solid curved black arrows = $connectedTo$, dashed arrows = $flink$, brown arrows = $rlink$. Red nodes = Track 1, dark blue nodes = Track 2

where the "," symbol represents conjunction (AND), : $book$, : WS are unary typing predicates, m, the free variable designates the access right (e.g., read, write, execute), and $partOf^{+}$ is the transitive closure of the part of relationship. Thus, the access of u to p depends on the access of u to w. Dependency statements like Eq. 1 have practical consequences. If there is an access control policy that an undergraduate student has no access to a specific departmental web server, then no Jupyter Notebook for exercise problems can be placed on that server.

Equation 1 represents an access control rule with single-variable dependency (i.e., the access depends on the existence of just a web server with proper access right) which we call *node dependency*.

We can extend the notion of node dependency if we want to state that in addition to Eq. 1, a user u can execute a Jupyter Notebook (NB) on the web server w, if Python is installed on w and u has execute privileges for that Python installation. Equation 2 states this condition.

$$\forall u : user, w : WS, j : NB\ hasAccess(u, j, \texttt{execute}) \Rightarrow \exists w, p \mid installed(p, \texttt{Python}, w), runsOn(j, p), hasAccess(u, p, \texttt{execute}) \quad (2)$$

where predicate $installed(p, \texttt{Python}, w)$ states that p is an instance of Python that is installed on w. Equation 2 differs from Eq. 1 because the two existentially quantified variables w and p on which the u's execute access to j depends are constrained to satisfy a chain of relationships. We tend to call this form of multi-variable dependency as *chain dependency*. In general, a chain dependency relationship has the form

$$\forall vars1[: typeSpec], hasAccess(user, resourceVar1, mode) \Rightarrow \exists vars2 \mid chainExpression, hasAccess(user, resourceVar2, mode) \ldots \quad (3)$$

where $vars1$ and $vars2$ are sets of variables, $resourceVar1$ refers to the original resource whose access is being determined, and $resourceVar2$ refers to the resource on which the original access depends. The variable $resourceVar1$ belongs to the universally quantified variable set $vars1$ and $resourceVar2$ belongs to the existentially quantified variable set $vars2$. The chain expression is a conjunctive predicate over a set of participating relationships and their transitive closures and must include all existentially quantified variables. The semantic types of the variables are optionally specified as $typeSpec$ of unary atomic type specifiers for variables in $vars1 \cup vars2$. Although Eq. 3 has one $hasAccess$ predicate, a general chain dependency expression may have multiple intermediary access requirements (indicated by the dots) for the LHS of the implication to be satisfied.

2.1 Inferences from Node and Chain Dependencies

The implication in the dependency-based formulation leads to two kinds of inferences. The first category derives from *tuple-generating dependencies* (TGD) used in data management literature [2, 3], and the second relates to *implicit access-mode assignment* (IAM), a generalization from prior work by Dasgupta et al for ontological data access for digital libraries [7].

TGD: Node dependency is a form of TGD. In our example, if we have a ground fact like $hasAccess('Joe', 'Section : 3.2', 'read')$ Eq. 1 also asserts the existence of a tuple $hasAccess('Joe', w1, 'read')$ in the *hasAccess* table for some web server $w1$. If not, the system is inconsistent. For chain dependency, a ground fact like $hasAccess('Joe', 'Exercise : 3.2.14', \texttt{execute})$ implies a set of tuples

$installed(p1, \texttt{Python}, w1)$,
$runsOn('Exercise : 3.2.14', p1)$, and
$hasAccess('Joe', p1, \texttt{execute})$

where the predicate names map to table names in an access control database. However, by virtue of their generation rule, these three tuples are not independent of each

other and must be considered to be a group. We call this generalized form of TGD a *tuple-group generating dependency* (TGGD). Note that the tuples inferred from a TGGD belong to *multiple relations*.
IAM: The IAM problem can be illustrated by slightly modifying Eq. 1 as follows.

$$\forall u : user, p, b : book, m, m' \ user(u), hasAccess(u, p, m) \Rightarrow \exists w : WS, f : file, m' \mid$$
$$partOf^{+}(p, b), hosts(w, p), hasAccess(u, f, m),$$
$$contains(w, f), locatedIn(p, f), hasAccess(u, w, m') \quad (4)$$

In this case, we have added an existentially quantified variable f which represents a file such that p, the part of the book, is located in file f, which is contained in web server w. Now, if u has access to the file f in modality m, u must also logically have access to the web server w in some mode m'. However, the nature of m' is not specified, although it is clear that m' depends on the value of m. There are two issues to be resolved:

1. Why is m' different from m, the original access mode assigned to user u?
2. How is m' implicitly assigned for accessing web server w?

A similar situation was encountered in [7] where the bibliographic metadata of a digital library was represented by an ontology. A $read$ access by a user u to a document d containing in the concept c would implicitly provide an access to the concept c itself. However, it does not infer that all documents in c can be read by u. The authors have solved this problem by defining an access mode called $browse$ where the user can access concept c to reach the document d but cannot execute any operation related to other access modes like (*read, write, execute*). It makes authorization inference mechanism at the resource/object side different from the same at user/subject side.

To remove the ambiguity for m', we create a set of template IAM rules that apply to the right side of the dependency equation. To formulate the template IAM rule, we interpret the predicates of the dependency equation as a graph (Fig. 2), where a binary predicate like $contains$ is interpreted as a directed edge from the first argument to the second, and a ternary predicate like $hasAccess$ is interpreted as a directed edge from the first argument to the second, where the third argument is a property of the edge. A unary predicate like *book* is interpreted as the node type of its argument variable. We say that the LHS $hasAccess(m')$ edge is dependent on the RHS $hasAccess(m)$ edge, contingent upon the *constraining subgraph* that connects the resources w and f. From Fig. 2, the $partOf+$ edge does not participate in the constraint because removing it does not change the nature of the dependency. Now we can formulate our template rule as follows. Given the constraining subgraph of the form $hosts(A, B), locatedIn(B, C), contains(A, C)$, if m is the mode of access for the LHS $hasAccess(u, A, m)$ edge, then the m', the mode of access for all RHS $hasAccess(u, C, m')$ edges will be given by `intention-to-`m which is m'. Thus if m is `read`, m' is `intention-to-read`, an idea borrowed from multiple granularity locking protocol in DBMS systems [10].

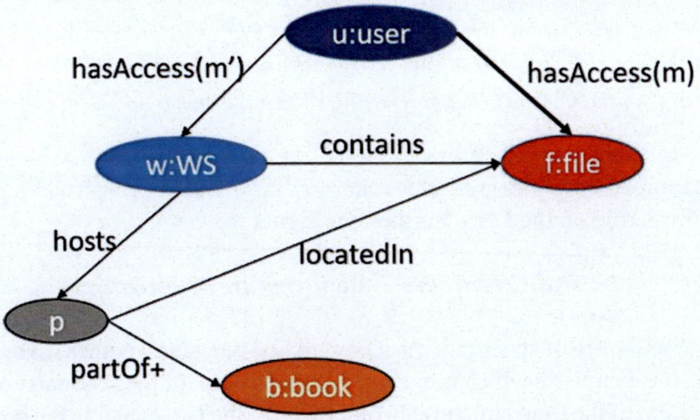

Fig. 2 A graph representation of dependency Eq. 4. The *hasAccess* edge from the LHS is made thicker

According to the locking protocol in a DBMS, if a table in a relational system is updated with a *write* lock, the entire database will have an `intention-to-write` lock indicating that down below in a more granular object actual write operation is in progress. The same concept can be used to infer access authorizations for objects which contain the actual object for which explicit authorization has been specified. Once again, access right of type `intention-to-`m assigned to work station w in Eq. 4 allows the concerned user to reach and access p and f by accessing work station w but no other operation is allowed on w.

To generalize this formulation, we recognize that like $partOf$, $hosts$, $contains$, and $locatedIn$ are transitive relationships. Now, we can write a more general form of the IAM dependency pattern, which applies to multiple dependency rules that have the same constraining subgraph.

IAM Template 1: Given an independent access relation $hasAccess(u, C, m')$, a dependent access relation $hasAccess(u, A, m)$ where m is a valid access mode, A, C are system resources, and a *constraining subgraph* pattern $hosts^{+}(A, B)$, $locatedIn^{+}(B, C)$, $contains^{+}(A, C)$ the implicit access mode m' is assigned as `intention-to-`m.

Using Chain Dependencies. There can be different use cases for chain dependency in our example application.

Example 1 sloppyA reader of the book has read permission to the pages of a section/subsection s if he has read all the prerequisite portions for s and completed their exercise problems (i.e., run the Jupyter Notebooks associated with the exercise problems).

$$\forall u : user, b : book, s, (typeOf(s) \textbf{ in } ('section','subsection')), hasAccess(u, s, \texttt{read})$$
$$\Rightarrow \exists s', e : exercise \mid (typeOf(s') \textbf{ in } ('section','subsection')), partOf^{+}(s, b),$$
$$partOf(e, s'), prereq^{+}(s', s), \{hasAccessed(u, s', \texttt{read}) \textbf{ then } hasAccessed(u, e, \texttt{execute})\} \quad (5)$$

In this example, $hasAccessed$ is a *state predicate* which serves as a precondition for the access rule on the LHS. Further, the signature

$$\{< statePredicate > \textbf{ then } < statePredicate >\}$$

indicates a sequence of states that must be satisfied one after another. To evaluate this signature, the exercise e which is executed has to be part of the section/subsection s', the antecedent of the **then** structure. In other words, the (s', e) pair in the precondition is constrained so that $partOf(e, s')$ holds.

Example 2 A user enrolled in a track T *does not* have read access to any subsection S of a book or write/execute access to the exercises if S is not in T.

$$\forall u : user, b : book, s : subsection, partOf^{+}(s, b), hasAccess(u, s, \texttt{read}) \Rightarrow$$
$$\neg\exists t : track \mid enrolledIn(u, t), inTrack(s, t) \quad (6)$$

$$\forall u : user, e : exercise, b : book, hasAccess(u, e, \{write, execute\}) \Rightarrow$$
$$\neg\exists t : track, s : subsection \mid partOf^{+}(s, b), contains(s, e),$$
$$enrolledIn(u, t), inTrack(s, t) \quad (7)$$

where the predicate $contains$ is the inverse of predicate $partOf$. In other words, $partOf(a, b)$ means a is $partOf$ b, whereas $contains(a, b)$ means b contains in a. In this pair of conditions, the chain dependency is purely *structural* because the access does not depend on any implicit or past access conditions.

Example 3 A student can create a new Jupyter Notebook page on a web server only if he falls within the reporting hierarchy of any of the authors and has the explicit permission from her supervisor to upload data to the web server.

$$\forall s : student, j : NB, w : WS, b : book, canCreate(s, j, w) \Rightarrow$$
$$\exists p : professor, d : dataFile, l : permissionToken, x \mid$$
$$author(p, b), supervises^{+}(p, s), supervises(x, s),$$
$$hasPermission(s, l, canUpload(s, d, w)) \quad (8)$$

We use $canCreate(s, j, w)$ as a specialization of the more standard form of $hasAccess(user, resource, accessMode)$ used so far. We can rewrite the $canCreate(s, j, w)$ predicate as $hasAccess(s, w, \texttt{create}(j))$ where the access mode `create` is parameterized by the object to be created. The permission token l is similar to an API key used for accessing web and mobile services. To apply the token, we use $hasPermission(user, token, accessPredicate)$ as a *second-order predicate* that enables an access pattern via an explicit permission condition. The permission condition can be viewed as a type of eventive precondition that must be satisfied for the LHS $hasAccess$ to take effect.

Incidentally, this chain dependency also takes care of provisions (pre-conditions) and obligations (post-conditions) involved in specifying access control rules [4].

2.2 From Chain Dependencies to Bilateral Path Dependencies

In the previous subsection, we used the term chain dependency to refer to a series of connected predicates on the RHS of the dependency rule. We can think of the aforementioned chains as "RH paths" that exist only on the right-hand side of the implication symbol. A more general form of dependency can be defined by having a path expression on both sides of the implication. We call this form a *bilateral path dependency* rule.

We initiate our approach with a simple implication rule that does not have a dependency formulation but has a LH path expression. In our example situation, such a rudimentary (but unrealistic) relationship-based access control statement can be "if g, a graduate student, has write access to a book chapter authored by p who is g's faculty advisor, then so does g's undergraduate advisee u".

$$\begin{aligned}\forall u : ugStudent, g : gradStudent, p : professor, b : book, c : bookChapter,\\ authorOf(p, b), partOf(c, b), advisor(p, g), supervises(g, u),\\ hasAccess(g, c, \texttt{write}) \Rightarrow hasAccess(u, c, \texttt{write})\end{aligned} \tag{9}$$

We can make this statement more realistic by adding the additional condition that u has this write access only if u writes an exercise for some portion of the chapter. Now Eq. 9 becomes

$$\begin{aligned}\forall u : ugStudent, g : gradStudent, p : professor, b : book, c : bookChapter,\\ authorOf(p, b), partOf(c, b), advisor(p, g), supervises(g, u),\\ hasAccess(g, c, \texttt{write}) \Rightarrow \exists e : exercise \mid hasAccess(u, e, \texttt{write}),\\ partOf(e, c), hasAccess(u, c, \texttt{write})\end{aligned} \tag{10}$$

With this extension, we have path expressions on both sides of the implication in a dependency rule, making this constraint an example of a bilateral path dependency.

The rule is shown as a graph in Fig. 3, where existence of the subgraph with heavy edges depends on the existence of the subgraph with light edges. We make the following observations about Eq. 10:

(a) One prerequisite for the ugStudent to get write access to the exercise is that he or she is an advisee of the author of the book. This encodes an ReBAC-style condition within the fold of the bilateral path dependencies. Thus, the language of bilateral path dependencies has the expressivity to capture both the subject-side ReBAC criteria and its object-side extensions.

(b) The ugStudent primarily has write access to the exercise and consequently also has write access to the book chapter. However, this is not explicitly captured in the equation. One way to capture this "derived" mode of write access is to change the access mode to the book chapter to `intention-to-write` as we did in the IAM case. The second, more direct way to represent this is to add a secondary implication in the RHS. In this case, the RHS is

$$\exists e : exercise \mid hasAccess(u, e, \texttt{write}), partOf(e, c), \longrightarrow hasAccess(u, c, \texttt{write})$$

We prefer to use this secondary implication notation to make the dependency expression more precise.

(c) In this case, the undergraduate student gets write access to the exercise of the chapter that the graduate student has write access to primarily because the graduate student supervises the undergraduate student. We can write a companion, non-dependency-generating rule that the graduate student who has access to a resource r has the capability to grant access of r or a part thereof, to the undergraduate student that he(she) supervises, is

$$\begin{aligned} &\forall u : ugStudent, g : gradStudent\ \exists_0 r : resource \\ &\quad \mid hasAccess(g, r, m), grantsAccess(g, u, r', m'), partOf^*(r', r) \end{aligned} \tag{11}$$

We use the symbol $\exists_0$ to denote that while g has the capability of granting access, in reality there may be no resources for which the capability is exercised. Further, in reality, the modality term $grantsAccess$(grantor, grantee, resource, modality) predicate can be more nuanced. For example, if g has read access, he cannot grant a write or execute access to u. This consideration directly shows that the well-known Discretionary-Based access control (available even in SQL) can also be mapped to our model.

Coupled and Independent Bilaterality. In Eq. 10, the variables used on the LHS of the implication are $V_L = (u, g, p, b, c)$ and those on the RHS are $V_R = (u, c, e)$. Thus, $V_L \cap V_R = (u, c)$, i.e., $V_L \cap V_R \neq \emptyset$. In this case, we call the dependency rule as *coupled bilateral path dependency* (or *coupled bilaterality*); on the other hand, a bilateral path dependency rule where $V_L \cap V_R = \emptyset$ is called *independent bilateral path dependency*. We make the following assertion.

Assertion 1 *An access control rule with an independent bilateral path dependency is unsatisfiable.*

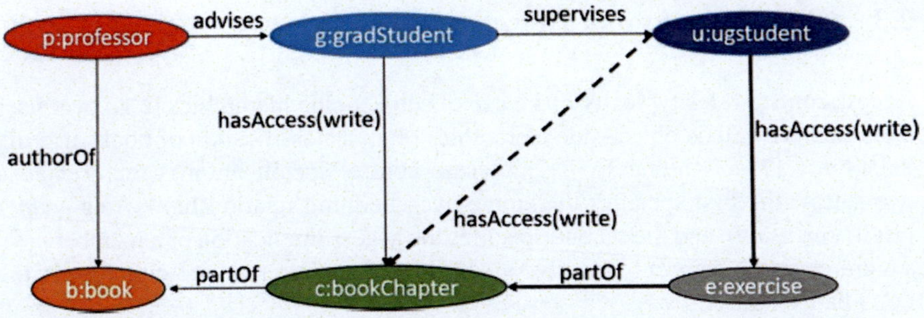

Fig. 3 A graph representation of dependency equation Eq. 10. The edge from the RHS is made thicker. The dashed edge reflects the secondary implication

Justification: Instead of a formal proof, we prefer to present a logical justification in support of the assertion made.

1. As shown in Eq. 10 and in other equations earlier, the complete set of chains on both sides of an equation is formed by considering the transitive closure of all the relationships connecting both the subject side and the object side. Thus $advises/supervises^+$ connects ugstudent to gradstudent to professor and $partOf^+$ connects exercise to book chapter to book as shown in Fig. 3. Thus chain of inferred authorizations connects different users to the initial user-group and the authors of the book and offers access to different parts of the initial object, the book.
2. If $V_L \cap V_R = \emptyset$ then it implies that even after considering the transitive closure of all relationships on both subject side and object side, some subjects/users cannot reach to initial set of subjects and/or objects where explicit authorizations were specified. Thus, such subjects/users will not have proper authorizations inferred to access the required objects. From a graphical point of view, no path will be available for any such access.

So, the assertion stands and only *coupled bilateral path dependency* are allowed.

In light of Assertion 1, we only focus on rules with coupled bilateral path dependency. Recall from Eq. 10 that nodes u, c are in the intersection of the LHS and RHS—we call them the *anchor nodes* of the dependency graph (Fig. 3). Justification given for Assertion 1 also asserts that coupled bilateral path dependency graphs (CBPD graphs) are always connected. Further, the anchor nodes of CBPD graphs can have outgoing edges only to other anchor nodes or to variables on the RH side of the implication.

2.3 CBPD Graphs and Hierarchies

Access control rules typically make use of subject-side hierarchies (e.g., over users and roles) as well as object-side hierarchies (e.g., a classification of books in a digital library [7]). A hierarchy-based access control specification is represented as two rules—the first specifies the hierarchy-generating relationship r_h (e.g., *supervises(user, user)*), and the second specifies the access implication of a member m_1 of a hierarchy with respect to member m_2 if they are in the same tree-path (sometimes DAG path) induced by r_h.

To express the interaction between a CBPD graph and hierarchies, we extend our example in Eq. 10. We discuss a scenario where professors have senior postdoctoral students, who supervise junior postdoctoral students, postdoctoral students (junior and senior) supervise senior graduate students, who in turn supervise the work of junior graduate students. Here, we treat the $supervises$ relationship as transitive and notice that it creates a DAG because a senior graduate student may be supervised by a junior or a senior postdoc. Now, we would like to still use the rule in Eq.10 with the following differences:

(a) A subgraph S depicting the $supervises$ hierarchy must be created.
(b) Following the same notation as Eqs. 7 and 8, the $advises$ edge is replaced with a $supervises+$ edge in Fig. 3.
(c) The variables (nodes) p and g should now be connected to the corresponding nodes of S by an explicit $instanceOf$ edge. This presents a slight expressivity problem if the node marked g can be any direct or indirect supervisee of p (i.e., a senior/junior graduate student or postdoc). We accomplish this by

  - creating a subgraph S' of S where S' does not include professors;
  - creating a *type-predicate* where instead of making the type assertion $g : gradStudent$, we write $g : memberOf(nodes(S'))$.

  The modified part of the graph is shown in Fig. 4

Thus, the inclusion of hierarchies changes the nature of the CBPD graphs because it introduces the need to create *named subgraphs* which can be referenced by type resolution logic at a node. To test whether a specific undergraduate student has write access to the exercises, the eligibility of her supervisor must resolve to a node type within subgraph S'.

2.4 Toward a BiReBAC Graph Model

From the example of book writing project and gradual development of a new relationship-based access control model through Eqs. 1–10, we possibly could establish that an extension of ReBAC model is necessary where object/resource-side and user/subject-side relationships and hierarchies need to be incorporated. We call this

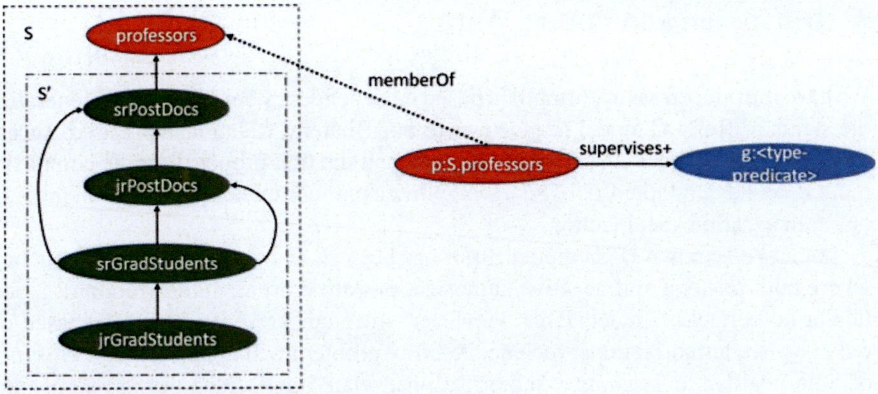

Fig. 4 The hierarchy and the affected nodes from Fig. 3 are shown. Note that the $memberOf$ edge is equivalent to the type assertion $p : S.professors$ made in the node

model the Bilateral Relationship-Based Access Control model (BiReBAC). While developing a full translation of the dependency-based formalism is beyond the scope of this paper, we outline below a rough sketch of how a property graph corresponding to Figs. 3 and 4 may be constructed from our rules, based on Eq. 10.

The general scheme of a BiReBac graph construction follows:

1. Both objects and subjects are represented as typed nodes such that node types are entity classes defined by the problem space. If these classes are enumerated over a hierarchy as in Fig. 4, nodes in the hierarchy are represented as node types in a BiReBAC graph. As defined in the explanation of the figure, a node may have multiple types specified through a *type-predicate*.
2. There are three types of edges—$hasAccess$ edges that specify access rights; relationship edges between subject pairs, object pairs, and subject-object pairs; and $subclass/membership$ edges. Traditional authorization relationships such as inheritance through user-group/subgroup or super-role/sub-role are expressed directly as edges or as predicates over these edges.
3. Node properties of BiRBAC graphs need to specify if a node is on the antecedent side, the consequent side, or is an anchor node. Similarly, a node property also captures if a node is existentially qualified on the consequent side (otherwise, all nodes are universally quantified).
4. When multiple dependency rules apply, our strategy is to create multiple graphs, and using the principles of the web ontology language (OWL), declare comparable nodes as equivalent by creating an `equivalent` edge between them. For example, the node type `Book` is used in graphs of Figs. 2 and 3—these nodes will be connected through the equivalence edge.

   In future work, we will present a formal construction for BiReBAC graphs and prove that the construction will produce unambiguous graphs.

3 Discussion and Future Work

We have introduced a new formalism based on dependency constraints that generalize the standard ReBAC model to give rise to our Bilateral Relationship-Based Access Control model (BiReBAC). However, this being the first proposal for the composite model, we have adopted *Closed Policy* to ensure completeness and soundness of our authorization specification.

In the well-known HRU model [8], it has been shown that in a network structure where both positive and negative authorizations are present, authorization of a particular node may be undecidable. However, since our present proposal is based on only positive authorizations, such decidability problem will not be present. Presence of both positive and negative authorizations, related undecidability problems, and possible mitigation is part of future work.

In Sect. 2.4, we qualitatively argued that the dependency-based formalism lends itself well to a property graph model. We will develop a provably sound and correct bidirectional translation from the dependency rules to an extended version of the property graph model. We also expect to develop a modified query engine that will accept this extended property graph. We will prove that the implication rules presented in the paper will be appropriately translated into a combination of query operations and inferencing over these property graphs. Finally, we expect to develop a policy server that will use this extended property graph-based model for its operations.

References

1. Abiteboul S, Hull R, Vianu V (1995) Foundations of databases, vol 8. Addison-Wesley Reading
2. Baudinet M, Chomicki J, Wolper P (1999) Constraint-generating dependencies. J Comput Syst Sci 59(1):94–115
3. Beeri C, Vardi MY (1984) Formal systems for tuple and equality generating dependencies. SIAM J Comput 13(1):76–98
4. Bettini C, Jajodia S, Wang XS, Wijesekera D (2003) Provisions and obligations in policy rule management. J Netw Syst Manag 11(3):351–372
5. Crampton J, Sellwood J (2014) Path conditions and principal matching: a new approach to access control. In: Proceedings of the 19th symposium on access control models and technologies, pp 187–198. ACM
6. Das SK, Bagchi A (2018) Representation and validation of enterprise security requirements, a multigraph model. In: Advanced computing and systems for security, vol 6. Springer, Berlin, pp 153–167
7. Dasgupta S, Pal P, Mazumdar C, Bagchi A (2015) Resolving authorization conflicts by ontology views for controlled access to a digital library. J Knowl Manag 19(1):45–59
8. Harrison MA, Ruzzo WL, Ullman JD (1976) Protection in operating systems. Commun ACM 19(8):461–471
9. Jajodia S, Samarati P, Subrahmanian V (1997) A logical language for expressing authorizations. In: Proceedings of IEEE symposium on security and privacy (Cat. No. 97CB36097). IEEE, pp 31–42
10. Molina HG, Ullman JD, Widom J (2002) Database systems the complete book

11. Rizvi SZR, Fong PW (2016) Interoperability of relationship-and role-based access control. In: Proceedings of the 6th international conference on data and application security and privacy (CODASPY). ACM, pp 231–242
12. Sellwood J (2017) RPPM: a relationship-based access control model utilising relationships, paths and principal matching. PhD thesis, Royal Hallway, University of London

A Robust Approach to Document Skew Detection

Barun Biswas, Ujjwal Bhattacharya, and Bidyut B Chaudhuri

Abstract This article presents a script-independent approach for the estimation of skew angle of the offline document image. It follows a certain divide and conquer strategy. A document image is segmented into several vertical strips of smaller and equal width and a simple strategy is applied to identify individual lines in each such vertical segment. Next, an association of horizontally adjacent lines in consecutive vertical strips takes place extracting individual text lines of the whole document image. Let P_{ij} and $P_{i(j+1)}$ be the center points of the baselines of two adjacent segments of the i-th text line of the input document. The angle of inclination θ_{ij} of the straight line segment $P_{ij} P_{i(j+1)}$ with the positive direction of x-axis is computed. The average of all θ_{ij}'s for the entire document image is used as the estimated skew of the document. This strategy of document image skew estimation has been tested on the DISEC'13 (ICDAR 2013 Document Image Skew Estimation Contest) database and its performance is comparable with the state-of-the-art results in terms of both accuracy and computation time. Additionally, this skew estimation scheme has been used on a wide variety of other documents of different scripts, mixed script documents, degraded quality documents, documents containing texts and doodles, etc. The results of skew correction in every such case have been observed to be satisfactory.

Keywords Document skew · Skew correction · Text line segmentation · Vertical projection profile · Optical Character Recognition (OCR)

B. Biswas (✉)
AKCSIT, University of Calcutta, Kolkata, India
e-mail: barun.isical@gmail.com

U. Bhattacharya
CVPR Unit, Indian Statistical Institute, Kolkata, India
e-mail: ujjwal@isical.ac.in

B. B. Chaudhuri
TECHNO INDIA UNIVERSITY, Kolkata, India
e-mail: bbc.isical@gmail.com

R. Chaki et al. (eds.), *Applied Computing for Software and Smart Systems*,
Lecture Notes in Networks and Systems 555,
https://doi.org/10.1007/978-981-19-6791-7_3

1 Introduction

Optical character recognition (OCR) systems are frequently used in the daily lives of the computer savvy population. Scanned documents which are subjected to OCR systems may be either printed or handwritten or a mixture of both of them. Input scanned documents to an OCR system may often be skewed due to either improper placement of the hard copy of the document on the scan bed or the original document may itself have some skew. Typical OCR systems usually include a skew correction module at the preprocessing stage and the performance of various other modules may get affected if the skew cannot be estimated sufficiently accurately. So, document skew correction has been studied [1] since the early days of OCR research, and studies of the same are still continued [2] to have a more robust approach capable of handling various complicated documents.

In the present article, we propose a divide and conquer strategy for the estimation of the skew in an offline document image. The present approach is computationally simple. It has been tested on a wide variety of documents of a large number of popular scripts which include Latin, Arabic, Devanagari, Bangla, Urdu, French, Tamil, Telugu, Oriya, Malayalam, Nepali, Gujarati, Greek, Japanese, etc. Also, it has been tested on multiscript documents and document pages containing doodles. It is capable of providing satisfactory results on both printed and handwritten documents. Moreover, its performance on degraded documents is acceptable.

The rest of this article is organized as follows. Section 2 gives a brief state of the art of some existing skew correction approaches. The details of the proposed algorithm have been described in Sect. 3. In Sect. 4, the results of our simulation have been presented. Section5 concludes the article.

2 Previous Work

A majority of existing document skew estimation studies have made use of one of the three main tools [3]: (i) projection profile (PP), (ii) Hough transform or (iii) nearest neighbor clustering. Also, a few studies involved more than one such tool. PP-based approaches [5, 6] for document skew estimation are perhaps the most popular ones and are simple in nature. However, such methods often fail in the presence of noise. The Power Spectral Density of horizontal PP was used in [4] for document skew estimation. Horizontal PP was used in [5] while vertical PP was used in [6] for this skew estimation problem.

Since the early days of OCR research for document skew estimation, Hough transform (HT) [7–9] has been used. A strategy based on Hough Transform for skew detection of Indian script documents was proposed in [10]. However, such methods are usually both computation and memory intensive.

A major advantage of nearest neighbor-based approaches is that it has no limitation in the range of skew angle. In these methods, connected components of the input document are extracted. For skew detection, O'Gorman [3] studied a generalized nearest neighbor (NN)-based approach. Liolios et al. [11] used a similar approach OF NN in which all the components belonging to the same text line of the document were grouped into a single cluster. An improved nearest neighbor-based skew estimation approach was proposed by Lu and Tan [12]. Usually, such an approach cannot estimate the skew of degraded documents efficiently.

Several other methods have been used for document skew estimation. These methods include cross-correlation [13, 14], principal component analysis (PCA) [15], Fuzzy run length [16], and some other example of skew detection are [22–25].

3 Proposed Methodology

The proposed scheme is based on dividing the preprocessed document into several vertical strips of certain small width. Our algorithm includes a step for the computation of strip width specific to the input document. Text lines of each such strip are segmented based on its horizontal PP. This segmentation strategy includes specific strategies to tackle vertically overlapping and touching text lines. Next, the set of text line segments of all the vertical strips are clustered such that all the segments belonging to a cluster form a text line of the input document. Mid-point of the baseline of each text line segment within each vertical strip is computed. Mid-points of each pair of spatially adjacent line segments within each individual cluster are joined by a straight line and its angle with the positive direction of the horizontal axis is obtained. The final estimated skew angle is the average of these angles of the input document image. Figure 1 shows a block diagram of the proposed algorithm.

3.1 Preprocessing Stage

We used a 3 × 3 mean filtering followed by binarization using [17, 19], denoising using [20, 21] as a preprocessing operation. We further obtain the minimum bound-

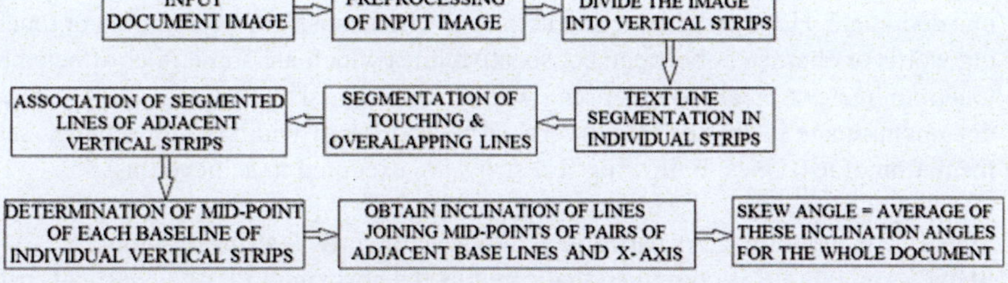

Fig. 1 Proposed skew detection method block diagram

ing rectangles containing text or non-text component of the binarized image. The foreground text pixels are given black color, while the background pixels are given white color. From the resulting image, we obtain connected components and the areas of the bounding rectangles of these components are sorted. Connected components corresponding to the upper 5% areas of such bounding rectangles, i.e., components with areas of their bounding rectangles greater than the 20th 20-quantile are removed. The resulting image is considered for further processing.

3.2 Division of Document Image Into Vertical Strips

Since the proposed methodology is developed for an arbitrary document, the consecutive text lines or the words in a text line may not be properly aligned. So, we divide the preprocessed document into several vertical strips of equal width (barring possibly the strip at the extreme right) within each of which words or lines are assumed to be properly aligned. The width of these strips is estimated as follows:

Step 1: The image is divided vertically into strips each of width $\max\{150, DW/10\}$, where DW = width of document.
Step 2: Obtain horizontal PP P_i, where $i = 1, 2, \ldots, n$, where n is the no. of strips.
Step 3: The connected components of P_i is obtained, for each $i = 1, 2, \ldots, n$.
Step 4: Obtain H = average height of all connected components (of the whole document) extracted in Step 3.
Step 5: Obtain $\overline{G}$ = average of all G_{ij}, where G_{ij} is the vertical gap between PP components C_{ij} and $C_{i(j+1)}$ for $i = 1, 2, \ldots,$ no. of strips, $j = 1, 2, \ldots, k_i$, where k_i = no. of PP components in i^{th} vertical strip-1.
Step 6: Obtain final strip width $S_w = 2 * (H + \overline{G})$.
Step 7: The input image is divided into S_w vertical strips each.

3.3 Initial Separation of Text Lines within Each Strips

In line segmentation, if the text lines of a document are free from touching, overlapping, skew, etc., then significant information can be obtained from the horizontal PP of a document. Handwritten document images often consist of overlapping or touching words or characters between consecutive lines which are sometimes affected by touching lines, vertically overlapping words, curved or skewed lines, etc. The input document image is divided into several vertical strips of width S_w for efficient segmentation of text lines. Following few strips are executed to achieve this:

Step 1: The input image is partitioned into several vertical strips of width S_w.
Step 2: An image P is computed that contains the horizontal PP of all vertical strips of the input image.

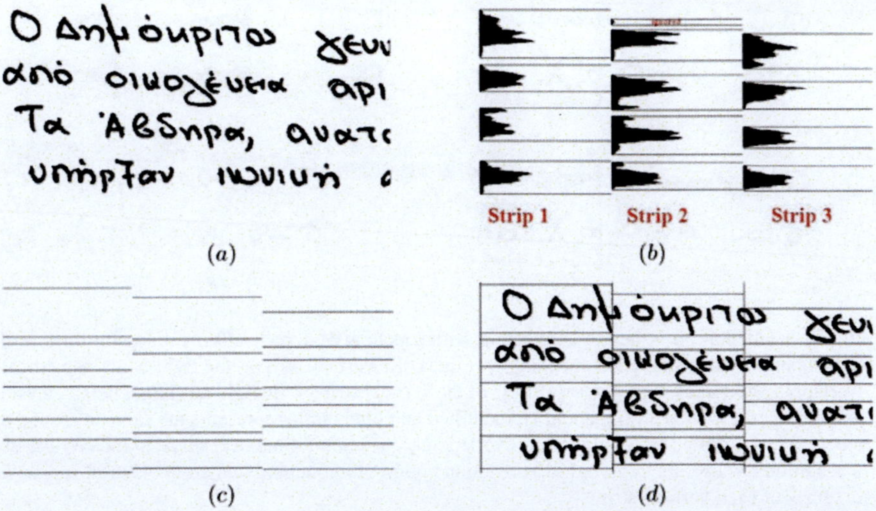

Fig. 2 Initial separation of text lines: **a** a part of a handwritten manuscript, **b** Inside each vertical strip of profile component, at the top and bottom horizontal line segments are drawn, **c** estimated line segments (barring the small height lines), **d** text lines are segmented inside individual strips

Step 3: From P we compute the average height (H_{avg}) of Connected Component Profile.
Step 4: The components of P having height less than $\frac{H_{avg}}{3}$ is ignored.
Step 5: For the remaining profile components, we obtain the top and bottom segments of horizontal lines within respective vertical strips.
Step 6: Inside individual strips of the image, the above segment provides an initial segmentation of text lines.

In Fig. 2a, a small part of a handwritten document is shown. In Fig. 2b–d, initial separation of text lines is illustrated. Next, we try to associate text lines in one strip with the text lines of the neighboring strips.

3.4 Segmentation of Touching and Overlapping Lines

In this section, we traced the strips from left to right. We start from the top strip and stop at the bottom strip in each vertical strip and perform the following segment on the strips:

Step 1: Until there is no more strips, we consider two consecutive lines from top to bottom.
Step 2: We consider the line component as single line if in the current strip the height of the next line is not more than $2H_{avg}$ and move to the next line. Move to step 3 if the component height exceeds the threshold.

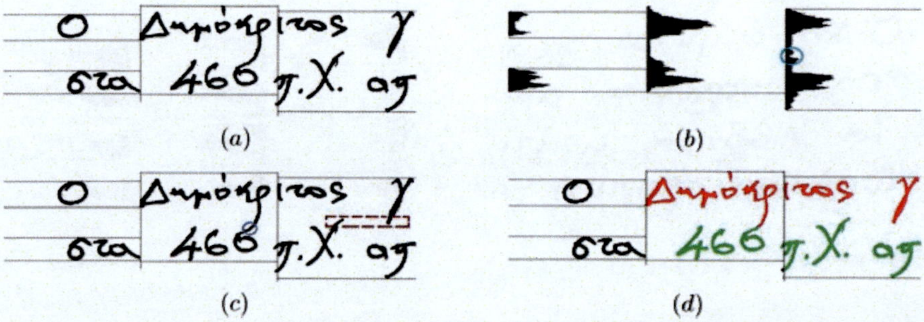

Fig. 3 Segmentation issues in connection with touching and vertically overlapping lines: **a** the vertical strip at the middle has touching text lines, the vertical strip at the right contains vertically overlapping text lines, **b** projection profile of the vertical strip of the right has larger height, a valley near the middle of this profile could be identified and blue circle represents the same, **c** around the valley of its PP dotted rectangle shows overlapping text region; blue oval shape in the vertical strip of the middle shows the required segmentation region of a touching text component, **d** segmented text lines of each vertical strip

Step 3: Components having the height greater or equal to $2H_{avg}$ consists of touching characters of two vertically consecutive lines. The middle point around the minimum valley of the PP component is considered to be the touching point. We horizontally segment at this point through the straight line. An illustration is provided in Fig. 3. We segment the initial line above and below of this horizontal line. Next, move to Step 2.
Step 4: As shown in Fig. 3a, vertically overlapped lines are considered here. In the current segmented line, valley region is found at the middle point of the PP segment. This is a similar situation, instead of single valley point, we find a small contiguous part of PP as valley. Through the middle of the valley region, we consider a horizontal line. We consider the upper line of which major parts are situated above this horizontal line and the rest part belongs to the lower line. An illustration is shown in Fig. 3c. The final segmentation within the strips is shown in Fig. 3d. Next, move to Step 2.

3.5 Association of Segmented Lines of Adjacent Vertical Strips

In Sect. 3.4, we have already identified the lines of the individual strips. In this step, corresponding lines of the adjacent vertical strips are associated, if any. Two adjacent strips are considered and we start from two leftmost strips and continue until the rightmost pair. Followings are the segment used here:

Step 1: Set $i = 1$
Step 2: The ith and $(i + 1)$th strips are considered until there are no more strips.
Step 3: In the $(i + 1)$th strip, we consider the next line. Increase i by 1 if it is end line and go to step 2, else go to step 4.

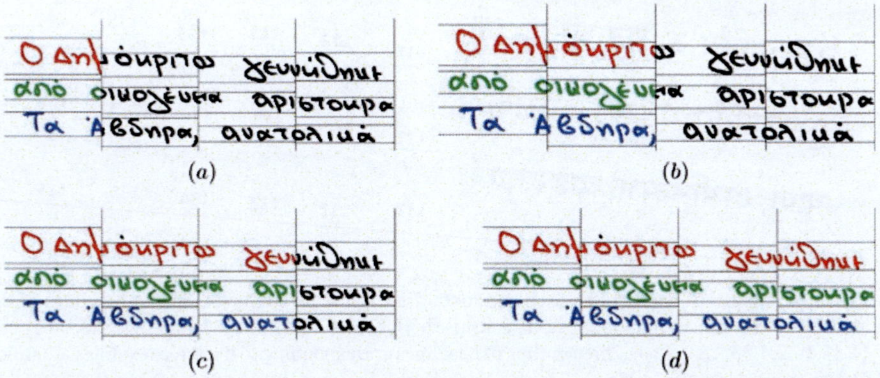

Fig. 4 Text lines of adjacent vertical strips association: **a** initial segmentation of text lines in individual vertical strips, **b** Respective segmented text lines of 1st strip and 2nd strip have been associated, **c** The lines of 3rd strip have been associated with the first two strips, **d** initially segmented lines of 4th vertical strip are associated with the text lines of the first three strips

Step 4: Move to step 5 if there is no component in the ith strip. Otherwise, the line component of ith strip is associated with the line of $(i + 1)$th strip which shares a part of its components. In the situation of more than one such common component present in both the strips and the strips are from different lines of ith strip, then the line with a larger component of ith strip is associated with the present line. Move to Step 3.
Step 5: We associate the current line with the maximum vertical overlapped line of ith strip. The line is considered to be a new line if ith strip contains the leftmost component, otherwise we search for similar overlap with the adjacent strips at left. We associate such two lines if we find such a strip at the left. Move to Step 3.

Figure 4 illustrates the above strategy using the document shown in Fig. 2d.

3.6 Skew Angle Computation

The baseline (lower boundary) of each part of all the text lines belonging to different vertical strips is computed. In Fig. 5, such baselines for a piece of text have been shown. In this Figure, the text is divided into six vertical strips. Its three lines have been identified by the above line segmentation strategy and the corresponding three groups of the baselines are {L11, L12, ..., L16}, {L21, L22, ..., L26}, and {L31, L32, ..., L36}.

Middle points P_{ij}'s of respective baselines L_{ij} are obtained for all $i = 1, 2, 3$ and all $j = 1, 2, \ldots, 6$. Now, the angle θ_{ij} of each line joining P_{ij} and $P_{i(j+1)}$ for all $i = 1, 2, 3$ and all $j = 1, 2, \ldots, 5$ with the positive direction of the horizontal

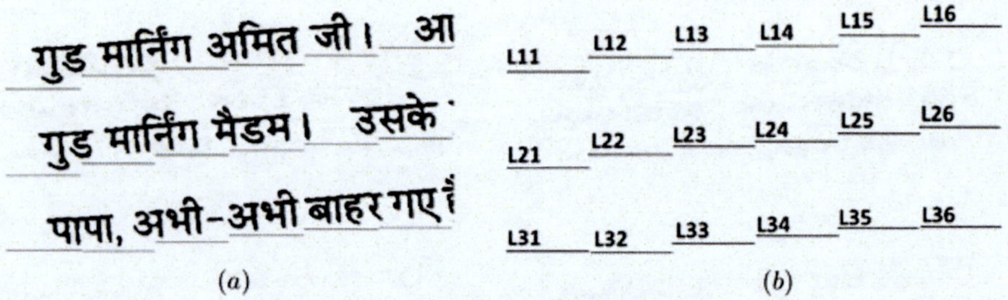

Fig. 5 **a** A piece of handwritten text; horizontal line segments are the computed baselines of the identified text lines in individual vertical strips. **b** {L11, L12, L13, …}, {L21, L22, L23, …} and {L31, L32, L33, …} are the three groups of baselines corresponding to the three text lines identified by the above described strategy

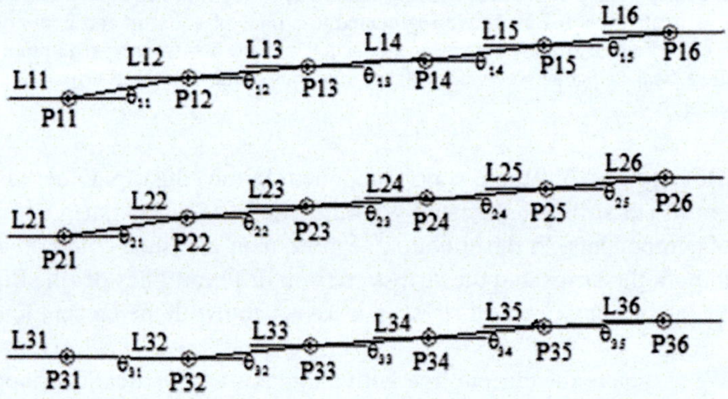

Fig. 6 Calculation of the skew angle: P_{ij} is the middle point of L_{ij}, the ith baseline of the jth vertical strip and θ_{ij} is the inclination of the line joining P_{ij} and $P_{i(j+1)}$ with the horizontal line. Skew is estimated as the average of all such θ_{ij} of the entire document image

axis (as shown in Fig. 6) is computed. The average (θ) of all such θ_{ij} for the entire document is the estimated skew. Now the entire document is rotated at an angle $-\theta$ to deskew the input document.

4 Experimental Results

The proposed algorithm has been simulated on the following two databases—(1) a document image database consisting of samples of 15 different scripts collected by us from different sources and (2) ICDAR2013 Document Image Skew Estimation Contest (DISEC'13) database. We term the first database as Multi-script Document Image Database. Simulation results on these two databases are detailed below.

4.1 Results on Multi-script Document Image Database

For the evaluation of the proposed skew estimation method, we have developed a database of 200 binarized document image samples collected from different sources. It consists of 10 printed document images of 15 different scripts which include Latin, Arabic, Devanagari, Bangla, Urdu, Japanese, French, Tamil, Telugu, Oriya, Malayalam, Nepali, Greek, Gujarati, and Sindhi. These printed pages have been collected from the Digital Library of India archive. Additionally, it includes 10 sample pages of degraded quality documents and another 10 printed samples containing texts mixed with a sketch or art design both being collected from the same archive. Also, it contains 10 samples of handwritten documents of each of the Devanagari and Bangla scripts. Finally, it contains 5 printed and another 5 handwritten samples of mixed scripts. These later two types of samples have been added to our database from various personal collections. All the 200 samples of this database have been manually checked so that they do not have any visually recognizable skew. Each of these samples has been rotated at 6 random angles in the range –45° to +45°. For each of these 1200 rotated samples, the difference between the estimated skew by the proposed algorithm and the amount of rotation is less than 0.1°. Some results of skew correction by the proposed method are shown in Figs. 7, 8, 9, 10, and 11.

4.2 Results on Disec'13 Database

This database was developed for the organization of the 1st International Skew Estimation Contest (DISEC'13) [18] held in conjunction with the 12th International Conference on Document Analysis and Recognition (ICDAR 2013). It consists of 1750 document images with different amounts of skew in the range –15° to +15°. These image samples were obtained by rotating 175 binarized images of various types of documents each of which was manually verified to have no skew. These 175 images form a representative set of various realistic document categories containing figures, tables, diagrams, etc. These were collected from different types of sources which include newspapers, literature, scientific books, dictionaries, etc., of several scripts such as English, Chinese, Japanese, and others.

For the purpose of evaluation, we compute $E(j)$ = the absolute difference between the estimated skew and the ground truth of the jth sample document image. Here, we consider three criteria similar to the evaluation strategy of DISEC'13—(1) Average Error Deviation (AED), (2) Average Error Deviation of the Top 80% best-performed documents (TOP80), and (3) Correct Estimation percentage (CE). These

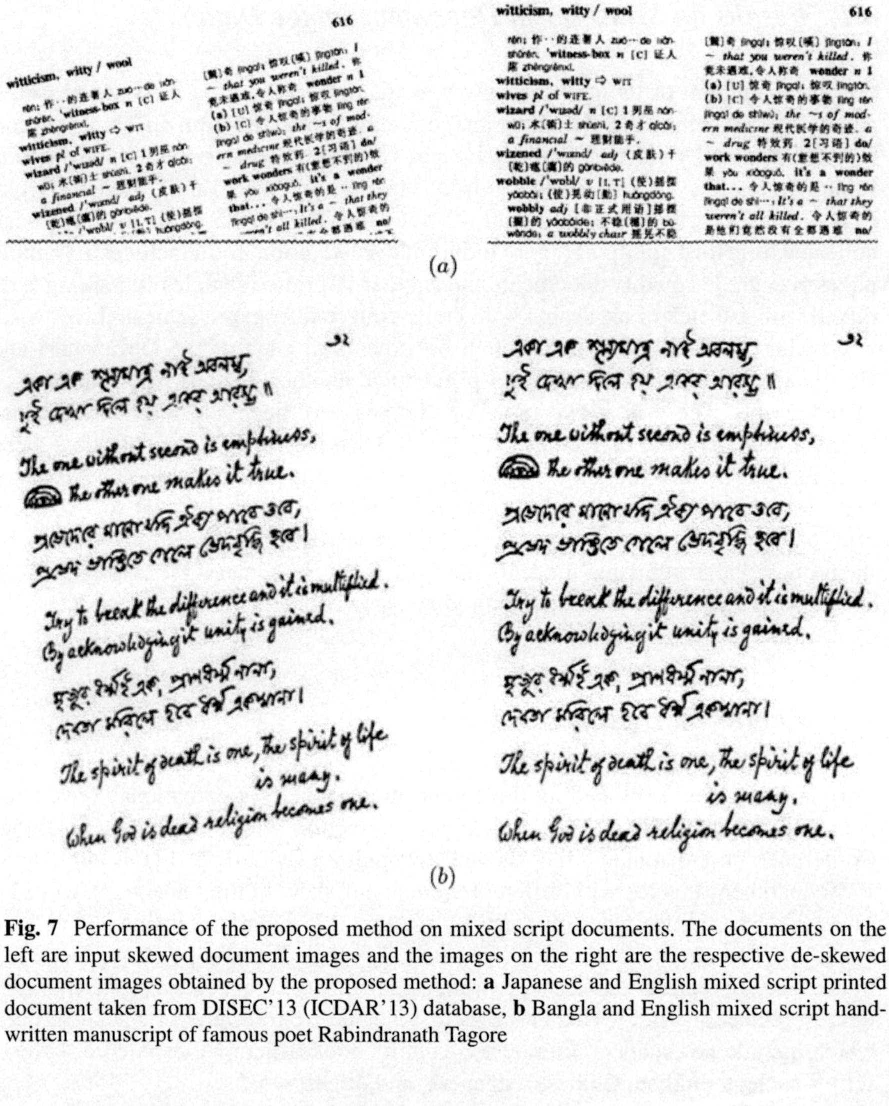

Fig. 7 Performance of the proposed method on mixed script documents. The documents on the left are input skewed document images and the images on the right are the respective de-skewed document images obtained by the proposed method: **a** Japanese and English mixed script printed document taken from DISEC'13 (ICDAR'13) database, **b** Bangla and English mixed script handwritten manuscript of famous poet Rabindranath Tagore

$$\text{AED} = \frac{1}{N}\sum_{j=1}^{N} E(j) \tag{1}$$

$$\text{TOP80} = \frac{1}{M}\sum_{i=1}^{M} E_s(i) \tag{2}$$

$$\text{CE} = \frac{1}{N}\sum_{j=1}^{N} K(j), \tag{3}$$

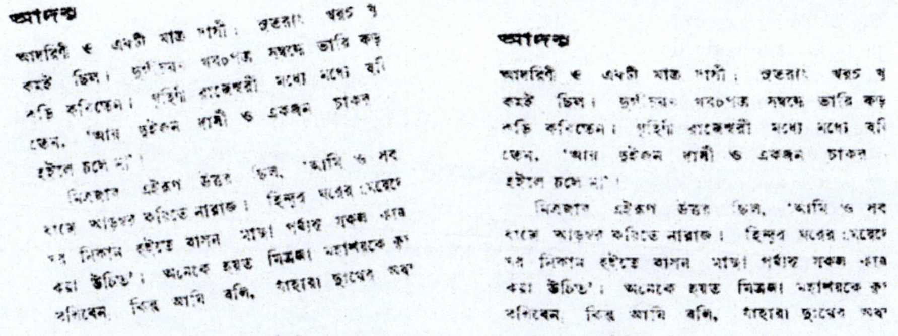

(a) (b)

Fig. 8 Performance of the proposed method on a degraded document image: **a** Original skewed degraded document image, **b** De-skewed degraded document image

where $M = 0.8N$, N is the total number of samples in the document image database, E_s is obtained by sorting the list E in ascending order and

$$K(j) = \begin{cases} 1 & if E(j) \leq 0.1, \\ 0 & otherwise. \end{cases} \tag{4}$$

DISEC'13 competition was participated by 12 different skew estimation algorithms and their evaluation results based on its database have been reproduced in Table 1. The evaluation results of the proposed method on this competition database are also presented in the last row of this Table. Following the protocol of DISEC'13, we have also ranked all 13 algorithms (including the proposed one) based on the above three criteria and the sum of these three ranks (S) for individual algorithms is shown in the last column of Table 1. Toward effective visualization of this evaluation result, we have presented a vertical bar chart for the values of 'S' corresponding to different algorithms in Fig. 12. It is to be noted that the smaller the value of 'S', the better the performance of the algorithm.

5 Conclusions

The proposed skew estimation strategy works equally efficiently on both printed and handwritten documents. It is independent of the script and layout of the document. It is suitable for both single and multi-column documents. Usually, projection profile-based skew correction approaches fail in the presence of graphics. However, since the preprocessing module of the proposed algorithm filters out 5% of the largest

Fig. 9 Performance of proposed skew correction approach on handwritten document samples: input skewed documents are shown on the left and respective output skew corrected documents are shown on the right. **a** Devanagari script, **b** Bangla script

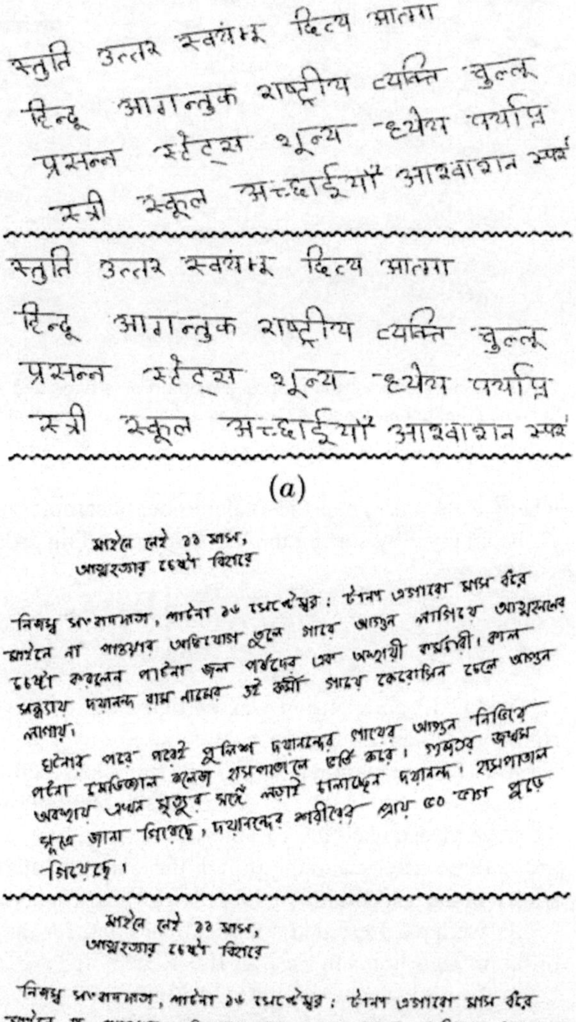

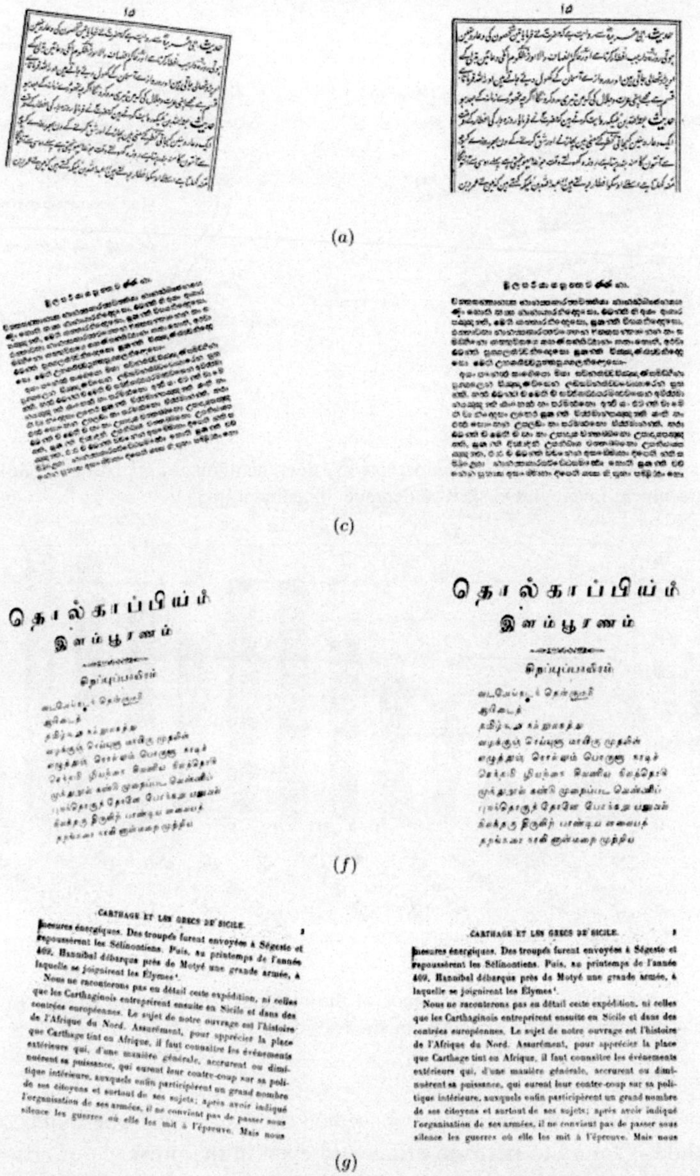

Fig. 10 Performance of proposed skew correction approach on printed document samples collected from Digital Library of India archive: input skewed documents are shown on the left and respective output skew corrected documents are shown on the right. **a** Urdu script, **c** Oriya script, **f** Tamil script, **g** French script

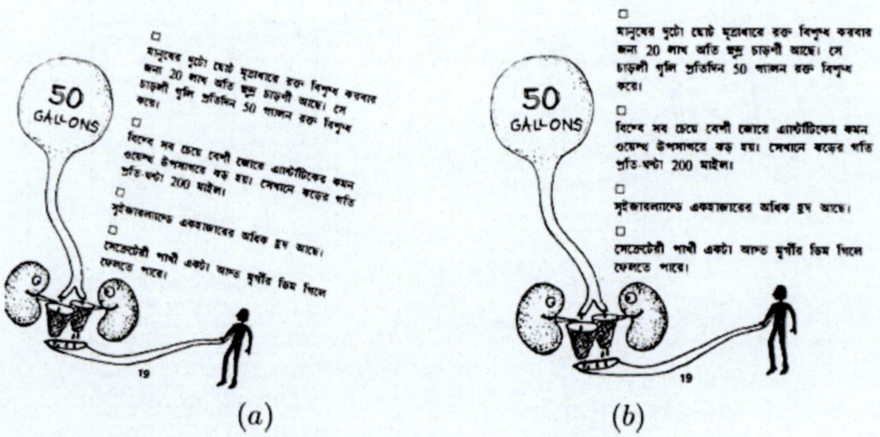

Fig. 11 Performance of the proposed method on a document image containing a Diddle: **a** Original skewed document image, **b** De-skewed degraded document image

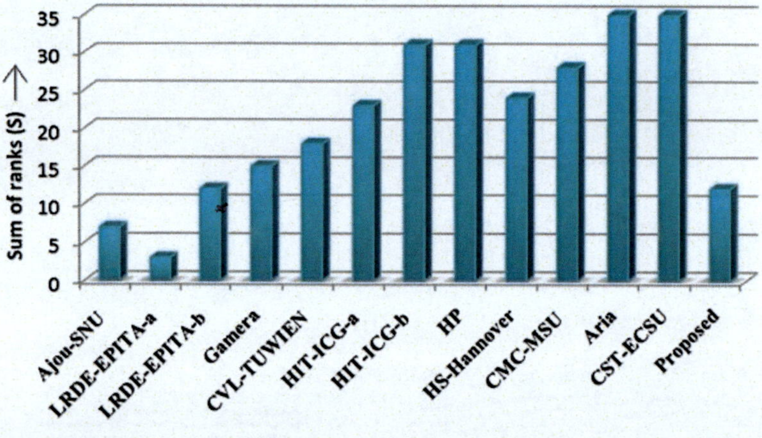

Fig. 12 Comparative overall performance of the methods submitted to DISEC'13 competition versus the performance of the proposed method

components and estimates the skew based on the rest of the foreground components, the method is found to perform efficiently even in the presence of artworks in the document. However, if the artworks of large sizes are more than 5% of the foreground components, then the proposed algorithm may not provide the desired result. Also, the proposed algorithm cannot be used on documents with multi-skew.

Table 1 Results of DISEC'13 competition versus the result of the proposed method

Methods	AED (Rank)	TOP 80 (Rank)	CE (Rank)	Sum of ranks (S)
Ajou-SNU	0.085 (2)	0.051 (2)	71.23 (2)	6
LRDE-EPITA-a	0.072 (1)	0.046 (1)	77.48 (1)	3
LRDE-EPITA-b	0.097 (3)	0.053 (4)	68.32 (5)	12
Gamera	0.184 (6)	0.057 (5)	68.9 (3)	14
CVL-TUWIEN	0.103 (4)	0.058 (6)	65.42 (7)	17
HIT-ICG-a	0.73 (10)	0.061 (7)	65.74 (6)	23
HIT-ICG-b	0.75 (11)	0.078 (10)	57.29 (10)	31
HP	0.768 (13)	0.073 (9)	58.32 (9)	31
HS-Hannover	0.227 (8)	0.069 (8)	58.84 (8)	24
CMC-MSU	0.184 (6)	0.089 (11)	50.39 (11)	28
Aria	0.473 (9)	0.228 (13)	19.29 (13)	35
CST-ECSU	0.75 (11)	0.206 (12)	28.52 (12)	35
Proposed	**0.104 (5)**	**0.052 (3)**	**67.56 (4)**	**12**

References

1. Hull JJ (1998) Document image skew detection: survey and anotated bibliography. In: Hull JJ, Taylor SL (eds) Document analysis systems II. World Scientific, Anal, pp 40–64
2. Mascaro AA, Cavalcanti GDC, Mello CAB (2010) Fast and robust skew estimation of scanned documents through background area information. Pattern Recognit Lett 31:141–1403
3. O'Gorman L (1993) The document spectrum for page layout analysis. IEEE Trans Pattern Anal Mach Intell 15(11):1162–1173
4. Liolios N, Fakotakis N, Kokkinakis G (2002) On the generalization of the form identification and skew detection problem. Pattern Recognit 35:253–264
5. Kavallieratou E, Fakotakis N, Kokkinakis G (2002) Skew angle estimation for printed and handwritten documents using the wigner—ville distribution. Image Vis Comput 20:813–824
6. Papandreou A, Gatos B (2011) A novel skew detection technique based on vertical projections. In: Proceedings of international conference on document analysis and recognition, pp 384–388
7. Srihari SN, Govindaraju V (1989) Analysis of textual images using the hough transform machine vision and applications. Image Vis Comput 2:141–153
8. Min Y, Cho SB, Lee Y (1996) A data reduction method for efficient document skew estimation based on hough transformation. In: Proceedings of the 13th international conference on pattern recognition, pp 732–736
9. Yin P-Y (2001) Skew detection and block classification of printed documents. Image Vis Comput 19(8):567–579
10. Chaudhuri BB, Pal U (1997) Skew angle detection of digitized Indian script documents. IEEE Trans Pattern Anal Mach Intell 19(2):182–186
11. Liolios N, Fakotakis N, Kokkinakis G (2001) Improved document skew detection based on text line connected- component clustering. In: Proceedings of international conference on image processing, pp 1098–1101
12. Lu Y, Tan CL (2003) Improved nearest neighbor based approach to accurate document skew estimation. In: Proceedings of the 7th international conference on document analysis and recognition, pp 503–507 (2003)
13. Chaudhuri A, Chaudhuri S (1997) Robust detection of skew in document images. IEEE Trans Image Process 6(2):344–349

14. Chen M, Ding X (1999) A robust skew detection algorithm for grayscale document image. In: Proceedings of international conference on document analysis and recognition, pp 617–620 (1999)
15. Paunwala CN, Patnaik S, Chaudhary M (2010) An efficient skew detection of license plate images based on wavelet transform and principal component analysis. In: Proceedings of international conference on signal and image processing, pp 17–22
16. Shi Z, Govindaraju V (2003) Skew detection for complex document images using fuzzy run length. In: Proceedings of the 7th international conference on document analysis and recognition, pp 715–719
17. Biswas B, Bhattacharya U, Chaudhuri BB (2014) A global-to-local approach to binarization of degraded document images. In: Proceedings of of the 22nd international conference on pattern recognition, vol 22, pp 3008–3013
18. Papandreou A, Gatos B, Louloudis G, Stamatopoulos N (2013) ICDAR2013 Document Image Skew Estimation Contest (DISEC'13). In: Proceedings of the 12th international conference on document analysis and recognition, pp 1444–1448
19. Hadjadj Z, Cheriet M, Meziane A et al (2017) A new efficient binarization method: application to degraded historical document images. SIViP 11:1155–1162
20. Shreyamsha Kumar BK (2013) Image denoising based on gaussian/bilateral filter and its method noise thresholding. Signal, Image Video Process 7:1159–1172
21. Shreyamsha Kumar BK (2013) Image denoising based on non-local means filter and its method noise thresholding. Signal, Image Video Process 7:1211–1227
22. Junjuan L, Guoxin T (2008) An efficient algorithm for skew-correction of document image based on cyclostyle matching. In: 2008 international conference on computer science and software engineering, pp 1267–1270
23. Tang Y, Wu X, Bu W, Wang H (2013) Skew estimation in document images based on an energy minimization framework. In: Proceedings of the international conference on image processing, computer vision, and pattern recognition (IPCV), pp 1267–1270
24. Guan YP (2012) Fast and robust skew estimation in document images through bilinear filtering model. IET Image Process 6(6):761–769
25. Li C, Wu G (2017) A novel fast hierarchical projection algorithm for skew detection in multimedia big data environment. Int J Mobile Comput Multimed Commun (IJMCMC) 8(3):44–65

Smart Systems and Networks

MBLEACH: Modified Blockchain-Based LEACH Protocol

Shubham Kant Ajay, Rishikesh, and Ditipriya Sinha

Abstract One of the primary clustering-based routing protocols in the wireless sensor network is the LEACH (Low-Energy Adaptive Clustering Hierarchy) protocol. In the LEACH protocol, cluster head plays an important role as entire data transmission is performed through cluster head nodes, and if any one of the cluster heads is corrupted, the entire network will be affected. Hence, the cluster head dysfunction must be discarded and addressed. To handle the above-mentioned issue this paper proposes a secure blockchain-based security mechanism for the detection and removal of malicious cluster head nodes in LEACH protocol. In the proposed work, a hybrid blockchain model is designed which is consisting of public and private blockchains. The base station joins the public blockchain as a miner node and monitors the cluster head nodes which join the local blockchain and are known as validators. Whenever the Base station monitors that some cluster head nodes are behaving maliciously (i.e., does not mine transactions and does not produce new blocks or does not sign the transactions), a smart contract on the public blockchain is triggered and stores the node information of the malicious cluster head node. After that, the Base station triggers an event to notify the cluster head nodes in the local blockchain and that cluster head node validates the information among themselves through PoA (Proof-of-Authority) consensus to remove these malicious nodes from the network.

Keywords Wireless sensor network · Blockchain · LEACH

S. K. Ajay (✉) · Rishikesh · D. Sinha
National Institute of Technology Patna, 800005 Patna, Bihar, India
e-mail: shubhama.pg20.cs@nitp.ac.in

Rishikesh
e-mail: rishikesh.ph21.cs@nitp.ac.in

D. Sinha
e-mail: ditipriya.cse@nitp.ac.in

R. Chaki et al. (eds.), *Applied Computing for Software and Smart Systems*,
Lecture Notes in Networks and Systems 555,
https://doi.org/10.1007/978-981-19-6791-7_4

1 Introduction

Designing a secure clustering protocol which is energy-efficient as well as provides security is one of the burning demand in WSN paradigm. All the communications in a clustering-based protocol are performed through the cluster head nodes and if the cluster head node starts behaving maliciously, the performance of entire cluster system will be affected. Therefore, the identification and removal of such malicious cluster head nodes is an important aspect of the smooth functioning of the entire network. One of the primary clustering-based routing protocols for the wireless sensor networks is the LEACH (Low-Energy Adaptive Clustering Hierarchy) [1] protocol. There is no security-based mechanism associated with the LEACH protocol and also it does not provide any mechanism for the identification of the malicious nodes. In order to remove the lack of security mechanism in the LEACH [1] protocol SLEACH (Secure low-energy adaptive clustering hierarchy protocol for wireless sensor networks) [3] is proposed. MSLEACH (a routing protocol combining multi-hop transmissions and single-hop transmissions) [2], is created to increase the security of SLEACH [3]. However, the above-mentioned versions of LEACH protocol do not address the problem of malicious cluster head node detection which is highly important for the smooth and secure working of the network.

To address this problem, a blockchain-based malicious cluster head node detection mechanism is designed which identifies as well as removes the malicious cluster head nodes from LEACH protocol. Novel blockchain technology can assist to secure sensors in the WSN paradigm. In a decentralized system, it provides a high level of security, transparency, immutability, and consensus during data transmission from one node to another. These properties of blockchain make it robust against different security attacks and motivate the authors to apply blockchain in LEACH protocol for the detection of malicious cluster head nodes. In the proposed work, we have designed a hybrid blockchain model which comprises of public and private blockchains for the detection and removal of malicious cluster head nodes from the network. The base station which is a high-energy node joins the public blockchain as a miner node and monitors all the cluster head nodes joined on the local blockchain network as validator nodes. The base station on observing any malicious activity from cluster head nodes stores the malicious cluster head node information on public blockchain and sends a signal to the cluster head nodes on the local blockchain so that they can verify the transaction and remove the malicious node from the network permanently through PoA (Proof-of-Authority) [4] consensus algorithm. This modified blockchain-based LEACH model is compared with other existing secured LEACH protocols such as SLEACH [3] and MSLEACH [2] in terms of packet drops and throughput in the presence of malicious cluster head nodes and it is observed that the proposed LEACH protocol outperforms than SLEACH and MSLEACH. It is also worth noting that the transaction message sizes of the proposed model are not particularly large.

2 Background

The following two subsections discuss the literature survey.

2.1 Blockchain Technology in WSN

Blockchain is likewise a terrific fit for the wireless sensor networks [5, 6] because of its distributed nature. Many researchers are creating an edge between IoT and blockchain as IoT infrastructures are susceptible to various attacks. However, it is difficult to acquire the blockchain deployment criterion on resource constraints IoT devices. To enhance the power barrier of IoT devices a blockchain-based virtual identification among different IoT devices and applications is hooked up. This is possible through a sequence of edge devices. A blockchain-based multi-WSN authentication mechanism is recommended in [7]. In this work, based on the capability, IoT nodes are labeled as base station, cluster head, regular node, and connected in hierarchical fashion. Public and private blockchains are incorporated to prefabricate a hybrid blockchain network. In this hybrid architecture of blockchain network, identity of ordinary nodes is checked out by private blockchain, identity of cell is mutually controlled and validated through communication range and the public blockchain controls the identity of cluster head nodes. Nowadays, we cannot constitute a WSN network which is devoid of security features. Traditional WSNs do not ensure fairness and traceability. Malevolent node identification is investigated in [8] using the association of blockchain with WSN (BWSN). This paper mainly focused on (1.) BWSN architecture for identification of malicious node and (2.) malicious node identification for smart contract components. The authors are also focused on traditional WSNs solutions along with the blockchain-based solutions for WSNs. Blockchain and WSNs combinedly do the data administration that includes online data aggregation, data storage for analytics, and also offline query processing. The authors [9] developed a malicious node detection mechanism in WSNs using the blockchain trust model (BTM). The simulation results state that the model can effectively discover the malicious nodes in WSN and make sure the traceability of WSN technique.

2.2 Cluster Head Selection in WSN

In this section, LEACH extensions are addressed alongside their advantages and downsides. MSLEACH [2] is the succeeder of SLEACH [3] that improves the safety and gives records confidentiality and node to cluster head (CH) authentication making use of paired keys shared by means of CHs and their cluster participants. In comparison to current secured LEACH protocol solutions, the proposed MSLEACH [2] has efficient security mechanisms and meets all WSN security goals. The authors of ME-LEACH [10] decrease the distance between nodes to BS and balance node load. TL-LEACH [11] is proposed by the authors of [11]. To separate the community into degrees: upper and lower degree clusters, a two-level LEACH is designed. Security component to the LEACH protocol [1] is given in SLEACH [3]. This model uses lightweight protection mechanisms for reliable communication. The drawback of this model is that no security is given for the duration of cluster formation and key updates also are now not approved here. One of the major concerns needs to be addressed is the security of intra-clustering. A much efficient version of LEACH is multi-hop LEACH (MH-LEACH) [12]. In advanced-LEACH [13], the authors choose CH (cluster head) based on general probability and current state probability. In this work, the most energy-efficient node is selected as the cluster head in every round. The cluster head node is responsible to collect the data from all the nodes present in the cluster and send it to the BS (base station). Authors in [11] suggested that minimizing the number of nodes communicating to the BS leads to less energy requirement.

State of the art shows that very few works are considering on detection of malicious cluster head in LEACH [1] protocol. This paper designs a blockchain-based novel cluster head selection protocol of LEACH [1].

3 Methodology

3.1 System Model

The system model details are described below. This section describes the network model for the blockchain-based LEACH protocol. The work of this paper is based on some reasonable assumptions.

Assumptions For the wireless sensor network, the malicious cluster head detection mechanism is proposed with the following assumptions:

- Each sensor node existing in the network has a unique Ethernet address.
- Base station has computational and storage capabilities to monitor the cluster head nodes and trigger Smart Contract on the public blockchain network.
- Cluster heads have an adequate amount of computational capabilities and storage to validate and remove the malicious cluster head node through PoA (Proof-of-Authority) [4] consensus.
- The base station is trusted by all the nodes in the network as it is the node manager for the entire network.

Nodes Description Based on the different functionalities of the sensor nodes in the network the nodes in the proposed model can be divided into the following types: base station, cluster head nodes, and ordinary nodes. Figure 1 describes the network model of MBLEACH. These nodes can be described as follows:

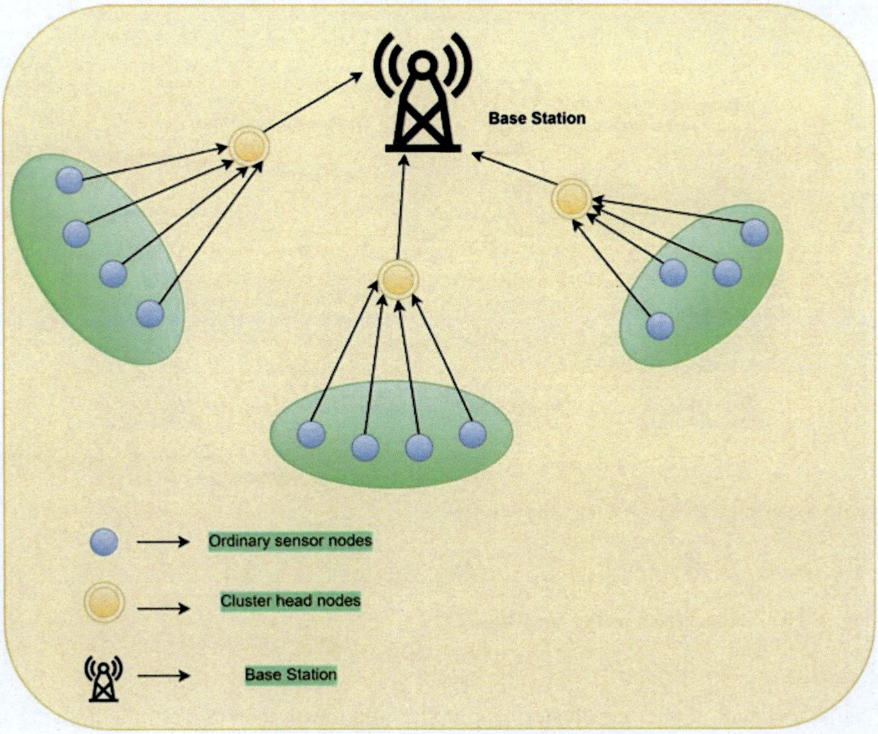

Fig. 1 Network model for proposed MBLEACH

- **Base Station**: The major responsibility of the base station is to monitor the network's nodes, collect sensory information from nodes, process, store, and analyze that data. The base station also initializes all the sensor nodes in the network by generating a unique ID for all the nodes in the network.
- **Cluster Head Node**: The cluster head node has sufficient computing power and memory to validate the transaction sent by Base station, performs basic processing and sends the data received from all the ordinary nodes to the base station.
- **Ordinary Node**: Ordinary nodes are the sensor nodes with low computational power and storage which are typically found at the edge of the network and their prime functionality is to sense the data from the environment. Each regular node belongs to only one cluster and under a single WSN network. Due to the limitations of their computing and memory capacity, as well as their energy constraint, these nodes are unable to perform complex operations and data processing.

Hybrid Blockchain Model In this paper, a hybrid blockchain model is designed which is a combination of both public and private blockchains. In the traditional LEACH [1] protocol, if any malicious cluster head node enters the network, it will affect the entire cluster and the functioning of the entire network will be disrupted. So, to avoid this issue, a hybrid blockchain model is designed in this paper which

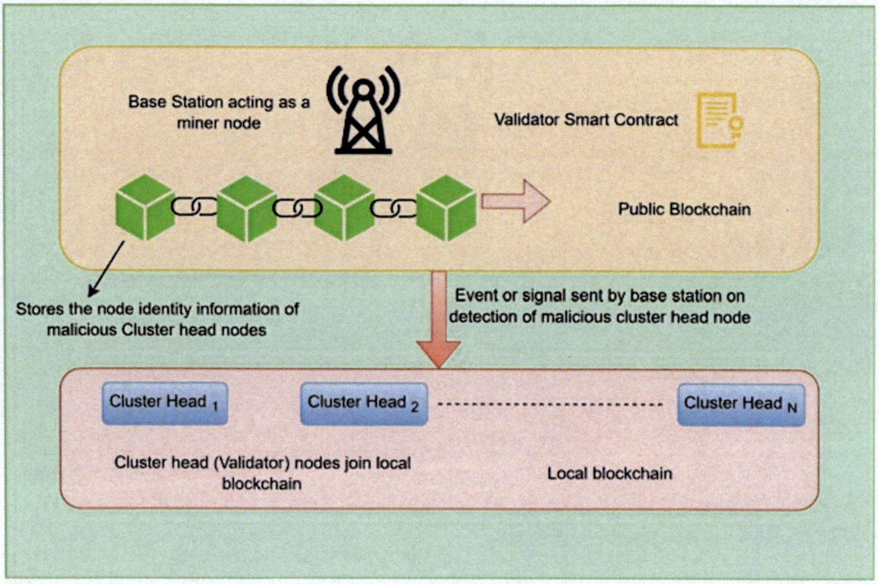

Fig. 2 Hybrid blockchain model of MBLEACH

can detect such malicious cluster head nodes and remove them from the network to enhance the performance of LEACH protocol in terms of energy dissipation, network lifetime, and packet drops. In the public blockchain network, nodes initially join to an unauthenticated network and with the help of consensus protocol create a decentralized trust network. Regular validation processes will cost a large amount of resources and time if all nodes of wireless sensor networks are joined to the public blockchain network. As a result, meeting the IoT's real-time requirements become unfeasible. Furthermore, because cluster head nodes have limited processing capabilities, they are unable to solve the computationally difficult mathematical puzzle each time a block is mined and added to the blockchain network. As a result, this work proposes a hybrid blockchain architecture that consists of two parts: Public blockchain and private blockchain, in order to effectively satisfy the needs of network. Figure 2 depicts the hybrid blockchain model which consists of the following two parts such as

- **Public Blockchain**: The base station joins the public blockchain as a miner node and monitors the cluster head nodes which are joined to the private/local blockchain as validator nodes. If the base station observes any malicious behavior (such as not mine transactions, or not produce new blocks) from any of the cluster head nodes then it triggers a Validator smart contract and stores the identity information of the malicious cluster head node on the public blockchain so that it cannot act as a validator on the private blockchain network for further rounds. In this way, the identity information of the malicious cluster head node is stored in a secure and tamper-proof way on the public blockchain.

- **Private Blockchain**: The private blockchain consists of the cluster head nodes which join to the blockchain as validator nodes (or signers) and using the PoA [4] consensus validate and remove the malicious cluster head node from private blockchain so that it cannot participate in future rounds.

3.2 *Flow Dagram of Modified Blockchain-Based LEACH Protocol (MBLEACH)*

The overall flow diagram of the modified blockchain-based SLEACH protocol is described in Fig. 3 below. The flow diagram of the modified blockchain-based LEACH protocol (MBLEACH) is as follows:

1. **Setup Phase**: In this phase, the network is set up, i.e., the sensor nodes elect among themselves cluster head node for the current round and the ordinary nodes join the cluster heads to form clusters.
2. **Initialization Phase**: The base station generates a unique ID for all nodes during this phase. The base station accomplishes this by hashing each node's Ethernet address. For example, $ID_i = Hash(E.A.i)$, where IDi is the unique identification for the ith sensor node and E.A.i is the ethernet address for the ith sensor node.
3. **Detection and Removal Phase**: This phase comprises of the detection of the malicious cluster head node and removing it from the local blockchain network. It is described by Algorithm 1 in Sect. 4.3.
4. **Resetting of Cluster head nodes**: After the identification and removal of malicious cluster head node for the current round. After that, for the next round, the cluster head nodes are again selected.

– **Setup Phase**: In this phase, similar to the traditional LEACH [1] protocol, all of the sensor nodes in the network elect who will be the cluster head nodes for the current round. For this, the sensor nodes choose a random number between 0 and 1. The node is designated as the cluster head for the current round if the chosen number is less than the required threshold T(n) requirement. Following that, all of the elected cluster head nodes broadcast an advertisement message to all of the regular nodes. Ordinary nodes chose their clusters based on the strength of the received broadcasted message.

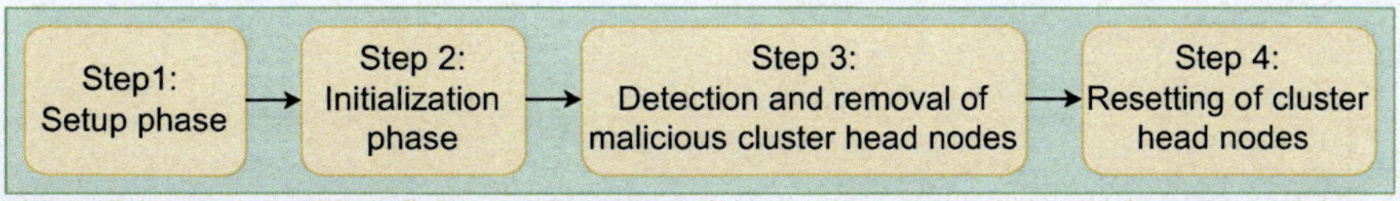

Fig. 3 Overall proposed model of MBLEACH

- **Initialization Phase**: After the cluster formation, the base station generates a unique identifier (ID) for each of the nodes, including itself. The base station hashes the Ethernet address and creates a unique identifying ID (hash value) for each node n_i in the network because each node has a distinctive Ethernet address E_i. The unique identifiers for the various nodes generated are as follows: BSID, which denotes the hash value of the base station's Ethernet address, CID, which signifies the hash value of the cluster head node's Ethernet address, and OID, which is the hash value of the Ethernet address of the ordinary node. Table 1 below depicts the table for unique ID generation.
- **Detection and removal of malicious cluster head node phase**: The detection and removal of malicious cluster head nodes are divided into the following two phases:

1. **Malicious Cluster Head Information Storage**: This paper designs a malicious cluster head information storage structure that primarily comprises of attribute and value pairs, with the attribute component storing the malicious cluster head node ID, the round number, and the block number at which the node started behaving maliciously. Table 2 below shows the format in which the malicious Cluster head node information will be stored where,
   - $r_n o$: indicates the round number in which the cluster head node behaved maliciously.
   - **MHCID**: Unique identification of the cluster head node which behaved maliciously in the round rno, which is generated by hashing the Ethernet address of the malicious cluster head node given by Eq. (1)

$$MCHID_i = SHA256(E.A._{MCH_i}) \tag{1}$$

   - b_{no}: indicates the block number at which the cluster head started behaving maliciously.
   - $addr_{MCH}$: 32 bytes address to which the malicious cluster head node ID is mapped to.
2. **Identification of Malicious Cluster head node** The cluster head nodes that have joined the local blockchain network are monitored by the base station for any malicious activity. Malicious action means that the malicious cluster head node will either stop mining transactions or stop signing transactions or stop producing new blocks. If any of the cluster head nodes behaves maliciously,

Table 1 Table of Unique ID generated

Unique Id Generation
BSID = SHA256 (Ethernet Address of Base Station)
CHID= SHA256 (Ethernet Address of Cluster Head)
ORDID = SHA256 (Ethernet address of ordinary nodes)

Table 2 Malicious cluster head node information storage structure

Attribute	Value
Round number	r_{no}
Malicious cluster head node ID	MCHID
Block number	b_{no}
Address of malicious cluster head	add_{MCH}

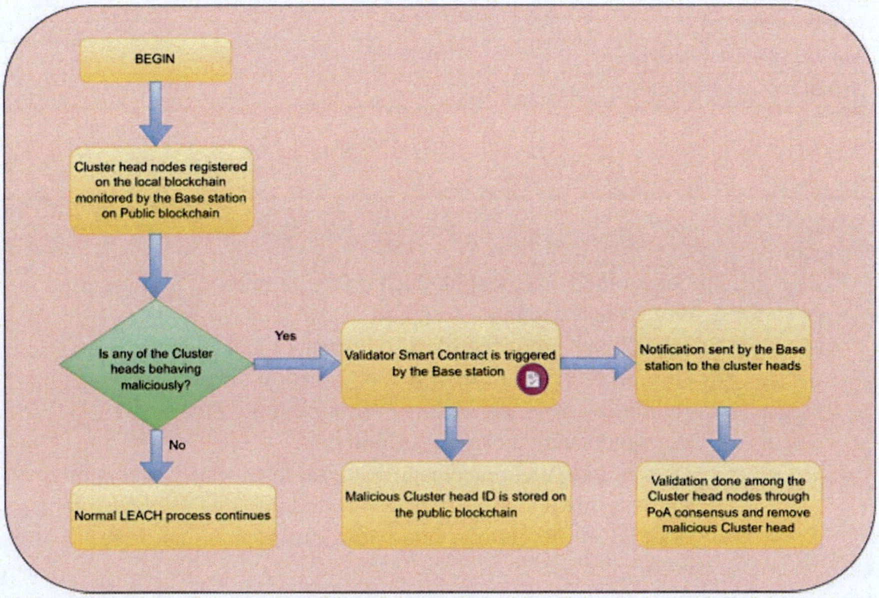

Fig. 4 Malicious cluster head node identification steps

a Validator Smart Contract will be triggered by the base station on the public blockchain network, and details about the malevolent cluster head node will be stored on the global blockchain network. An event will be triggered by the base station to notify the cluster head nodes. The cluster head nodes on receiving this event/signal validate among themselves and then discard or remove the particular malicious cluster head node from their network table using the PoA (Clique) [4] consensus mechanism. Consequently, those nodes will be discarded for participation in further rounds. The complete detection mechanism of malicious Cluster head node is described by Algorithm 1 and Fig. 4 below.

Table 3 Experimental setup

Components	Tools required
Smart contract	Solidity programming language
Public blockchain	Ethereum main network
Private blockchain	Go-Ethereum Client (Geth)
Integrated development environment	Remix IDE (For building and testing SmartContracts)
Simulation of network	Python language
Number of nodes	100
Number of cluster head nodes	35
Number of malicious cluster heads	1–11

Algorithm 1

```
Begin
Cluster head nodes joined on Local blockchain are monitored by Base Station
      for (each Cluster head node CH_i) do
        if (time since last produced block on private blockchain >10 sec)
          if ( current_block_number = = last_mined_blocknumber)
              Not mining new new blocks or dropping the blocks
              BS triggers validator Smart Contract
              BS stores (malicious_clusterhead_id, last_mined_blocknumber,
               address_malicious_clusterhead,round_no) on Public Blockchain
              Emit (Event(malicious_node_id,address_maliciousnode,
              round_no))
      each Cluster head node using PoA consensus:
         Verify malicious_node_address
            clique.discard ("Address_of_malicious_node", "False")
End
```

4 Simulation Results

A software implementation has been performed in order to simulate the performance of the proposed blockchain-based LEACH mechanism. The Solidity programming language is utilized for the development of the Smart Contracts. Remix IDE has been used to write and test the Smart Contract functionalities before deploying it on the Ethereum main network. The Geth (Go-Ethereum) private blockchain network is used to construct the local or private blockchain network, and all Cluster head nodes join the Geth network. The entire simulation procedure is carried out in the presence of 100 nodes, with 35 cluster heads and 11 malicious cluster heads (Table 3).

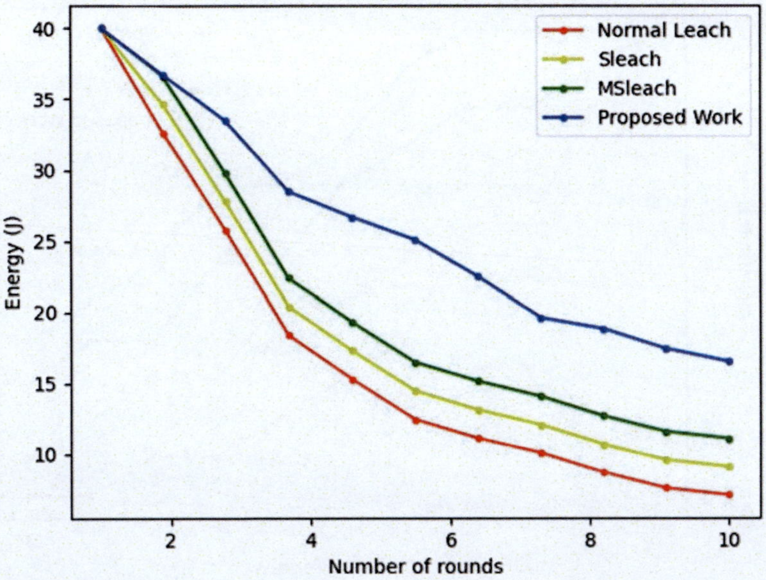

Fig. 5 Energy dissipation comparison for Normal LEACH [1], SLEACH [3], MSLEACH [2], and MBLEACH in the presence of malicious cluster head nodes.

4.1 Experimental Setup

4.2 Performance Analysis

Energy Dissipation and Network Lifetime in the presence of malicious CH nodes: In the traditional LEACH protocol [1], the elected Cluster head nodes may start behaving maliciously by receiving the data packets from all the ordinary nodes within in its cluster but instead of forwarding it to the Base Station drop these packets. Therefore, the malicious Cluster head nodes dissipate the energy of all the sensor nodes within its cluster thereby decreasing the overall network lifetime which is evident from Figs. 5 and 6. Moreover, our proposed model detects as well as removes the malicious Cluster head nodes which dissipates the energy by dropping the packets or by injecting false packets in the network or by changing the destination address, etc. Therefore, the proposed work increases the network lifetime by saving energy.

Packet Drops in the presence of malicious nodes: Because there are malicious cluster head nodes in the normal LEACH protocol, more data packets are dropped, whereas in our model, packets are dropped only if the nodes are dead. Figure 7 illustrates this point. Malicious cluster head nodes are recognized in the proposed model using Algorithm 1. For this reason, performance of the proposed model is better. Compared to other approaches such as LEACH [1], SLEACH [3], and MSLEACH [2].

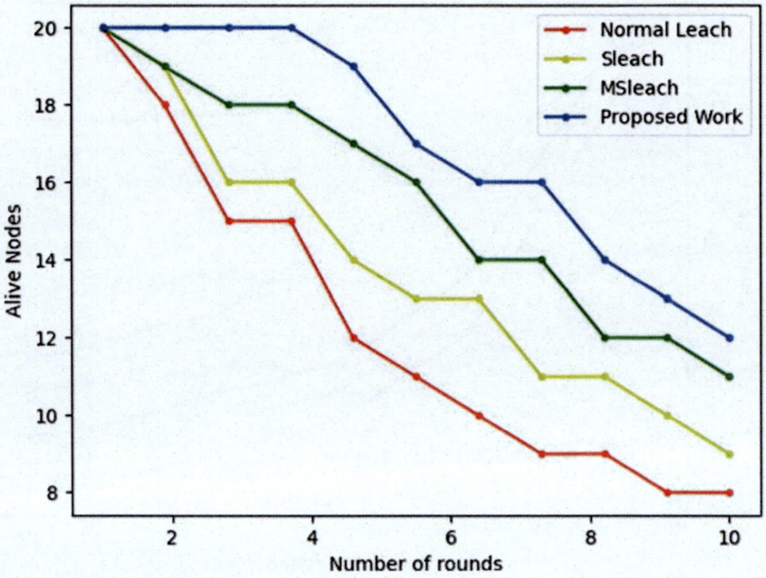

Fig. 6 Network Lifetime comparison for Normal LEACH [1], SLEACH [3], MSLEACH [2], and MBLEACH in the presence of malicious cluster head nodes

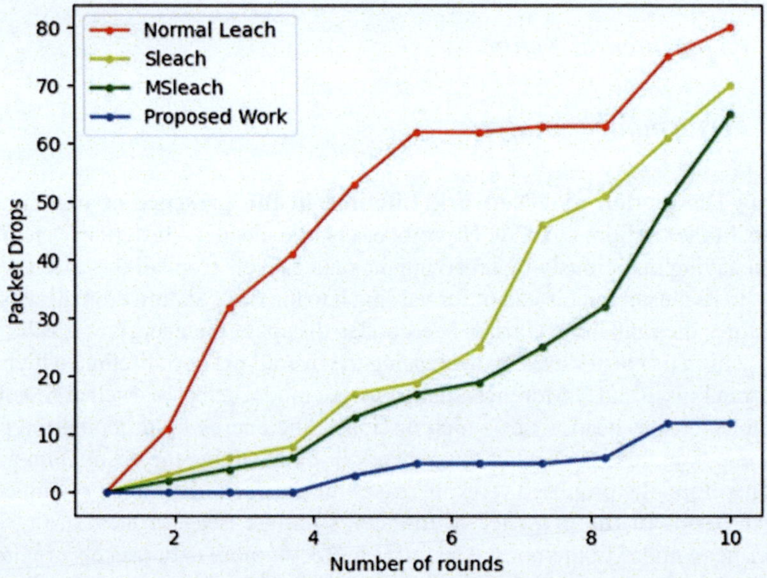

Fig. 7 Packet Drops comparison for Normal LEACH [1], SLEACH [3], MSLEACH [2] and MBLEACH in the presence of malicious Cluster head nodes.

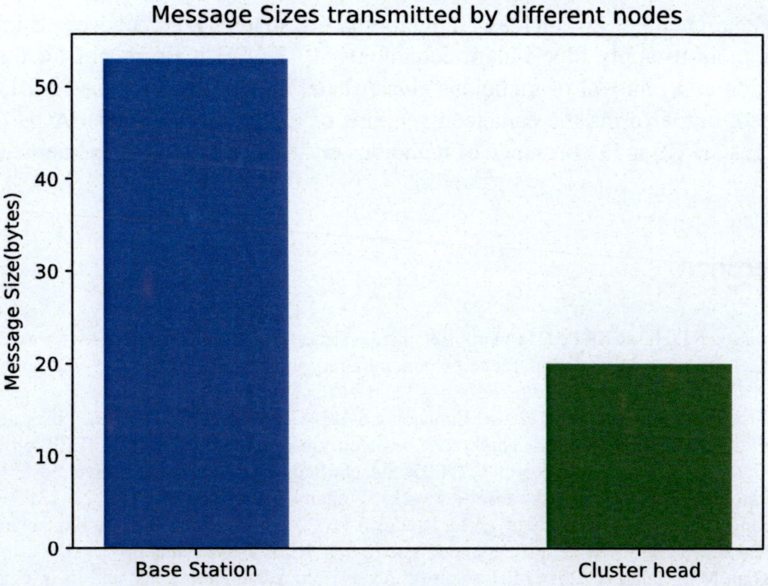

Fig. 8 Comparison of message sizes transmitted by different nodes

Comparison of Message Sizes transmitted by different nodes: We compare the message interaction between nodes and the size of the blockchain transaction submission message during the execution of the proposed model, and the result is shown in Fig. 8. The message size transmitted by base station is more because, after the detection of malicious cluster head, it emits an event containing the complete identity information of malicious node, whereas, in order to validate and remove the malicious nodes through PoA [4] consensus, the cluster head nodes only propagate the address of malicious node.

5 Conclusion

Cluster head selection and cluster formation play an important role in wireless sensor networks. In this scenario, LEACH [1] is the conventional cluster-based protocol that establishes milestones. However, the major drawback of this protocol is that it does not have any security policies associated with it. Consequently, any malicious node can act as cluster head which lowers the performance of LEACH protocol in terms of packet drops, throughput, etc. Various versions of LEACH protocol such as SLEACH [3] and MSLEACH [2] have already been built to enhance the security aspect of LEACH [1] protocol. However, they are not able to deliver outstanding performance in the presence of malicious nodes. Blockchain, whereas is a new decentralized technology that ensures the security of smart sensors. Features of the blockchain technology such as immutability, transparency, and consensus motivate the authors to apply blockchain technology in LEACH [1] protocol for the identification and removal of malicious cluster head nodes. The proposed MBLEACH protocol outperforms the enhanced versions of LEACH such as SLEACH [3] and MSLEACH [2] in the presence of malicious cluster head nodes in the network.

References

1. Hasson ST, Hasan SE (2021) an improvement on leach protocol for wireless sensor network. In: 2021 7th international conference on contemporary information technology and mathematics (ICCITM), pp 130–135. https://doi.org/10.1109/ICCITM53167.2021.9677653
2. ElSaadawy M, Shaaban E (2012) Enhancing S-LEACH security for wireless sensor networks. In: 2012 IEEE international conference on electro/information technology. IEEE, pp 1–6
3. Xiao-yun W, Li-zhen Y, Ke-fei C (2005) Sleach: Secure low-energy adaptive clustering hierarchy protocol for wireless sensor networks. Wuhan Univ J Nat Sci 10(1):127–131
4. Yang J, Dai J, Gooi HB, Nguyen H, Paudel A (2022) A proof-of-authority blockchain based distributed control system for islanded microgrids. IEEE Trans Indus Inf
5. Khan MA, Salah K (2018) IoT security: review, blockchain solutions, and open challenges. Futur Gener Comput Syst 82:395–411
6. Reyna A, Martín C, Chen J, Soler E, Díaz M (2018) On blockchain and its integration with IoT. ChallS Oppor, Futur Gener Comput Syst 88:173–190
7. Cui Z, Fei XUE, Zhang S, Cai X, Cao Y, Zhang W, Chen J (2020) A hybrid blockchain-based identity authentication scheme for multi-WSN. IEEE Trans Serv Comput 13:241–251
8. Ramasamy LK, KP FK, Imoize AL, Ogbebor JO, Kadry S, Rho S (2021) Blockchain-based wireless sensor networks for malicious node detection: a survey. IEEE Access 9:128765–128785
9. She W, Liu Q, Tian Z, Chen J-S, Wang B, Liu W (2019) Blockchain trust model for malicious node detection in wireless sensor networks. IEEE Access 7:38947–38956
10. Chen J, Shen H (2008) MELEACH-L: More energy-efficient LEACH for large-scale WSNs. In: 2008 4th international conference on wireless communications, networking and mobile computing. IEEE, pp 1–4
11. Zhixiang D, Bensheng Q (2007) Three-layered routing protocol for WSN based on LEACH algorithm
12. Lei Y, Shang F, Long Z, Ren Y (2008) An energy efficient multiple-hop routing protocol for wireless sensor networks. In: 2008 first international conference on intelligent networks and intelligent systems. IEEE, pp 147–150
13. Ali MS, Dey T, Biswas R (2008) ALEACH: advanced LEACH routing protocol for wireless microsensor networks. In: 2008 international conference on electrical and computer engineering. IEEE, pp 909–914

Community Detection in Large and Complex Networks Using Semi-Local Similarity Measure

Saikat Pahari, Anita Pal, and Rajat Kumar Pal

Abstract Community detection is essential to know the fundamental property of a large and complex network. The majority of the existing methods available in the literature either employ a global approach that requires structural information about the entire network which is computationally expensive for a large network or a local approach which has less accuracy. To create a balance between these two objectives, we propose a semi-local approach. A semi-local similarity index, *expanded overlap coefficient,* is presented, which considers not only the immediate common neighbor between two nodes in a network, but also considers second level common neighbors between them. The proposed method consists of three key steps: Detecting central seeds, expanding central seeds using the proposed semi-local centrality index and finally merging small communities to form a stable community. A local metric, *normalized local modularity* is proposed to expand the central seeds. Experimental results illustrate that the proposed method exhibits excellent results on real-world and computer generated artificial networks and is superior to the state of the art approaches.

Keywords Semi-local Similarity Measure · Community detection · Expanded Overlap Coefficient · Normalized Local Modularity · LFR networks

---

S. Pahari
Omdayal College of Engineering and Architecture, Howrah 711316, India

A. Pal (✉)
National Institute of Technology, Durgapur 713209, India
e-mail: amitmandal.nitdgp@gmail.com

R. K. Pal
University of Calcutta, Kolkata 700106, India

R. Chaki et al. (eds.), *Applied Computing for Software and Smart Systems*, Lecture Notes in Networks and Systems 555,
https://doi.org/10.1007/978-981-19-6791-7_5

1 Introduction

Various difficult systems in our daily life can be described using complex networks [1] like technological networks [2], information networks [3], biological networks [4, 5], social networks [6] etc. Technological networks include the transport system network, electrical grid, telephone network which are important in our daily life. Information networks consisting of web and citation networks help us to retrieve information easily. Biological networks like protein–protein interaction, neural networks, and gene-regulatory networks represent functions and interactions among neural entities. Due to the diversity of the context and extent in which complex networks appear, study of network science has attracted great attention from researchers of various disciplines. A wide range of studies have been organized to achieve deep understanding of network structures and uncover the underlying functions or characteristics of complex networks.

One common property of any real world network is that it should not be classified as a random network [7]. The degree distributions of nodes are skewed and that follows power–law distribution [8], small average distance between nodes [9], edges are not evenly distributed resulting in various clusters of nodes with high internal edge density and low inter cluster density [10]. These high edge density groups are generally referred to as communities and are of vast interest in the field of real-world networks. Typically a community is considered as a cluster of nodes which are closely connected to each other compared with other nodes in the network. The closeness among the nodes is based on a similarity measure that is defined over them. In the areas of data mining the task of clustering can be considered as unsupervised learning where the objective is to find the clusters of similar nodes without any prior information about the clusters [11]. Over the past decades researchers have conducted various studies on discovering communities using different ways. Broadly these approaches can be classified as global and local community detection methods. Global methods can be broadly categorized into three groups: sub-graph optimization [12], hierarchical [13], modularity optimization [14], random walk [15], and matrix factorization [16]. Main drawbacks of global methods are they require information about the entire network which is computationally intensive in today's large network with millions and billions of nodes. To overcome this problem, recently several local approaches are introduced which require local information to discover communities. In local approaches a seed set (containing one or more nodes of interest) are selected first, and then expand these nodes through optimizing an objective function. A widespread approach in seed expansion is probability distribution from the seeds [17]. Label propagation algorithm (LPA) [18] is another local method which detects communities by propagating the community labels to their neighborhood nodes. Though simple and fast, LPA suffers from instability due to its random behavior. Another drawback of LPA is ignoring small communities due to label propagation which affects the detection of local communities. Local similarity-based community detection has drawn a great attention in recent times. Qian et al. [12] proposed an overlapped community discovery method by maximizing the similarities between cliques.

All of these methods exploit either global approaches with high computational time or local methods with less accuracy.

In this paper, we propose a semi-local similarity-based community discovery method that can mitigate the gap between extensive computation time of global approach and lower accuracy of local approach. We provide a new similarity method and expanded overlap coefficient to capture the semi-local similarity among nodes in a network. The algorithm works in three phases: first a central seed is detected using the expanded overlapping coefficient, and the second seed is expanded based on whether its neighbors can satisfy a predefined condition. To do this we define local normalized modularity, which is a modification of the local modularity proposed by Clauset [19]. If the obtained local normalized modularity of the partial community after inclusion of a neighbor of the seed is above the threshold then the neighbor will be included. Finally small communities are merged to get a stable community. Our contributions in the paper are manifold:

- Proposed method can mitigate the gap between the extensive computation time of global approach and lower accuracy of local approach.
- Most existing methods in literature exploit any of these two objective functions, either maximizing internal density or maximizing structural density. There is no requirement of optimizing any objective function here.
- Time complexity of the proposed algorithm is near linear, which is acceptable for large networks.

Remaining part of the paper is prepared as follows: In Sect. 2 literature review and related work is discussed. Section 3 illustrates the proposed method with algorithm analysis. Next Sect. 4 shows the experimental results on real-world and artificial networks. Finally Sect. 5 summarizes the paper.

2 Related Work

Global methods can be broadly categorized into three groups: sub-graph optimization, hierarchical, modularity optimization and random walk-based methods. Hierarchical approaches find communities at different levels based upon similarities between nodes. There are two main categories of hierarchical methods: divisive and agglomerative. Divisive methods are basically top-down approaches where a network is decomposed iteratively until a strong partition is achieved. Girvan and Newman [20] have introduced a divisive algorithm that starts with considering the whole network as a single community and then iteratively removing the edges based on edge-betweenness score. Edge-betweenness is a measure of each edge which calculates the number of shortest paths an edge is part of. It is obtained by computing the sum of the fraction of the shortest path going through the edge between any two vertices. Ni et al. [21] proposed a community detection method using geometric decomposition of a network. The method utilizes the principle of Ricci flow process which is used iteratively to find the heavily traveled edges. On the other hand agglomerative methods are bottom up where initially every node is considered as a single node community and then based on optimization of some objective function they are merged iteratively to expand the community. Sharma et al. [13] used complete linkage hierarchical clustering which is an example of agglomerative clustering method to cluster the biological data. The authors started the method assuming all nodes are in its own community and then they are combined with other nodes with the shortest distance. The distance between communities equals the distance between those two nodes (one in each cluster) that are farthest away from each other. The minimum distance of these links that remains at any step causes the union of the two clusters. Another agglomerative approach proposed by Gorke et al. [22] is a cut-based approach to maintain the communities in a dynamic network. This method maintains the communities by updating the minimum cut tree continuously.

Another approach of discovering communities is modularity optimization where communities are found by maximizing the modularity measure. Louvan [23] proposed a greedy method where each node is considered as a single community. Next each node is replaced from its community to its neighboring community till there is no increase in modularity. Next all nodes of one community are merged together to form one node with self a loop and weighted edge and form a new network. This process continues till all nodes converge to a single community. The computational time of this method is O $(n \log n)$ where n represents the total number of nodes in the network. However, as modularity optimization is a NP-hard [24] problem, it is inappropriate for large networks. The third group of methods used for community detection achieves the global network information using random walk. It increases the overall accuracy using local heuristic search. Pon and Latapy [25] proposed a method, named Walktrap, which is a hierarchical clustering approach that employs random walk. The main idea of this method is that if we compute short distance random walks then the traversed nodes will always stay in the same community. Initially it assumes that there are no communities at all, then using random walk calculate the distance between all adjacent nodes. Next phase is the community merging phase where two adjacent communities are merged and the distance between them is modified. This process continued for $(n-1)$ times, where n is the number of nodes. Time complexity of this algorithm is $O(en^2)$. If the network is a sparse network the time complexity will be $O(n^2 \log n)$. Rosvall et al. [26] proposed Infomap which is based on the concept of information theory. Here a random walk-based shortest description length is computed. The description length is computed using map equation where a vertex has to encode the path traversed by a random walk. Time complexity of this method is $O(e)$.

Recently a growing interest is observed in the literature where researchers are adopting various methods to discover local community around a set of nodes, often referred to as seed set. One such method is the Spectral method where eigen vectors of the graph Laplacian are computed first. Then the leading eigenvectors determine the communities in a network. Michael et al. [27] utilized local properties of a network around an input seed which uses the second eigenvector to discover the local communities. Finding sparse vectors from local spectral subspace using l_1

norm optimization is used by many authors for local community mining [28]. Here the authors have modified the network with a self loop in each node and employed a conventional random walk-based power method for subspace iteration. Coscia et al. [29] proposed a similar to LPA approach, called DEMON, where authors defined that each node is equally responsible in the process of community expansion and merging. Though this method takes the whole network along with a threshold value as input, it discovers communities locally. The start nodes are selected randomly and for each node an ego network is obtained. To do this, the authors have defined two graph operations: ego network extraction and graph vertex difference to find the ego network which they call as 'EgoMinusEgo'. DEMON takes $O(n + m)$ time to discover the communities where n is the number of vertices and m is the number of edges in the network.

Structural similarity plays a vital role in detecting communities in large networks. Qian et al. [12] proposed a structural similarity based overlapping community detection method by maximizing the similarities between cliques. This method uses the similarity measure of connection in a community by finding the maximum clique. Finally overlapping communities are obtained by merging the closely connected communities. Peixoto [30] proposed a stochastic block model approach where a group of nodes are sifted simultaneously instead of individual nodes. To implement this authors have used Markov Chain Monte Carlo (MCMC) where more than one node can be shifted in every step. There are three main methods that are used to shift the nodes: reshuffling, splitting and merging a group of nodes. A core node expansion method named as ECES is proposed by Berahmand et al. [31] where core nodes of a network are obtained using an extended similarity measure. First they found the core nodes which are the central nodes of a dense area. This is done by getting weight of the edges of the graph using extended Jaccard similarity where not only the first degree, but even the second degree neighbors are also considered for more accuracy. Total time complexity of this method is $O(E)$ where E is the total number of edges in a network.

3 Preliminaries

Before entering into the proposed method, let us discuss some important concepts and definitions which are required for our method.

Background and notations: Let a network is represented by an unweighted graph G = (V, E) where V is a set of vertices, E is a set of edges. Now n = |V| is number of vertices and m = |E| is the number of edges. $A_{i,j}$ is adjacency matrix of graph G.

Definition 1

(*Semi-local similarity measure*) Global similarity measure requires prior knowledge about the entire network which is inappropriate for large a network due to its high complexity. On the contrary, local similarity considers only nearest nodes as a similarity index which has less accuracy [32]. To create a balance between these two objectives semi-local indices are gaining popularity nowadays. A semi-local similarity index considers not only the immediate common neighbors between two nodes, but also considers the second hop common neighbors, i.e. neighbors of neighbors between them.

Definition 2
(*Overlap Coefficient*) Overlap coefficient [33] is a similarity measure that measures similarity between two finite sets. For a given network the overlapping coefficient between two nodes i, j it can be denoted as:

$$OC(i, j) = \frac{(|\Gamma(x) \cap \Gamma(y)|)}{\min|\Gamma(x), \Gamma(y)|} \tag{1}$$

where $\Gamma(x)$ and $\Gamma(y)$ are the number of neighbors of X and Y, respectively. If $\Gamma(x)$ is a subset of $\Gamma(y)$ or converse the *OC(i, j)* will be 1. So it measures the ratio of common neighbors to minimum neighbors of the two nodes.

Definition 3
(*Expanded Overlap Coefficient*) While Overlap Coefficient considers the immediate neighbor, the expanded overlap coefficient extends the neighborhood area of distance *L*. Here instead of considering only immediate neighbors, neighbors up to distance *L* are considered for better accuracy and can be represented as:

$$OC_L(i, j) = \frac{\left(\left|\Gamma(x)_L \cap \Gamma(y)_L\right|\right)}{\min\left|\Gamma(x)_L, \Gamma(y)_L\right|} \tag{2}$$

where $\Gamma(x)_L$ defines the number of neighbors of distance *L*. It sets a good example of semi-local similarity measure. In this paper we set $L = 2$ i.e. along with immediate neighbors of a node and the next level of neighbors are also considered. Figure 1. Shows that the expanded overlap coefficient provides better similarity degree compared with the overlap coefficient.

3.1 Community Measuring Functions

Here we provide two well defined community measure functions: modularity and conductance. Also, we define a new measuring function, normalized local modularity (LQ_N) to estimate the goodness of a detected community. The most popular community measuring function that assesses the quality of a community is the modularity of Newman and Girvan [34], which is based on

comparison between real density of edges within a subgraph with probable density assuming the nodes were attached randomly. This can be represented as:

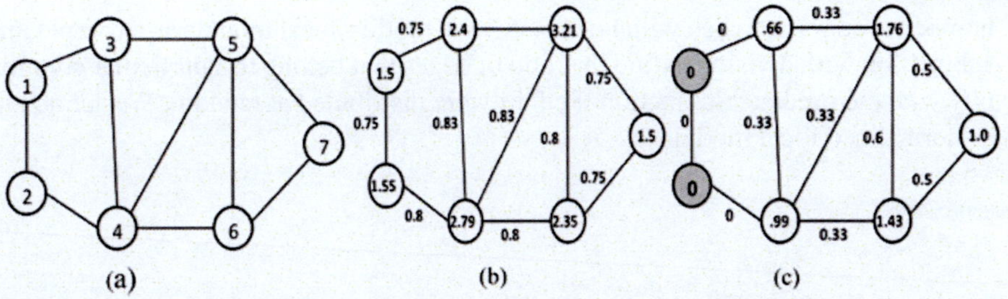

Fig. 1 Similarity measure of a network. **a** A graph with 7 vertices and 10 edges. **b** Similarity measure with expanded Overlap coefficient and **c** Same with Overlap coefficient, Edges 1–3, 1–2 and 2–4 have no similarity values. Centrality score of vertices obtained using Eq. (8)

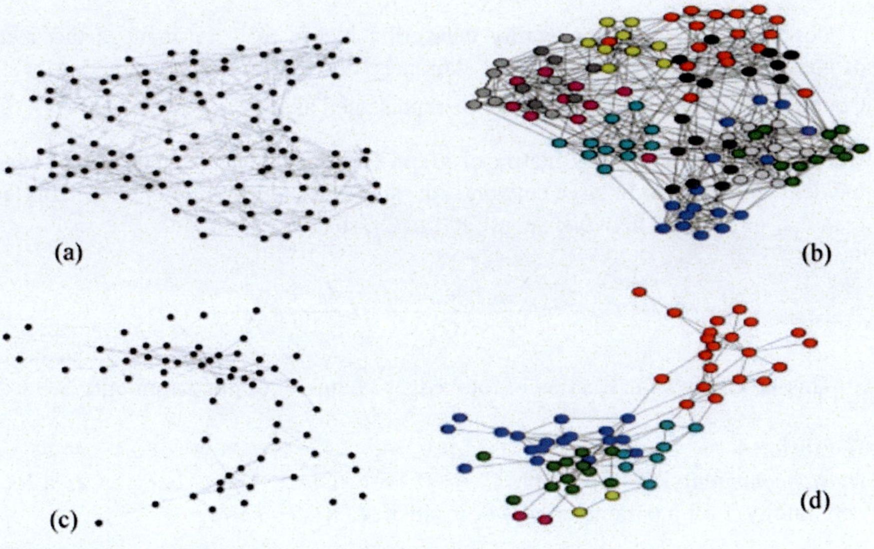

Fig. 2 Original **a**, **c** Football and Dolphin network, **b**, **d** communities obtained by proposed algorithm. Different communities are indicated with different colors

$$Q = \frac{1}{2m} \sum_{i \cdot j} \left(A_{i,j} - \frac{k_i k_j}{2m} \right) \partial (C_i C_j) \quad (3)$$

Here m denotes the number of edges, k_i, k_j are degree of vertex i and j, C_i is a community that contains vertex i, and $\partial(C_i C_j) = 1$ if i, j belong to the same community, otherwise 0. The network with a high modularity has dense connection between nodes of a community and spares connection with other nodes in other communities. But global community measures require all edges to be known which is not suitable for large networks. Clauset [19] proposed local modularity LQ by modifying global modularity Q as follows: $LQ = \frac{e_c}{S} - (\frac{d_c}{2S})^2$. where e_c and d_c are the number of edges within community C and total degree of community C respectively, S represents the number of edges attached with nodes in C. We modify local modularity by replacing subtraction with division to find the ratio of edges that belong to a particular community with the predictable fraction if edges were distributed at random. We define this as normalized local modularity:

$$LQ_N = \frac{4Se_c}{dc^2} \tag{4}$$

Now we can simplify this by ignoring the coefficient 4 as this is constant in the same sub-network, then we simply define

$$\overline{LQ_N} = \frac{Se_c}{dc^2} \tag{5}$$

Conductance is another quality measuring metric that can manage the speed of convergence of a network to the stationary distribution during a random walk. Conductance of a community C can be represented as $\varphi_C = \frac{\sum_{\mathrm{i}\in \mathrm{C},\mathrm{j}\in\overline{\mathrm{C}}} \mathrm{A}_{\mathrm{i,j}}}{\min(\mathrm{a(C)},\mathrm{a}(\overline{\mathrm{C}}))}$

where $\mathrm{A}_{\mathrm{i,j}}$ is the adjacency matrix of graph G, a(C) is the total number of edges incident with C. Now in large networks, the size of a single community remains less than half of the total network i.e. a(C) < $a(\overline{\mathrm{C}})$, so

$$\varphi_C = \frac{\sum_{i\in C, j\in\overline{C}} A_{i,j}}{a(C)} = \frac{d_c - 2e_c}{d_c} \tag{6}$$

This φ_C denotes the fraction of total edges going out of the community S.

Definition 4
(*Local community*) A sub-graph $C \subset V$ with $|C| \ll |V|$ is denoted as a local community if for a parameter $\sigma > 0$, it satisfies $\overline{LQ_N}(C) > \sigma$.

Theorem 1
Normalized local modularity LQ_N of a network is equal to $1 - \varphi_C^2$, where φ_C is the conductance of that network.

Proof We have already discussed conductance φ_C can be expressed as follows:
$\varphi_C = \frac{d_c - 2e_c}{d_c}$ where $d_c = 2e_c + e_{out}$. Therefore φ_C can be represented as:
$\varphi_C = \frac{2e_c + e_{out} - 2e_c}{2e_c + e_{out}} = \frac{e_{out}}{2e_c + e_{out}}$
Alternatively, $1 - \varphi_C = 1 - \frac{e_{out}}{2e_c + e_{out}} = \frac{2e_c}{2e_c + e_{out}}$
So,

$$\frac{1 - \varphi_C}{2} = \frac{e_c}{2e_c + e_{out}} \tag{7}$$

We have defined Normalized local modularity LQ_N as follows: $LQ_N = \frac{4Se_c}{d_c^2}$.

where $S = e_c + e_{out}$.
So, $LQ_N = \frac{4(e_c+e_{out})e_c}{(2e_c+e_{out})^2} = \frac{4(e_c+e_{out})}{(2e_c+e_{out})} \cdot \frac{e_c}{(2e_c+e_{out})} = 4.\left[\frac{e_c}{2e_c+e_{out}} + \frac{e_{out}}{2e_c+e_{out}}\right].\left[\frac{e_c}{(2e_c+e_{out})}\right]$
Substituting equations (vi) and (vii).
$= 4[\frac{1-\varphi_C}{2} + \varphi_C].[\frac{1-\varphi_C}{2}] = 4\left[\frac{1+\varphi_C}{2}\right].\left[\frac{1-\varphi_C}{2}\right] = 1-{\varphi_C}^2$
Hence it is proved that $LQ_N = 1-{\varphi_C}^2$.

4 Proposed Method

Local community detection method works in three phases: finding central seeds, expansion of central seed and community merging. In the first phase it finds central seeds in a weighted graph using a semi-local similarity measure. With an intuition that a community grows around a central node, we select the seed from a denser part of the network. To implement this, we introduce a new similarity evaluate index named as the expanded overlap coefficient that gives a higher similarity measure compared with other similar types of indexes. We use expanded overlapping coefficient $OC_{L=2}(i, j)$ between two nodes i and j to obtain the weight of the edge $e_{i.j.}$ Then the centrality score of vertex i is obtained by adding all weights of the edges connected with it:

$$\text{Centrality score (i)} = \sum_{j=1}^{k} A_{ij} OC_{L=2}(i, j) \tag{8}$$

where A_{ij} is the adjacency matrix of the graph. A queue is used to store the centrality score of all the nodes. The element with the highest score is chosen as a central seed. The seed node along with neighbors of the seed forms an initial local community which will be expanded next. In the next step, the initial community expands according to the definition (1) of the local community. All neighbors of the initial community form the expansion set. A node j will be added to the initial community if its inclusion in the initial local community satisfies the condition $\overline{LQ_{N}}_j(C) > \sigma$. We have tested σ in a range 0.3 to 0.7. Based upon the test result, the value of σ is set to 0.4 as more than this value will make the expansion process very strict. The neighbors of j will also be added to the expansion set. This step is continued till there does not exist any nodes to satisfy the condition. First community of the network is obtained. Now the nodes belonging to the detected community will be removed from the queue. The remaining nodes are searched to find the node with the highest centrality score and that is selected as the seed of the next community. The expansion process of the newly found seed is followed as discussed. In the last stage small communities are merged together depending upon satisfying a condition. At this point each community is treated as a single node whose weight is obtained by adding the weights of each edge within the community. For all edges between any two communities X and Y, we create one single weighted edge by adding the weights of the edges that attach between nodes of X and Y. If the weight of any node X is less than the weight of any of the edges of the neighboring node Y, then X will be merged with Y. Also the community list will be updated. This process continues until no more community remains for merging. In this way very small communities will be merged together to form a more stable community. Here is a pseudo code of the proposed algorithm.

Algorithm 1: LocalCommunityDetection

Input: Network G(V, E)

Output: All local communities LC

1. **while** (NodeList = empty)
2. v_i = CentralSeedSelection(G, N)
3. S = SeedExpansion(vi, G, N)
4. C_{com}= C_{com} ∪ S
5. **end while**
6. LC = MergeCommunities(C_{com})
7. **Return** LC

Algorithm 2: CentralSeedSelection

Input: Network G(V, E)

Output: A node with highest semi-local similarity

1. **for** each edge e_{ij} ∈ E **do**
2. Compute similarity measure $OC_{L=2}(i, j)$ using equation (3)
3. **end for**
4. **for** each vertex v_i ∈ V**do**
5. Compute centrality score using equation (8)
6. Store into a queue
7. **end for**
8. **sort** the queue in descending order
9. **Return** front node

Algorithm 3: SeedExpansion

Input: Network G(V, E), seed v_i, local com C_{loc}, initial community C_{ini}, NodeList N

Output: Local community C_{loc}

1. Initialize $C_{loc} = \{v_i\}$
2. **for** all j ∈ V do
3. **if** $A_{ij} = 1$ **then**
4. $C_{loc} = C_{loc} \cup j$
5. **for** all k ∈ V do
6. **if** $A_{jk} = 1$ **then**
7. $C_{ini} = C_{ini} \cup k$
8. **for** all j ∈ C_{ini} **do**
9. compute $\overline{LQ_N}$ of $C_{loc} \cup j$ using equation (5)
10. **if** $\overline{LQ_N} > \sigma$ **then**
11. $C_{loc} = C_{loc} \cup j$
12. **for** all k ∈ V **do**
13. **if** $A_{jk} = 1$ **then**
14. $C_{loc} = C_{loc} \cup k$
15. Remove j from C_{ini}
16. Remove i from N
17. **Return** C_{loc}

Algorithm 4: MergeCommunity

Input: Set of small local community S

Output: Stable Local community C_l

1. Wi = ø
2. **for** each i ∈ S do
3. **for** each node v_i **do**
4. $Wi = W_i + e_{ii}$
5. **for** all j **do**
6. $e_{ij} = e_{ij} + OC_{L=2}(i, j)$
7. **if** $Wi < e_{ij}$ **then**
8. $S_i = S_i \cup S_j$
9. **Return** C_l

Overall algorithm analysis: The proposed method has three key phases. First edge weights of the graph are determined using the expanded overlap coefficient with O (nk^2) time. All edge weights are summed up to obtain the node weight in O (nk) time. Nodes are sorted to find the central seed in O (n log n) time. The second phase deals with the expansion of the central seed up to community diameter d, and most of the cases d lie within 4. Time complexity of this phase is O (nk^d). Finally small communities are merged O (n) time. On the whole time complexity T(n) of the proposed method is as follows:

$$T(n) = O\left(nk^2 + nk + n\log n + nk^d + n\right) \approx O(n\log n).$$

5 Experiments and Results

5.1 Datasets

To illustrate the performance and effectiveness of the proposed method, we have used both real-world and computer generated networks. Details of the dataset are given here.

Real-world networks: Eight real world networks which comprise a good combination of small and large networks are used to evaluate the proposed method (Table 1).

The Zachary Karate club (ZKC) [35] is a social network of friendship among 34 members in a US university karate club in the 1970s. Zachary observed that in two years the network parted into two different groups due to some disagreement between the club administration and the instructor. Dolphin [36] is a network of 62 dolphins observed between 1994 and 2001 in New Zealand. Dolphins are represented as nodes and an edge between the two dolphins signifies that they are seen together "more than expected". American college football (ACF) [37] is a games network of division 1 colleges in America. Colleges (nodes) are grouped into conferences and each conference is treated as a community. An edge between two nodes represents a match between them. E-mail network [38] is constructed using the email transferred between two users in a research institution in Europe. An edge between node u and v in the network means at least one mail is transferred between u and v. DBLP computer science bibliography [38] is a network of co-authorship with communities consisting of authors published in the same conference or journal. Amazon [38] is a network where communities signify the group of products that are often purchased together. Youtube [38] is a video sharing website and LiveJournal is an online community [38] that allows members to maintain journal, group or personal blogs. It allows users to create groups where other members can connect. These groups can be treated as ground truth communities.

Table 1 Dataset contains eight real-world networks with ground truth communities

Network	No. of nodes (n)	No. of edges (m)	Ground truth communities (k)	Refs.
Zachary Karate Club	34	78	2	[35]
Dolphin	62	159	4	[36]
Football	115	613	10	[37]
E-mail	1,005	15,571	42	[38]
DBLP	317,080	1,049,866	13,477	[38]
Amazon	334,863	925,672	75,149	[38]
Youtube	1,134,890	2,987,624	8385	[38]
LiveJournal	3,997,962	34,681,189	287,512	[38]

Artificial networks: Apart from real-world networks, we also used computer generated artificial networks to evaluate our algorithm. LFR benchmark is a widely used artificial network for community detection. It is proposed by Lancichinetti et al. [39] with an objective to generate a benchmark network where the degree of nodes and the size of the communities follow power–law distribution with exponents γ and β, respectively. Total numbers of nodes are represented with n and average degree k. The mixing parameter μ defines the ratio of the external degree of a node to the total degree of that node. So $\mu = 0$ indicates that all edges lie within a community and $\mu = 1$ indicates that all links are connected with the nodes in other communities. Minimum and maximum degree of a node is represented with d_{min} and d_{max},respectively. Minimum and maximum community size denoted with C_{min} and C_{max}.

As one strong community contains higher inter-cluster edges than intra-cluster therefore $\mu \leq ½$ is a good choice. We create four networks with varying number of nodes and are denoted as LFR1 to LFR4 with a diverse range of mixing parameter, $\mu = 0.1$ to 0.8 as follows:

LFR1 : $n = 5, C_{\min} = 10, C_{\max} = 40, d_{\min} = 50, d_{\max} = 80, \gamma = 2, \beta = 1, \mu = 0.1 - 0.8$

LFR2 : $n = 1,000, C_{\min} = 10, C_{\max} = 40, d_{\min} = 50, d_{\max} = 80, \gamma = 2, \beta = 1, \mu = 0.1 - 0.8$

LFR3 : $n = 10,000, C_{\min} = 20, C_{\max} = 60, d_{\min} = 60, d_{\max} = 80, \gamma = 2, \beta = 1, \mu = 0.1 - 0.8$

LFR4 : $n = 100,000, C_{\min} = 20, C_{\max} = 80, d_{\min} = 60, d_{\max} = 80, \gamma = 2, \beta = 1, \mu = 0.1 - 0.8$

Performance metric: We have used *normalized mutual information,f1-score* and *slandered internal density* to evaluate the result of the proposed method.

Normalized mutual information (NMI) [40]: Let N be a confusion matrix, whose rows and columns represent 'real' (ground truth) and 'obtained' communities, respectively. N_{ij} is the element of N which stands for the number of nodes deviated from the real community i to the detected community j. Then the NMI is $\frac{-2\sum_{i=1}^{c}\sum_{i=1}^{\overline{C}} N_{ij}\log(N_{ij}N/N_{io}N_{oj}}{\sum_{i=1}^{c} N_{io}\log(N_{io}/N)+\sum_{i=1}^{\overline{C}} N_{oj}\log(N_{oj}/N)}$ where C denotes the number of communities indicated by the LFR model and $\overline{C}$ denotes the number of obtained models using any algorithm. N_{io},N_{oj} are the sum of the elements of the i-th row of N and that over the j-th column respectively. In case the obtained communities are the same as the real one then NMI will be 1, and if both are totally different NMI vanishes.

F1—score [41]: To evaluate the goodness of a community with given ground truth data, *F1—score* is a widely used metric. It can be represented as fusion of precision and recall:

$$F1 - \text{score} = 2\frac{\text{precision} * \text{recall}}{\text{precision} + \text{recall}}, \text{precision} = \frac{|\overline{C} \cap C|}{\overline{C}} \text{ and recall} = \frac{|\overline{C} \cap C|}{C}$$

where $\overline{C}$ is the detected community by the proposed algorithm and C is the actual ground truth community. *F1—score* is the amalgamation of recall and precision. Its value lies between 0 and 1, more score indicates a better community.

Omega Index [42]: To evaluate the cohesiveness of a community based on its connection we have used *Omega Index.* It uses Rand Index for correcting chance agreement by considering the total communities that contain the same pair of nodes.

To evaluate the performance of the proposed method, we compared it with the existing state of the art methods like LPA [18], Infomap [26], Markov Chain Monte Carlo (MCMC) [30], DEMON [29] and ECES [31] on both real-world and artificial network. Infomap and MCMC consider a global approach while DEMON and ECES consider a local approach. ECES is a local similarity-based approach which is nearly similar to this paper.

Experiments on artificial dataset: The proposed method is applied over the class of LFR benchmark network. Figure 3 depicts the quality of different compared methods with respect to its NMI values. It is clear for LFR1 that LPA performs well for lower values of μ, but the proposed method outperforms others for $\mu > 0.4$. For LFR2, Infomap keeps steady the results throughout. In LFR3 our method shows good results in the range $0.1 > \mu > 0.3$ whereas in LFR4 initially MCMC performs well but the proposed method exhibits good results with a higher μ value.

Figure 4a shows composite performance of various artificial networks. We have considered three measures for composite performance, namely NMI, F1 score and omega value. The proposed is the efficient one with a composite score 2.65 compared with others.

Experiments on real-world dataset: Our proposed algorithm is evaluated by applying it over eight real-world networks (Table 1). Figure 4b shows the performance of the ZKC network. It is clear that in the NMI metric both the Infomap and the proposed one yields good results, whereas for F1, LPA is better. For omega, MCMC

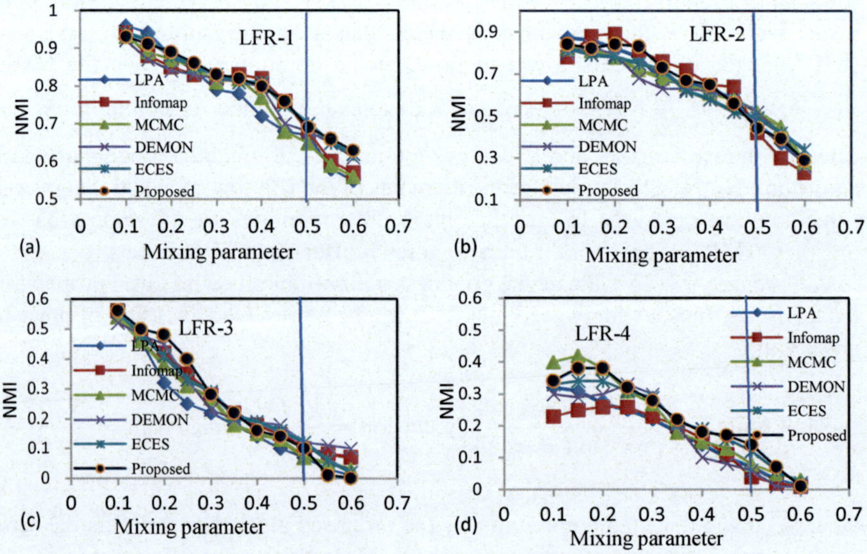

Fig. 3 Performance of different artificial networks with respect to its NMI values

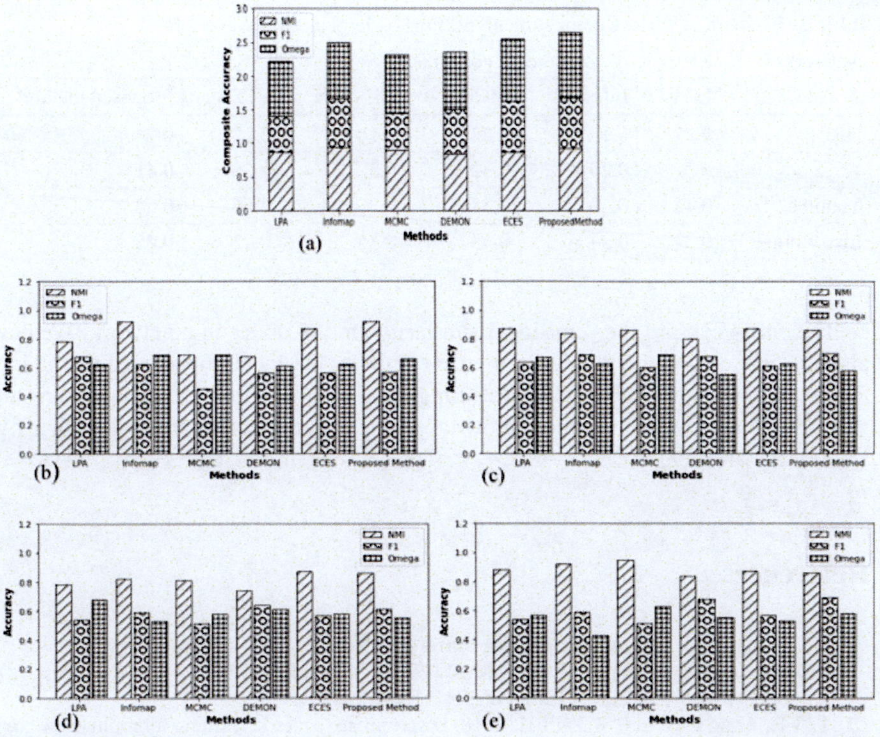

Fig. 4 **a** Shows the composite score over various artificial networks. **b**, **c**, **d** and **e** shows the NMI, f1 and Omega values of Karate Club, Dolphin, Football and E-mail networks with different mixing parameter

and the proposed method show satisfactory results. Figure 4c depicts the accuracy of the dolphin network. The F1 score of the proposed algorithm is best compared with others. Football network shows satisfactory NMI and F1 values (Fig. 4d) and E-mail network shows the highest F1 score for the proposed method (Fig. 4e). For large datasets like DBLP, Amazon, Youtube and LiveJournal, NMI and omega values cannot be obtained. F1 values of compared algorithms are presented in Table 2. For DBLP, LPA yields the best result. For Amazon and LiveJournal both the proposed method and MCMC algorithm perform well compared with others.

6 Conclusion

Most of the methods that exist in literature deal with either local or global approaches. Here we have presented a semi-local similarity-based community discovery method that will work as a tradeoff between the expensive global approach and the less accurate local approach. We provide a new similarity method, the expanded overlap

Table 2 F1 score of various competing algorithms on large networks

Networks	F1 score of compared algorithms					
	LPA	Infomap	MCMC	DEMON	ECES	Proposed method
DBLP	**0.27**	0.15	0.22	0.18	0.22	0.24
Amazon	0.39	0.20	0.32	0.32	0.36	**0.41**
Youtube	0.42	0.23	0.38	0.26	**0.66**	0.62
LiveJournal	0.32	0.34	**0.45**	0.35	0.42	**0.45**

coefficient to capture the semi-local similarity among nodes in a network. We have also introduced a new modularity metric named as the local normalized modularity to expend the central seeds while detecting local communities. Experimental results show that the proposed methods show favorable accuracy compared with other five state of the art methods.

References

1. Newman M (2018) Networks. Oxford University Press
2. Mishra S, Hota C, Kumar L, Nayak A (2019) An evolutionary GA-based approach for community detection in IoT. IEEE Access 7:100512–100534
3. Li J-H, Wang C-D, Li P-Z, Lai J-H (2018) Discriminative metric learning for multi-view graph partitioning. Pattern Recognit 75(C):199–213
4. Menche J, Sharma A, Kitsak MSD, Ghiassian M, Vidal M, Loscalzo J, Barabási A-L (2015) Uncovering disease-disease relationships through the incomplete interactome. Science 347(6224):1257601
5. Ma X, Sun P, Zhang Z-Y (2018) An integrative framework for protein interaction network and methylation data to discover epigenetic modules. IEEE/ACM Trans Comput Biol Bioinf 16(6):1855–1866
6. Havens TC, Bezdek JC, Leckie C, Ramamohanarao K, Palaniswami M (2013) A soft modularity function for detecting fuzzy communities in social networks. IEEE Trans Fuzzy Syst 21(6):1170–1175
7. Erd¨os P, R´enyi A (1960) On the evolution of random graphs. Publ Math Inst Hung Acad Sci 5:17–61
8. Barabasi AL, Albert R (1999) Emergence of scaling in random networks. Science 286(5439):509–512
9. Albert R, Jeong H, Barabasi A-L (1999) The diameter of the world wide web. Nature 401:130–131
10. Girvan M, Newman ME (2002) Community structure in social and biological networks. Natl Acad Sci 99(12):7821–7826
11. Jain AK, Murty MN, Flynn PJ (1999) Data clustering: a review. ACM Comput Surv 31(3):264–323
12. Qian X, Yang L, Fang J (2018) Heterogeneous network community detection algorithm based on maximum bipartite clique. In: ICPCSEE, pp 253–268
13. Sharma A, Boroevich K, Shigemizu D, Kamatani Y, Kubo M, Tsunoda T (2016) Hierarchical maximum likelihood clustering approach. IEEE Trans Biomed Eng 64(1):112–122
14. Traag VA, Waltman L, van Eck NJ (2019) From Louvain to Leiden: guaranteeing well-connected communities. Sci Rep 9:5233

15. Chen J, Xu G, Wang Y, Zhang Y, Wang L, Sun X (2018) Community detection in networks based on modified pageRank and stochastic block model. IEEE Access 6:77133
16. Kamuhanda DW, M. and He, K. (2020) Sparse Nonnegative Matrix Factorization for Multiple-Local-Community Detection. IEEE Trans Comput Soc Syst 7(5):1220–1233
17. Li Y, He K, Kloster K, Bindel D, Hopcroft J (2018) Local Spectral Clustering for Overlapping Community Detection. ACM Trans Knowl Discov Data 12(2):1–27
18. Raghavan UN, Albert R, Kumara S (2007) Near linear time algorithm to detect community structures in large-scale networks. Phys Rev E 76:036106
19. Clauset A (2005) Finding local community structure in networks. Phys Rev E 72(2):026132
20. Girvan M, Newman ME (2002) Community structure in social and biological networks. Proc Natl Acad Sci 99(12):7821–7826
21. Ni C-C, Lin Y-Y, Luo F, Gao J (2019) Community detection on networks with Ricci flow. Sci Rep 9(1):1–2
22. Görke R, Hartmann T, Wagner D (2012) Dynamic graph clustering using minimum-cut trees. J Graph Algorithms Appl 16(2):411–446
23. Blondel VD, Guillaume JL, Lambiotte R, Lefebvre E (2008) Fast unfolding of communities in large networks. J Stat Mech Theory Exp 10
24. Brandes U, Delling D, Gaertler M, Gorkee R, Hoefer M, Nikoloski Z, Wagner D (2008) On modularity clustering. IEEE Trans Knowl Data Eng 20:172–188
25. Pons P, Latapy M (2005) Computing communities in large networks using random walks. In: Computer and information sciences-ISCIS. Springer, Berlin, pp 284–293
26. Rosvall M, Axelsson D, Bergstrom CT (2009) The map equation. Euro Phys J Special Top 178(1):13–23
27. Michael WM, Orecchia L, Vishnoi NK (2012) A local spectral method for graphs: with applications to improving graph partitions and exploring data graphs locally. J Mach Learn Res 13(1):2339–2365
28. He K, Shi P, Bindel D, Hopcroft JE (2019) Krylov subspace approximation for local community detection in large network. ACM Trans Knowl Discov Data 13(5):1–30
29. Coscia, M. Rossetti, G. Giannotti, F. Pedreschi, D.: DEMON: A local-first discovery method for overlapping communities. In: Proceedings of 18th ACM SIGKDD international conference of knowledge discovery data mining (KDD), pp 615–623
30. Peixoto TP (2020) Merge-split Markov chain Monte Carlo for community detection. Phys Rev E 102(1)
31. Berahmand K, Bouyer A, Vasighi M (2018) Community detection in complex networks by detecting and expanding core nodes through extended local similarity of nodes. IEEE Trans Comput Social Syst 5(4):1021–1033
32. Broder A, Kumar B, Maghaoul F, Raghavan P, Rajagopalan S, Stata R, Tomkins A, Eiener J (2000) Graph structure in the Web. Comput Netw 33:309–320
33. Vijaymeena MK, Kavitha K (2016) A survey on similarity measures in text mining. Mach Learn Appl 3(1):19–28
34. Newman MEJ, Girvan M (2004) Finding and evaluating community structure in the networks. Phys Rev E 69(2):026113
35. Zachary WW (1977) An information flow model for conflict and fission in small groups. J Anthropol Res 33(4):452–473
36. Lusseau D, Schneider K, Boisseau OJ, Haase P, Slooten E, Dawson SM (2003) The bottlenose dolphin community of doubtful sound features a large proportion of long-lasting associations. Behav Ecol Sociobiol 54(4):396–405
37. Girvan M, Newman MEJ (2002) Community structure in social and biological networks. Proc Natl Acad Sci USA 99(12):7821–7826
38. Yang J, Leskovec J (2015) Defining and evaluating network communities based on ground-truth. Knowl Inf Syst 42(1):181–213
39. Lancichinetti A, Fortunato S, Radicchi F (2008) Benchmark graphs for testing community detection algorithms. Phys Rev E 78(4)

40. Danon L, Díaz-Guilera A, Duch J, Arenas A (2005) Comparing community structure identification. J Stat Mech Theory Exp (9):09008
41. Chen Q, Wu T-T, Fang M (2013) Detecting local community structures in complex networks based on local degree central nodes. Phys A 392(3):529–537
42. Murray G, Carenini G, Ng R (2012) Using the omega index for evaluating abstractive community detection. In: Proceedings workshop evaluation metrics and system comparison for automatic summarization, pp 10–18. Association of Computational Linguistics

Evaluation of SVM Kernels with Multiple Uncorrelated Feature Subsets Selected by Multiple Correlation Methods for Reflection Amplification DDoS Attacks Detection

Kishore Babu Dasari and Nagaraju Devarakonda

1 Introduction

A Distributed Denial of Service (DDoS) attacks [6] to prohibit legitimate users from accessing the network or system service by halting the system servers. The attackers employ a variety of compromised or controlled sources to generate enormous volumes of packets or requests in order to launch the attack. The target system becomes overloaded as a result of these requests, leading it to underperform and become unreachable to legitimate users.

DDoS attacks are divided into reflection-based attacks and exploit-based attacks [4]. Because legitimate third-party components are employed in reflection-based DDoS attacks, the attacker's identity is disguised. To overload the target victim with response packets, attackers send packets to reflector servers with the target victim's IP address as the source IP address. These attacks might leverage the Transmission Control Protocol (TCP), the User Datagram Protocol (UDP), or a mix of the two. SSDP, MSSQL, NTP, TFTP, CharGen, SNMP, NETBIOS, LDAP, and DNS are examples of Reflection-based DDoS attacks. Exploitation-based DDoS attacks use a number of vulnerabilities in TCP/UDP-based protocols. SYN flood is TCP-based, and UDP-Lag and UDP flood are UDP-based exploitation attacks.

A reflection amplification attack allows attackers to increase the quantity of harmful traffic they can generate while also hiding the source of the attack activity. This sort of distributed denial-of-service (DDoS) attack overwhelms the target, resulting in system and service interruption or loss. SNMP and DNS [2, 3] are reflection amplification DDoS attacks. The attacker sends a large number of SNMP

K. B. Dasari (✉)
Department of CSE, Acharya Nagarjuna University, Andhra Pradesh 522510, India
e-mail: dasari2kishore@gmail.com

N. Devarakonda
School of Computer Science and Engineering, VIT-AP University, Amaravati 522237, India

R. Chaki et al. (eds.), *Applied Computing for Software and Smart Systems*,
Lecture Notes in Networks and Systems 555,
https://doi.org/10.1007/978-981-19-6791-7_6

queries to a large number of connected devices, each of which responds with the fake address by using Simple Network Management Protocol. The attack volume increases as more devices answer, until the target network is forced down by the cumulative number of these SNMP responses. DNS amplification is a reflection-based DDoS attack that manipulates domain name systems and causes them to flood the target system with enormous amounts of UDP packets, bringing the target servers down.

The DDoS attack and its types are discussed in this section. In Sect. 2 of this paper, the methodology is explained, including proposed methodology, preprocessing, correlation methods, SVM classification algorithm and its kernels. The results and discussion are explained with experimental results in Sect. 3 of this paper. This study's conclusion and feature enhancement are found in Sect. 4 of this paper.

2 Methodology

Three Proposed feature selection methods are depicted from Figs. 1, 2 and 3. The SNMP and DNS Reflection Amplification DDoS attack dataset collected CIC-DDoS2019 datasets and performed operations on those to evaluate SVM classification algorithms different kernels with three different proposed uncorrelated feature subsets. The term "data preprocessing" refers to a set of processes for preparing data for machine learning algorithms. First, remove socket features which vary from one network to another. After that, clean the data by deleting missing and infinite values entries using 0 and 1 to represent the benign and DDoS attack target classes, respectively. Standardize feature values are used to improve the efficiency of the classification algorithms.

Machine learning classification techniques [1] use feature selection to reduce data computation and model training time. In this study, feature selection is utilized for filter-based feature selection methods such as variance threshold and correlation methods. The variance threshold is used to filter out features that vary by less than a specific threshold. Variance threshold takes the relationship of the feature itself in all records of the dataset. The link between the feature and the target label is ignored. Correlation [10] describes the relationship between two or more features. The values of the correlation coefficients vary from −1 to +1, showing the strength of the association between the features. When the coefficient value is ± 1, it means that the features have a strong correlation. When the coefficient value is 0, it means that the features are strongly uncorrelated. In this study, Pearson, Spearman, and Kendall correlation techniques are used to identify uncorrelation features.

The correlation coefficient calculated by Pearson is

$$r = \frac{\sum(x_i - \overline{x})(y_i - \overline{y})}{\sqrt{\sum(x_i - \overline{x})^2 \sum(y_i - \overline{y})^2}} \quad (1)$$

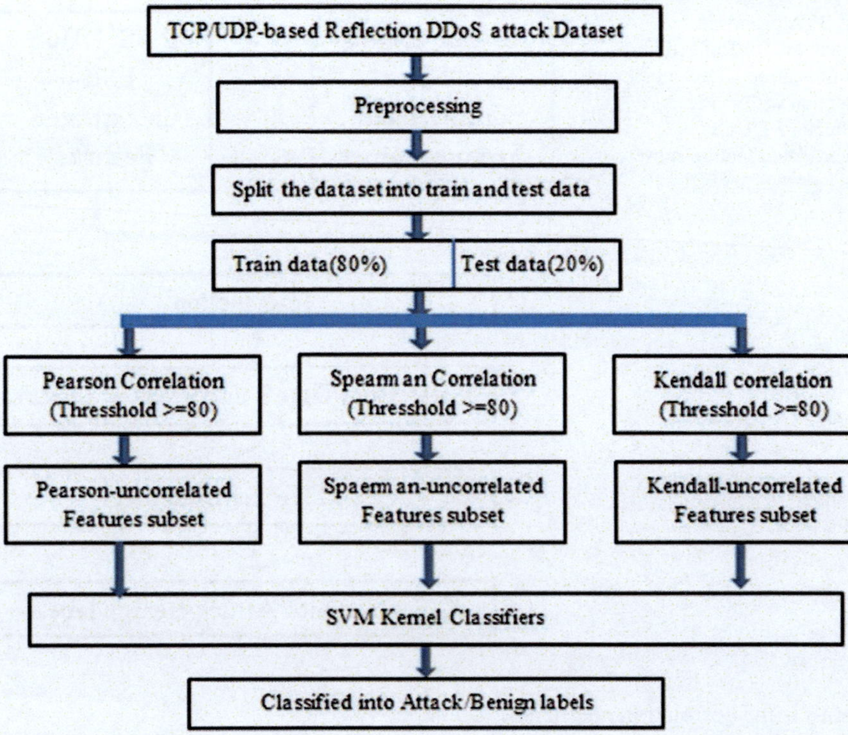

Fig. 1 A proposed framework for Reflection Amplification DDoS attack detection with Pearson, Spearman and Kendal uncorrelated feature subsets

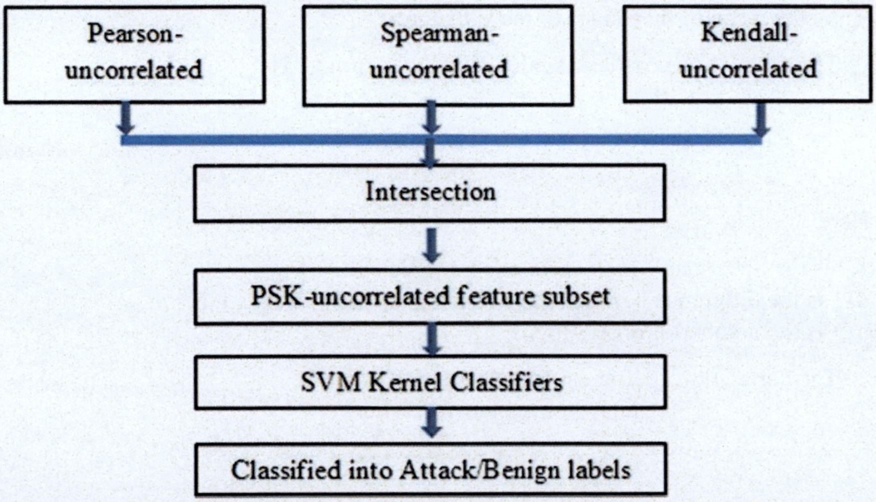

Fig. 2 A proposed framework for Reflection Amplification DDoS attack detection with PSK-uncorrelated feature subsets

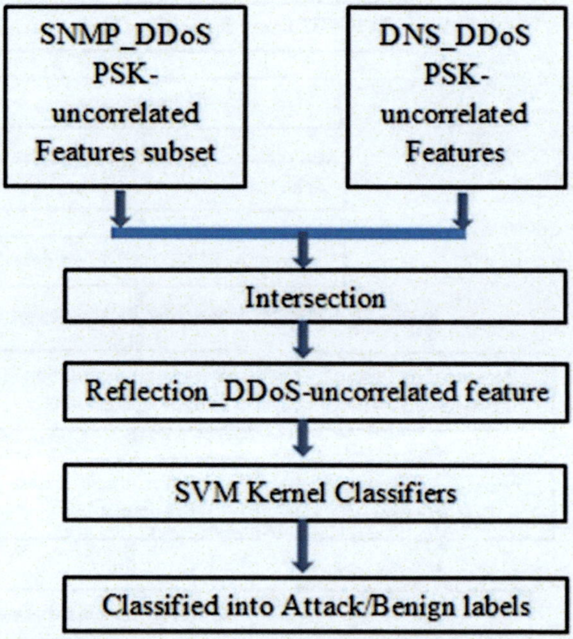

Fig. 3 A proposed framework for Reflection Amplification DDoS attack detection with Reflection_DDoS uncorrelated feature subsets

Here r is the correlation coefficient

x_i is the value of the x-feature in a sample.
$\overline{x}$ is the mean of the values of the x-feature.
y_i is the value of the y-feature in a sample.
$\overline{y}$ is the mean of the values of the y-feature.

The correlation coefficient calculated by Spearman is

$$\rho = 1 - \frac{6\sum d_i^2}{n(n^2 - 1)} \tag{2}$$

Here

ρ is the Spearman's rank correlation coefficient
d_i is the difference between the two ranks of each observation.
n is the number of observations.

The correlation coefficient calculated by Kendall is

$$\tau = \frac{N_c - N_d}{\frac{n(n-1)}{2}} \tag{3}$$

Here

τ is the Kendall rank correlation coefficient

N_c is the number of concordant.
N_d is the number of discordant.

In Machine Learning, classification challenges [7] are defined as the difficulty of learning to identify records in a dataset that correspond to two or more target class labels. The Support Vector Machine (SVM) [5, 9] is a powerful machine-learning technique that identifies label classes by finding a hyperplane in an N-dimensional space. Support Vectors are the data points with the shortest distance to the hyperplane. Because of the kernel functions that turn the input data space into a higher-dimensional space, SVMs are also known as Kernelized SVMs. The kernel functions used in this research are Linear, Polynomial, Radial Basis Function (RBF), and Sigmoid.

The Linear kernel function is written as follows:

$$k(x_i, x_j) = x_i * x_j \tag{4}$$

The Polynomial kernel function is written as follows:

$$k(x_i, x_j) = (1 + x_i * x_j)^d \tag{5}$$

The Radial Basis Function (RBF) kernel function is written as follows:

$$k(x_i, x_j) = \exp(-\gamma ||x_i - x_j||^2) \tag{6}$$

The Sigmoid kernel function is written as follows:

$$k(x_i, x_j) = \tanh(\propto x^T y + c) \tag{7}$$

Carry out the experiments in this study with the Python programming language using sklearn, pandas and numpy libraries for SVM classification algorithm processing and matplotlib and seaborn libraries for visualization of the ROC-AUC curve on a Google Collab with 25 GB of RAM and a TPU environment. CICFlowMeter is the network traffic flow generator tool used in this study to generate CSV files from extracted PCAP files which are network traffic packet capture files.

3 Results and Discussion

In this research experiments are performed on Reflection amplification DDoS attack dataset of SNMP and DNS datasets. After pre-processing, delete the data set's constant and quasi-constant features using a variance threshold filter-based feature selection method. Features which have variance-threshold as 0 are called constant features. Features which have variance-threshold as 0.01 as called quasi-constant features. Now, depending on a threshold value of >=80, identify correlated features of the dataset using correlation methods. By deleting correlation feature subsets from feature sets, you can select uncorrelation feature subsets of the correlation methods. This is shown in Fig. 1. Common uncorrelated features of Pearson, Spearman and Kendall correlation methods are called PSK-uncorrelated feature subsets which are shown in Fig. 2. This is the second proposal for the uncorrelated feature selection. Tables 1 and 2 show the list of Pearson, Spearman, Kendall and PSK uncorrelated features of SNMP and DNS DDoS attack datasets, respectively. Common uncorrelated features of PSK-uncorrelated subsets of SNMP and DDoS attack datasets is called the Reflection_DDoS uncorrelated feature subset which is shown in Fig. 3. This is the third proposal for the uncorrelated feature selection. Table 3 shows the list of features of the Reflection_DDoS uncorrelated feature subset.

This study evaluates SVM classification algorithm different kernels for DDoS attack detection on TCP/UDP based Reflection DDoS attacks with three proposed uncorrelated feature selections using accuracy, specificity, log loss, K-Fold cross validation and ROC-AUC score valuation metrics.

Table 1 List of uncorrelated feature subsets of SNMP DDoS attack

Pearson uncorrelated feature subset	Spearman uncorrelated feature subset	Kendall uncorrelated feature subset	PSK uncorrelated feature subset
Bwd Heade Length Bwd IAT Mean Bwd IAT Min Bwd Packet Length Min Bwd Packets/s CWE Flag Count Flow Duration Flow IAT Mean Flow IAT Min Fwd Header Length Fwd Packet Length Max Fwd Packet Length Std Idle Std Inbound Init_Win_bytes_backward Max Packet Length Protocol Total Fwd Packets Total Length of Bwd Packets Bwd IAT Total Flow Bytes/s Fwd PSH Flags Idle Mean Total Length of Fwd Packets	Active Std Bwd IAT Std Bwd Packet Length Min CWE Flag Count Flow Duration Flow IAT Min Flow IAT Std Fwd Header Length Fwd Packet Length Max Fwd Packet Length Std Inbound Protocol Total Backward Packets Total Fwd Packets Total Length of Bwd Packets Active Mean Fwd PSH Flags Total Length of Fwd Packets	ACK Flag Count Active Std Bwd IAT Std Bwd Packet Length Min Bwd Packet Length Std Flow Duration Flow IAT Min Fwd Header Length Fwd Packet Length Max Fwd Packet Length Std Inbound Protocol Total Backward Packets Total Fwd Packets Total Length of Bwd Packets min_seg_size_forward Active Mean Flow Bytes/s	Fwd Packet Length Max Fwd Header Length Total Length of Bwd Packets Fwd Packet Length Std Protocol Bwd Packet Length Min Total Fwd Packets Inbound Flow IAT Min CWE Flag Count Fwd PSH Flags Total Length of Fwd Packets Flow Duration

Table 2 List of uncorrelated feature subsets of DNS DDoS attack

Pearson uncorrelated feature subset	Spearman uncorrelated feature subset	Kendall uncorrelated feature subset	PSK uncorrelated feature subset
ACK Flag Count	ACK Flag Count	Active Std	Total Fwd Packets
Active Std	Active Std	Bwd IAT Std	Bwd Packet Length Min
Bwd Header Length	Bwd IAT Std	Bwd Packet Length Min	Total Backward Packets
Bwd IAT Mean	Bwd Packet Length Min	CWE Flag Count	Flow IAT Min
Bwd IAT Min	Bwd Packet Length Std	Flow Duration	ACK Flag Count
Bwd Packet Length Min	Flow Duration	Flow IAT Min	min_seg_size_forward
Bwd Packets/s	Flow IAT Min	Flow IAT Std',	Flow Duration
Down/Up Ratio	Fwd Header Length	Fwd Header Length	Protocol
Flow Duration	Fwd Packet Length Max	Fwd Packet Length Max	Active Std
Flow IAT Mean	Fwd Packet Length Std	Fwd Packet Length Std	Inbound
Flow IAT Min	Inbound	Inbound	Fwd Packet Length Max
Flow IAT Std	Protocol	Protocol	Fwd Packet Length Std
Fwd Header Length	Total Backward Packets	Total Backward Packets	Total Length of Bwd Packets
Fwd Packet Length Max	Total Fwd Packets	Total Fwd Packets	Active Mean
Fwd Packet Length Std	Total Length of Bwd Packets	Total Length of Bwd Packets	Fwd Header Length
Idle Std	min_seg_size_forward	Active Mean	
Inbound	Active Mean	Fwd PSH Flags	
Init_Win_bytes_backward		Total Length of Fwd Packets	
Protocol			
Total Backward Packets			
Total Fwd Packets			
Total Length of Bwd Packets			
URG Flag Count			
min_seg_size_forward			
Active Mean			
Bwd Packet Length Max			
Flow Bytes/s			
Init_Win_bytes_forward			
Total Length of Fwd Packets			

Table 3 List of uncorrelated feature subsets of Reflection-DDoS attacks

Pearson uncorrelated feature subset
Bwd Packet Length Min
Total Length of Bwd Packets
Flow Duration
Fwd Header Length
Inbound
Protocol

$$accuracy = \frac{TP + TN}{TP + TN + FP + FN} \tag{8}$$

$$Precision = \frac{TP}{TP + FP} \tag{9}$$

$$Recall = \frac{TP}{TP + FN} \tag{10}$$

$$F1score = \frac{2 * Precision * Recall}{Precision + Recall} \tag{11}$$

$$Specificity = \frac{TN}{TN + FP} \tag{12}$$

where TP is TRUE POSITIVE, TN is TRUE NEGATIVE, FP is FALSE POSITIVE, FN is FALSE NEGATIVE.

$$Log - loss = -\frac{1}{N}\sum_{i=1}^{N}[y_i \ln p_i + (1 - y_i)ln(1 - p_i)] \tag{13}$$

where ‘N’ is the number of observations, ‘p’ is the prediction probability and ‘y’ is the actual value.

K-Fold Cross (KFC) validation dataset divided into K folds, selects one fold as a test, the remaining folds as training and then evaluates the model. This process is continued until all folds have been selected as a test fold.

Area Under the Receiver Operating Characteristic Curve (AUC-ROC) is a probability curve that plots the True Positive Rate against False Positive Rate (FPR) at different threshold values for determining the efficiency of the classification models.

Tables 4, 5, 6, 7 and 8 show the results of evaluation metrics of the SVM classifier kernels on the SNMP DDoS attack dataset. The SVM Poly kernel gives the best accuracy with Spearman, Kendall and PSK uncorrelated feature subsets on SNMP dataset. The SVM Linear kernel gives better accuracy with Pearson uncorrelated feature subsets on SNMP dataset. The SVM RBF kernel gives better accuracy with Reflection_DDoS uncorrelated feature subsets on the SNMP dataset. Pearson uncorrelated feature subset gives good accuracy scores with Linear and RBF kernels and Spearman, Kendall uncorrelated feature subsets give good accuracy scores with Poly and Sigmoid kernels. The SVM Linear, RBF and Poly kernels produce the better K-fold cross validation accuracy score with all proposed uncorrelated feature subsets. The SVM Sigmoid kernel produces the poor K-fold cross validation accuracy score with all proposed uncorrelated feature subsets. Reflection_DDoS uncorrelated feature subset gives the better specificity values with SVM Linear, RBF, and Poly kernels. Sigmoid SVM kernel gives a poor specificity value with all proposed uncorrelated feature subsets especially the SNMP DDoS attack dataset. Pearson uncorrelated feature subset with SVM Linear and the remaining proposed uncorrelated feature subset with Poly kernel produces better Log-loss values on the SNMP DDoS attack. The SVM RBF kernel gives the best ROC-AUC scores with all proposed uncorrelated feature subsets. Figure 4 Shows ROC-AUC curves of SVM kernels on SNMP dataset.

Table 4 Overall model accuracy of the SVM kernels on an SNMP DDoS attack dataset with different uncorrelated feature subsets

SVM Kernels	Pearson	Spearman	Kendall	PSK	Reflection_DDoS
Linear	97.07	95.97	95.97	96.70	96.34
RBF	96.70	95.60	95.60	95.97	96.70
Poly	96.34	97.07	97.07	97.07	96.34
Sigmoid	95.24	96.34	96.34	93.77	95.24

Table 5 K-fold cross-validation accuracy scores (with a standard deviation) in % of the SVM kernels on a SNMP DDoS attack dataset with different uncorrelated feature subsets

SVM Kernels	Pearson	Spearman	Kendall	PSK	Reflection_DDoS
Linear	98.7156 (0.6742)	98.7156 (0.6742)	98.7156 (0.6742)	98.7156 (0.6742)	98.5321 (0.8895)
RBF	98.3486 (0.7998)	98.3486 (0.2247)	98.3486 (0.2247)	98.3486 (0.2247)	98.3486 (0.2247)
Poly	98.4404 (0.8508)	98.5321 (0.1835)	98.5321 (0.1835)	98.5321 (0.1835)	98.5321 (0.1835)
Sigmoid	97.6147 (0.9795)	96.3303 (1.0460)	96.3303 (1.0460)	96.3303 (1.0460)	96.3303 (1.0460)

Table 6 The specificity of the SVM kernels on the SNMP DDoS attack dataset with different uncorrelated feature subsets

SVM Kernels	Pearson	Spearman	Kendall	PSK	Reflection_DDoS
Linear	0.67	0.5	0.5	0.58	0.75
RBF	0.58	0.42	0.42	0.58	0.83
Poly	0.50	0.67	0.67	0.67	0.67
Sigmoid	0.42	0.67	0.67	0.42	0.42

Table 7 The log—loss value of the SVM kernels on an SNMP DDoS attack dataset with different uncorrelated feature subsets

SVM Kernels	Pearson	Spearman	Kendall	PSK	Reflection_DDoS
Linear	1.0121370313	1.39168988	1.391689882	1.1386556246	1.2651654312
RBF	1.1386556246	1.51820847	1.518208475	1.3916869535	1.1386468379
Poly	1.2651742180	1.01213703	1.012137031	1.0121370313	1.0121370313
Sigmoid	1.6447241403	1.26516836	1.265168360	2.1507867981	1.6447241403

Table 8 ROC-AUC scores of the SVM kernels on a SNMP DDoS attack dataset with different un-correlated feature subsets

SVM Kernels	Pearson	Spearman	Kendall	PSK	Reflection_DDoS
Linear	0.82758621	0.67401022	0.674010217	0.7886334610	0.7436143040
RBF	0.97988506	0.97669221	0.976692209	0.9811621967	0.8984674330
Poly	0.82567050	0.74968072	0.749680715	0.7496807152	0.7496807152
Sigmoid	0.74297573	0.96679439	0.966794381	0.9604086845	0.8888888889

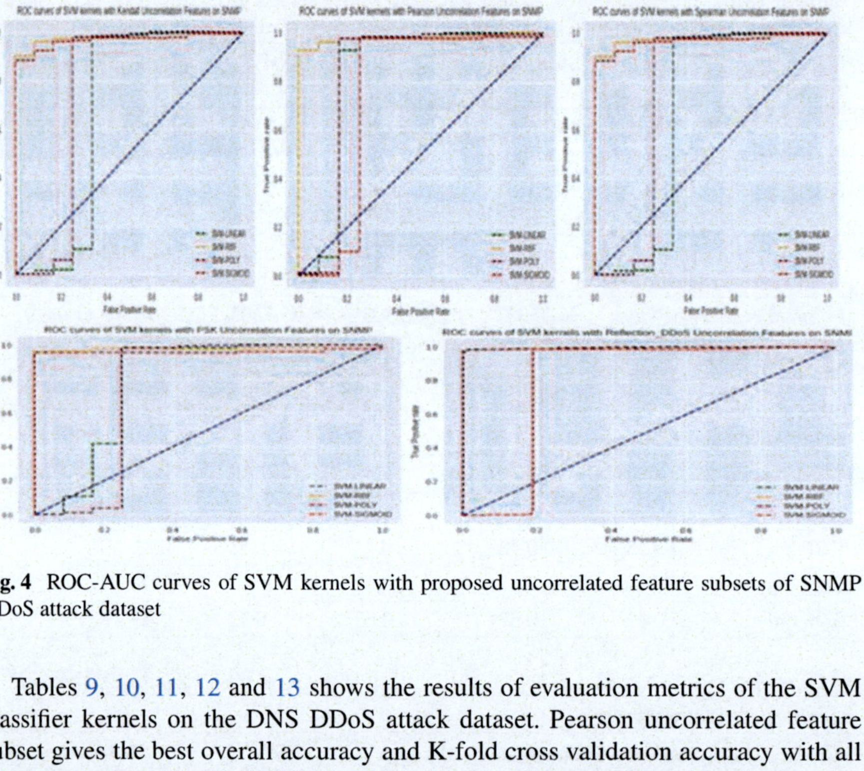

Fig. 4 ROC-AUC curves of SVM kernels with proposed uncorrelated feature subsets of SNMP DDoS attack dataset

Tables 9, 10, 11, 12 and 13 shows the results of evaluation metrics of the SVM classifier kernels on the DNS DDoS attack dataset. Pearson uncorrelated feature subset gives the best overall accuracy and K-fold cross validation accuracy with all SVM kernels on the DNS DDoS attack dataset. The poly SVM kernel gives the overall accuracy and K-fold cross validation accuracy with Pearson uncorrelated feature subset. Reflection_DDoS uncorrelated feature subset gives the better specificity values with SVM Linear, RBF, and Poly kernels on the DNS DDoS attack dataset. Sigmoid SVM kernel gives poor specificity value with all proposed uncorrelated feature subsets including the DNS DDoS attack dataset. Pearson uncorrelated feature with Poly kernel gives the best specificity value while the remaining proposed feature subsets give the best specificity value with RBF kernel. The SVM RBF kernel gives the best log-loss value with Pearson uncorrelated feature subsets. The SVM Poly kernel gives the best ROC-AUC score with Pearson uncorrelated feature subsets. The SVM RBF and Poly kernels give the better ROC-AUC scores with all proposed uncorrelated feature subsets. Figure 5 Shows ROC-AUC curves of SVM kernels on DNS dataset.

Table 9 Overall model accuracy of the SVM kernels of DNS DDoS attacks dataset with different uncorrelated feature subsets

SVM Kernels	Pearson	Spearman	Kendall	PSK	Reflection_DDoS
Linear	99.47	97.75	97.75	96.70	97.51
RBF	99.74	98.46	98.46	98.42	98.02
Poly	99.99	98.66	98.68	97.44	98.08
Sigmoid	98.95	95.10	95.23	93.77	97.27

Table 10 K-fold cross-validation accuracy scores (with a standard deviation) in % of the SVM kernels on a DNS DDoS attack dataset with different uncorrelated feature subsets

SVM Kernels	Pearson	Spearman	Kendall	PSK	Reflection_DDoS
Linear	99.5829 (0.1280)	98.0223 (0.2863)	98.0223 (0.2863	98.6239 (0.5802)	97.8345 (0.2676)
RBF	99.6487 (0.1613)	98.3932 (0.1425)	99.0903 (0.1264)	98.3486 (0.2247)	98.1064 (0.1781)
Poly	99.7146 (0.2560)	98.3684 (0.0842)	98.3833 (0.1246)	98.3684 (0.0842)	98.3833 (0.1246)
Sigmoid	99.3414 (0.0000)	95.2239 (0.4179)	95.3327 (0.3717)	95.9633 (0.7892)	97.1769 (1.0989)

Table 11 The specificity of the SVM kernels on a DNS DDoS attack dataset with different uncorrelated feature subsets

SVM Kernels	Pearson	Spearman	Kendall	PSK	Reflection_DDoS
Linear	0.33	0.83	0.83	0.58	0.81
RBF	0.67	1.00	1.00	1.00	1.00
Poly	1.00	0.99	0.99	0.58	1.00
Sigmoid	0.00	0.76	0.77	0.42	0.78

Table 12 The log—loss value of the SVM kernels on a DNS DDoS attack dataset with different uncorrelated feature subsets

SVM Kernels	Pearson	Spearman	Kendall	PSK	Reflection_DDoS
Linear	0.1819468456	0.77861762	0.778617616	1.1386556247	0.8605772985
RBF	0.0909734223	0.53273193	0.532731928	0.5463917167	0.6829896063
Poly	9.9920072216	0.46443330	0.457603246	0.8856242958	0.6624999229
Sigmoid	0.3638915831	1.69382743	1.646017378	2.1507867982	0.9425379297

Table 13 ROC-AUC scores of the SVM kernels on a DNS DDoS attack dataset with different uncorrelated feature subsets

SVM Kernels	Pearson	Spearman	Kendall	PSK	Relection_DDoS
Linear	0.88662734	0.99620240	0.996248883	0.7886334610	0.9942147923
RBF	0.99891839	0.99709220	0.997178020	0.9972757614	0.9947154181
Poly	1.0	0.99545920	0.998243042	0.7503192848	0.9955214852
Sigmoid	0.95329400	0.97182341	0.971293581	0.9604086845	0.8062280231

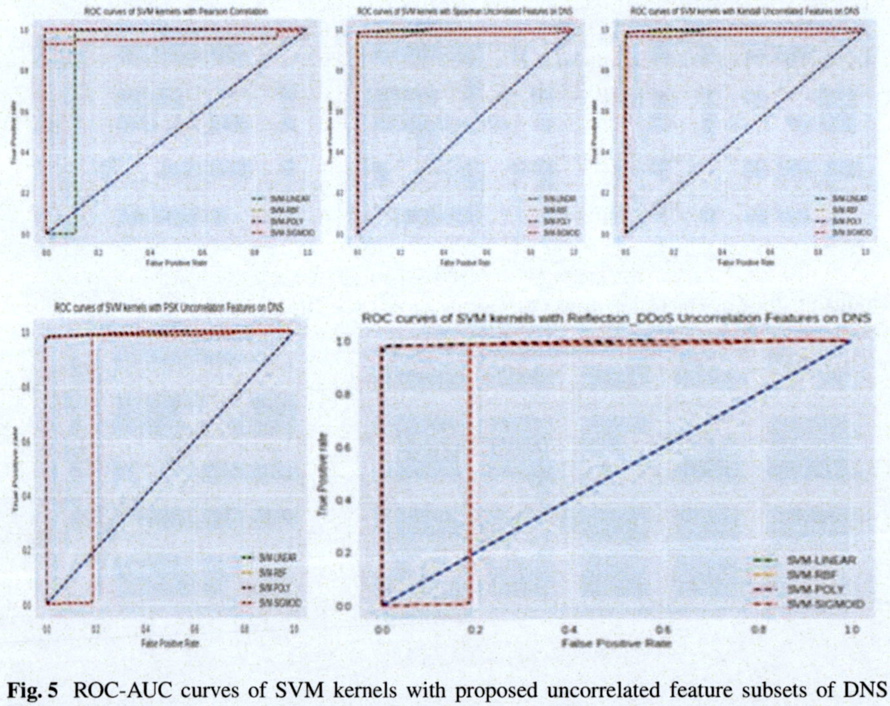

Fig. 5 ROC-AUC curves of SVM kernels with proposed uncorrelated feature subsets of DNS DDoS attack dataset

4 Conclusion

This research is a quantitative research on Reflection amplification DDoS attack datasets of SNMP and DNS with Support Vector Machine (SVM) classification algorithm using Linear, RBF, Poly and Sigmoid kernel functions with three proposed uncorrelated feature subsets selected based on Pearson, Spearman and Kendall correlation methods. Overall, the Pearson uncorrelated feature subset produces the best results, while other feature sets also produce good results. Among SVM kernels RBF and Poly kernels produce the better results. Sigmoid kernel produces the poor results. DDoS attacks detection research will be enhanced in the future using the SVM classifier with feature selection using Principal Component Analysis (PCA) [8].

References

1. Al-Naymat G, Al-Kasassbeh M, Al-Harwari E (2018) Using machine learning methods for detecting network anomalies within SNMP-MIB dataset. Int J Wireless Mobile Comput 15(1):67–76
2. Chen L et al (2018) Detection of DNS ddos attacks with random forest algorithm on spark. Procedia Comput Sci 134:310–315
3. Dasari KB, Devarakonda N (2021) Detection of different DDoS attacks using machine learning classification algorithms. Ingénierie des Systèmes d'Information 26(5):461–468. https://doi.org/10.18280/isi.260505
4. Dasari KB, Devarakonda N (2022) TCP/UDP-based exploitation DDoS attack detection using AI classification algorithms with common uncorrelated feature subset selected by Pearson, Spearman and Kendall correlation methods. Revue d'Intelligence Artificielle 36(1):61–71. https://doi.org/10.18280/ria.360107
5. Sahoo KS et al (2020) An evolutionary SVM model for ddos attack detection in software defined networks. IEEE Access 8:132502–132513. https://doi.org/10.1109/ACCESS.2020.3009733
6. Dasari KB, Devarakonda N (2018) Distributed denial of service attacks, tools and defense mechanisms. Int J Pure Appl Math 120(6):3423–3437. https://acadpubl.eu/hub/2018-120-6/3/247.pdf
7. Kwang P (2017) A countermeasure technique for attack of reflection SSDP in Home IoT. J Converg Inf Technol https://doi.org/10.22156/CS4SMB.2017.7.2.001
8. Mekala S, Padmaja Rani Supervisor B, Padmaja Rani B (2020) Article ID: IJARET_11_11_121 Kernel PCA Based Dimensionality Reduction Techniques for Preprocessing of Telugu Text Documents for Cluster Analysis. Int J Adv Res Eng Technol 11(11):1337–1352. https://doi.org/10.34218/IJARET.11.11.2020.121
9. Subbulakshmi T, BalaKrishnan K, Shalinie SM, AnandKumar D, GanapathiSubramanian V, Kannathal K (2011) Detection of DDoS attacks using enhanced support vector machines with real time generated dataset. Third Int Conf Adv Comput 2011:17–22. https://doi.org/10.1109/ICoAC.2011.6165212
10. Xiao P et al (2015) Detecting DDoS attacks against data center with correlation analysis. Comput Commun 67:66–74

BLRS: An Automated Land Records Management System Using Blockchain Technology

Swagatika Sahoo, Saksham Jha, Somenath Sarkar, and Raju Halder

Abstract Land registry and management system is a crucial component of every government system since they preserve records of land ownership in the country. Traditional land registration and transfer is a complex and arduous process that requires numerous intermediaries and is prone to fraud and forgery. One solution to this problem is to digitize land records, but even if the records are stored in a database, they can be altered due to its centralized nature and lack of adequate security. Blockchain technology has the potential to close all of these loopholes by increasing trust, security, transparency, and traceability of data shared across a network through the use of cryptographic primitives. This paper proposes a novel blockchain-based land records management system which, unlike existing approaches, considers several crucial factors as follows: (1) involvement of all existing stakeholders in the system, (2) land fragmentation and merging during transfer, (3) classification of lands like industrial, farming, barren, (4) marking lands that are under some court dispute, (5) overall Government monitoring, (6) automated pricing of lands with land search feature, and (7) support of proper payment channels. As a proof of concept, we provide a prototype implementation of the system written in solidity on the Ethereum platform with a user interface using ReactJS. We conduct an experimental evaluation to establish the feasibility and effectiveness of our proposed system using Hyperledger Caliper which demonstrates the system performance in terms of execution gas costs, CPU utilization, average latency, and transaction throughput.

Keywords Land records management · Blockchain technology · Smart contracts · Ethereum · Traceability

S. Sahoo (✉) · S. Jha · S. Sarkar · R. Halder
Indian Institute of Technology, Patna, India
e-mail: swagatika_1921cs03@iitp.ac.in

S. Jha
e-mail: 1801cs64@iitp.ac.in

S. Sarkar
e-mail: 1801cs50@iitp.ac.in

R. Halder
e-mail: halder@iitp.ac.in

R. Chaki et al. (eds.), *Applied Computing for Software and Smart Systems*,
Lecture Notes in Networks and Systems 555,
https://doi.org/10.1007/978-981-19-6791-7_7

1 Introduction

The advancement of blockchain technology has resulted in tremendous changes in the economic, industrial, academic, and administrative sectors, among many others [1, 17]. Its use is also essential in the real estate industry's land registry sector where a dispute-free transparent and verifiable land registration process is an essential component in case of land/property ownership transfer from one party to another [5, 8, 15]. The buyer and seller must complete numerous steps in this process, including stamp duty, sale agreement, sale deed, etc. Even though the land registration and transfer process is automated in many countries, such systems may suffer from a number of challenges and limitations, such as lack of accountability, transparency, tracing land ownership details, unprotected owner's rights, sale frauds, posing imposters, lengthy title registrations process, single-centered system failure, and so on.

Post the advent of Bitcoin by Satoshi Nakamoto in 2008 [12], blockchain technology has emerged as a ground-breaking disruptive technology showing its enormous potential for a wide range of applications. Following this, Buterin in 2013 [4] came up with Ethereum for making decentralized applications using Smart Contracts. Several use cases of smart contracts, including healthcare management, identity management, banking sector, supply chain, voting, etc., have been highlighted in [7]. The enticing characteristics, such as decentralization, immutability, transparency, and auditability, make blockchain technology a perfect choice to address the above-mentioned limitations in case land registration and transfer process, enabling to achieve the following goals: making land management system secure and tamper-proof, easy information accessibility, land record traceability, speedier tracing, system fairness, detecting fake title ownership, and elimination of fraud, corruption, unscrupulous land record manipulation, and multiple land sales. The benefits of blockchain as a smart technology for land management have been highlighted in many studies for its ability to increase transparency, trust, and security of data [16]. Additionally, it improves the quality, accuracy, and integrity of the data through appropriate consensus mechanism among the stakeholders [18].

These give us the motivation to propose a novel approach to the land registration process by leveraging blockchain technology, which addresses the above-mentioned challenges and limitations. In particular, unlike existing approaches, the proposed system considers several crucial factors as follows: (1) involvement of all existing stakeholders in the system, (2) land fragmentation and merging during transfer, (3) classification of lands like industrial, farming, barren, (4) marking lands that are under some court dispute, (5) overall Government monitoring, (6) automated pricing of lands with the land search feature, and (7) support of proper payment channels. As a proof of concept, we provide a prototype implementation of the system written in solidity on the Ethereum platform with a user interface using ReactJS. We conduct an experimental evaluation to establish the feasibility and effectiveness of our proposed system using Hyperledger Caliper which demonstrates the system performance in terms of execution gas costs, CPU utilization, average latency, and transaction throughput.

The structure of the paper is organized as follows. Section 2 describes the related works in the literature. The detailed description of our proposed approach is presented in Sect. 3. Security and privacy analysis is discussed in Sect. 4. Section 5 presents a proof of concept. Experimental results are shown in Sect. 6. Finally, Sect. 7 concludes our work.

2 Related Works

Owing to the attractive features of blockchain technology in almost every field, there have been few proposed blockchain-based solutions for land record management. Let us now provide a complete overview of the existed solutions for the land record management system. Nandi et al. [10] presented a framework for storing land as a digital asset on the blockchain and discussed the storage of land records and ownership history data. In [6], Eder et al. proposed the blockchain-based land titles approach specific to the countries of Ghana, Georgia, and Honduras. The paper discusses the policies, alliances, and strategies adopted to develop land ownership and registration systems using blockchain technology to accomplish technical advancement. However, these papers address some land registry issues but focus only on the theoretical benefits of blockchain technology, despite its lack of actual implementation. Khan et al. [9] presented a blockchain-based land registry system for India. They explored the possibilities and issues that could be resolved by using a blockchain-based system for land ownership transfer and have also implemented the system. However, the paper lacked a clear visualization of the proposed system and did not adequately cover each stakeholder involved in the system with their respective functionalities. In [2], the authors presented an Ethereum-based Land Registry system for Bangladesh. They included various stakeholders such as buyers, sellers, and banks. However, the authors have not addressed the issue of how the Government will be involved in the system and how will it monitor the system. Also, there is no explanation given for how exactly the details of lands will be stored or how the land transfer will happen. In [11], authors have discussed the issues faced in Land registry in India, like centralization, time delays, and fraud cases, and have come up with a solution based on Ethereum blockchain platform. However, they have not discussed the registration and authentication of stakeholders. They have also not discussed the scalable solution to store images and documents in the system. The system proposed in [13] uses Ethereum as its platform and has explored various issues present in the current system. They have used IPFS for decentralized storage and suggested a verification module using third parties to verify documents. However, they have not addressed various cases of land ownership transfer like partitioning of land, hereditary causes, or tracking the history of transactions for a particular land. In [14], shuaib et al. proposed a blockchain-based approach that attempts to incorporate stakeholders from the current system (registrar office, revenue Office, surveyor, and banks) while leveraging the benefits of blockchain. In their concept, there is a pre-agreement between buyer and seller, which is followed by the generation of a sale request that is then verified by each participant. Upon verification, the transfer of land occurs. However, the paper does not discuss how these concepts should be implemented or which frameworks should be utilized.

3 BLRS: Proposed Land Records Management System

In this section, we present our proposed land records management system by leveraging the power of blockchain technology, without changing existing practices followed in India. As mentioned earlier, unlike traditional system which is exceedingly time-consuming, less transparent, less synchronized among the discrete processes, and prone to compromise data integrity, the proposed system achieves transparency in management processes, reduces service delays, eases record maintenance and access in a tamper-proof manner, and provides a detailed audit trail.

The overall system components of the proposed system involve various stakeholders, such as Land Seller, Land Buyer, Lawyer, Land Inspector, Land Registry Officers, and five smart contracts RegistrationSc, LandSc, LISc, RegistrySc, and PaymentSc. The details about the stakeholders are discussed below:

1. **Land Sellers or Land Buyers**: They are the end-users who own or wish to purchase land. After logging onto the system, they can perform any land management-related tasks, including selling their land, purchasing new land, retrieving information about lands for sale, and registering new land.
2. **Lawyers**: Lawyers are responsible for producing Official Land Deeds for land transfer requests. They must perform data verification on transfer requests. Upon satisfactory verification, Lawyers prepare land deeds and forward the request to the respective Land Inspector.
3. **Land Inspectors**: Land Inspectors are accountable for the verification of land data and the cross-checking of that data with the lawyer's deed. After verifying that everything is in order, the request is forwarded to Land Record Officers for further processing.
4. **Land Records Officers (LRO)**: LROs conduct the final checks, and upon their permission, the land is finally registered or transferred.
5. **Govt. Agents**: The government officials are responsible to supervise the entire system. They have complete access to the system's data, but only in a read-only format. Additionally, they can purchase or acquire lands on the government's behalf for various development purposes.

The overall workflow of the proposed land record management framework is demonstrated in Fig. 1. Let us briefly discuss it. Initially, all land buyers, sellers, lawyers, land inspectors, and land registry officers of the system should undergo the registration process via RegistrationSc smart contract. This stakeholder registration process is depicted in steps (1) and (2). Prior to start any land management activity in the system, all the land owners are required to register their land using the RegistrySc

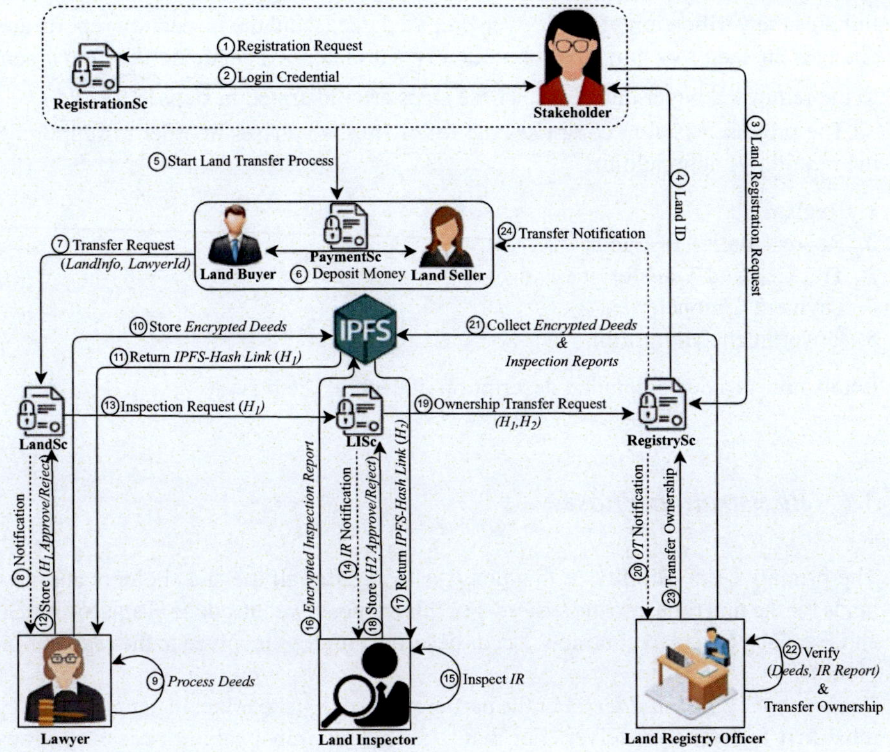

Fig. 1 High level flow diagram

smart contract, as shown in steps (3) and (4). In case an owner wishes to sell her land to a buyer, the buyer deposits funds into the PaymentSc in step (6). Then a land transfer transaction request is made by the seller to the LandSc smart contract to carry out the land transfer process between the seller and the buyer, as shown in step (7). Accordingly, in step (8), a notification is sent to the lawyer as specified in the request. On receiving the request, the lawyer starts preparing a deed taking the transfer requests into consideration and stores the encrypted deed in a private InterPlanetary File System (IPFS) via LandSc. These steps are depicted in steps (9)–(12). Next, LandSc submits an inspection request to the local Land Inspector (LI) via LISc smart contract, depicted in step (13) and (14), after which LI starts inspecting the land details in step (15) and accepts/rejects it accordingly. Once the inspection is successful, the land inspector stores the inspection result (approve/reject) in LISc and uploads the encrypted inspection reports in the IPFS. These processes are shown in steps (16)–(18). Finally, the ownership transfer request is forwarded to the land registry officer (LRO) via RegistrySc smart contract who initiates the verification of the lawyer-prepared deeds and the inspection reports and changes the land's ownership. Subsequently, a transfer ownership notification is sent to the requested buyer and seller. These steps are illustrated in steps (19)–(24).

The proposed system comprises the following five phases in order to support all the required functionalities:

1. Registration
2. Representation of Lands
3. Tracking and Transfer of Land
4. Payment Channel
5. Government Monitoring

Let us now provide a detailed description of each of the phases.

3.1 Registration Phase

The primary responsibility of this phase is to register all the stakeholders and new lands for the first time into the system. For this purpose, we introduce RegistrationSc and RegistrySc smart contracts. Let us describe the steps involved in the registration process.

Registration of Stakeholders: For the participation of a stakeholder in our system, they must first register themselves. For that, stakeholders must submit necessary information, such as their identities and addresses, along with other required information. These data are forwarded to the regulatory authorities for verification via RegistrationSc smart contract. Once the stakeholder registration verification is passed, the regulatory authorities record the verification result on the RegistrationSc. Then, a unique automated generated ID is assigned to each user enrolled in this system.

Registration of lands: In this phase, the registration of land is performed by the LRO using RegistrySc smart contract. A legitimate owner (seller) can request the land office to record a land against her name. However, she needs to provide the required detail (such as area, khaata number, proper address, landmark near the land, details of the area surrounding the land, required documents, etc.). Later, an LRO will verify the details and the documents and will assign ownership against the user account using her official account, and a unique land ID is generated corresponding to the land. By default, LRO will mark each land as non-sellable. If the land is not disputed, the owner may alter this status at their discretion.

The complete registration process of land is summarized in Algorithm 1. The owner of the land invokes `RegisterLand` function with all required input details of their owned land (like Owner ID OI, Land Details LD (area, proper address, landmark near land, details of the area surrounding the land, required documents, etc), Current Price CP). If the land details verification by the LRO is successful, its registration is initiated by invoking the function `createLandId` (Lines 4). This function generates a unique ID LID to be assigned to the land being registered and then increments it by 1 for the next land. Observe that the owner is granted access to the status change of the land depending upon whether the land is under some court dispute or not (Lines 5–8). Finally, the owner ID OI will be appended to the ownerHistory of the land and the land ID LID will be added to the list of lands owned by OI. The value 1 for splitParts attribute of the land indicates ‘no spilt of the land’. Observe that this algorithm has linear time complexity.

Algorithm 1: Land Registration

```
1  Function RegisterLand(Owner ID OI, Land Details LD, Current Price CP):
2      if Document Verification is correct then
3          landDet = land details struct with input parameters of LD;
4          LID = createLandId(landDet, OI, CP);
5          if Land is under some court dispute then
6              Remove Owner's Access to Status Change of Land
7          else
8              Grant Owner's Access to Status Change of Land
9          Append OI to ownerHistory[LID];
10         Append LID to ownerList[OI];
11         Return LID;
12     else
13         Return "Error in Documents"
14 Function createLandId(Land details struct landDet, Owner ID OI, Current Price
   CP):
15     LandID ← tokenID;
16     tokenID++;
17     landIDList[LandID].landDetails ← landDet;
18     landIDList[LandID].splitParts ← 1;
19     landIDList[LandID].owner ← OI;
20     landIDList[LandID].currentPrice ← CP;
21     Return LandID;
```

3.2 *Representation of Lands*

A single person can own single or multiple lands. When transferring a land or a group of lands to a buyer, tokens are used to represent them in order to enhance system scalability. Due to the lack of popular existing token standards that support the required features for our system, we introduce a new form of tokens to make the task easier. The primary benefit of tokenization is the reduced computation required to update the shared specifications of lands, such as owner, current price, and other land-specific information. With tokenization, we perform aggregation of multiple lands into a single land and splitting up a land into multiple parts. The splitting function will be invoked if the owner wants to sell a part of the land instead of the whole of it, which generates a new token for the part of the land. Similarly, the merging function helps to merge a set of lands and generates a single token to represent the set.

3.3 Tracking and Transfer of Land

This section explains the land transfer process and various features of the land incorporated into our system. The land transfer process (depicted in Fig. 2) includes all stakeholders working together, whose respective functions are explained below.
Citizen: Every citizen who wants to buy some land will have access to a list of all the lands that are marked as saleable. An advanced land search feature is also available based on certain parameters like state, city, pin code, price, type of land, etc. Once the buyer decides which land she wants to buy, she can contact the seller and start negotiations. After a successful negotiation, the seller creates a request containing the land id, buyer id, and seller id and sends the request to a particular lawyer.

Lawyer: The lawyer receives the incoming request and starts to verify all the details. Once everything is verified, the deed is created and stored on the LandSc smart contract. The request is then forwarded to Land Inspector, and the participating stakeholders are notified about the request status.
Land Inspector: The Land Inspector receives the request along with the deed. Her responsibility includes validation of land details, checking the authenticity of the

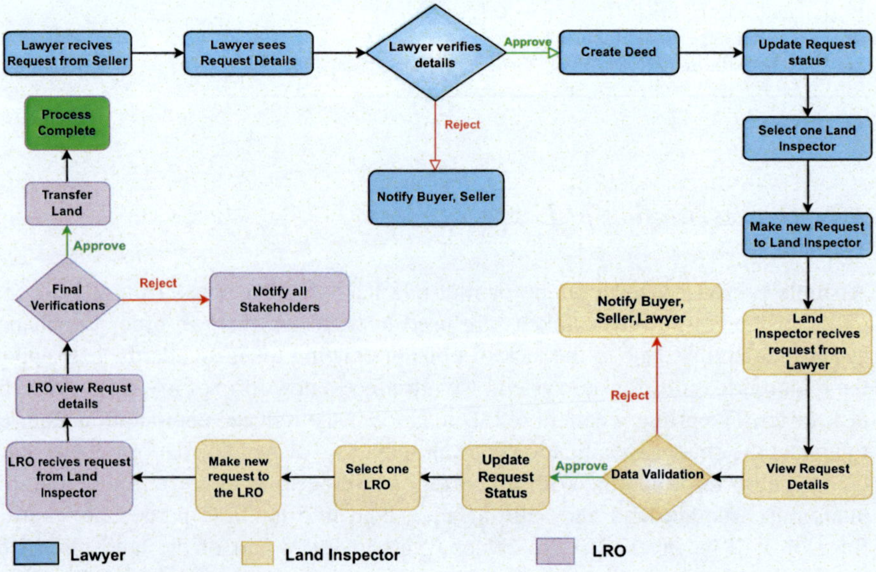

Fig. 2 Land transfer process

land owner, and reviewing deed correctness. If all checks are successful, the request is approved and sent to LRO for final steps.
LRO: Land Record Officer has the ultimate power to change ownership of land. The incoming request is minutely scrutinized and reviewed. Once the request is approved, a deadline is set for the buyer to complete the Payment of the land negotiated price. If the Payment is successful within the deadline, the land ownership change occurs, and all the participants involved in this request get notified. The owner history is also maintained and can be accessed easily.

Algorithm 2 describes the land transfer process from one owner to the next owner. If the entire land is being sold, the land ID LID of the transferred land will remain the same as that of the original land. But if the land has to be split, the new ID $newID$ of the part of the land will be obtained from the `splitToken` function (Line 14). Line 13 will extract the details of split land according to the information x provided as a parameter. Now for both the cases (full land sale or part of land sale), the new owner BI will be appended to the ownerHistory of the sold land (Lines 3 and 15). The current owner will be updated in the landIDList mapping for the sold land ID (Lines 4 and 16). If the full land is being sold, we can remove the LID from the ownerList[SI] because the seller SI no longer owns LID (Line 5). If part of the land is being sold, then we need to check if all the split parts have been sold or not. If all parts have been sold, then seller SI no longer owns that piece of land, so we remove LID from ownerList[SI] (Line 18). The ID of the sold land will be appended to the ownerList[BI] because buyer BI now owns that land (Lines 6 and 20). If the full land is being sold then ownerHistory access will be revoked from the seller (Line 8). If the land is being sold by parts, then the access will be removed from the seller only when the seller has no part of the original land LID left with him (Line 19). All the stakeholders will be informed about the transfer and the LandID LID being sold will be returned to the buyer (Lines 9–10 and 22–23).
Some additional features of this phase are discussed below.
Tracking history of previous owners of lands: This feature allows the present landowner to view the history of all previous landowners. Important land papers pertaining to this property are also accessible. Every land transfer dynamically updates the owner list history. With the help of this list, the current owner will know about all past transactions involving the land and the stakeholders involved.
Selling/Buying Part of a Land (Land Fragmentation): Land Fragmentation feature of the land transfer (depicted in Algorithm 2) permits a citizen to sell a part of the land. A unique land ID is assigned to the new land based on the split token. The owner can trace the history of a split land.

Marking a land non-sellable: All lands are categorized as either sellable or non-sellable according to the citizen's choice. The lands marked as sellable are visible during the land search. Disputed lands are marked as non-sellable till the court case is resolved.

Automated Calculation of Present Market Value of Land: The current market value of each land is automatically calculated using the cost inflation index. This helps in the price negotiations during the land transfer process as well as the land verification process.

Algorithm 2: Land Transfer

```
Function TransferLand( Land ID LID, Split Land Information x, Buyer ID BI,
  Seller ID SI, Transfer Price TP):
    if Full Land is being sold then
        Append BI to ownerHistory[LID];
        landIDList[LID].owner ← BI;
        Remove LID from ownerList[SI];
        Append LID to ownerList[BI];
        Grant Access of ownerHistory to Buyer;
        Remove Access of ownerHistory from Seller;
        Update all stakeholders about the transfer;
        Return LID;
    else
        LD = landDetails(LID);
        SL = splitLandDetails(LD, x);
        newID = splitToken(SL, BI, TP);
        Append BI to ownerHistory[newID];
        landIDList[newLandID].owner ← BI;
        if All split parts of LID has been sold then
            Remove LID from ownerList[SI];
            Remove Access of ownerHistory from Seller;
        Append newID to ownerList[BI];
        Grant Access of ownerHistory to Buyer;
        Update all stakeholders about the transfer;
        Return newID;
```

3.4 Payment Channel

A payment channel is introduced in our system to facilitate the stakeholders' payment method. This is made possible through the PaymentSc smart contract. After a successful negotiation between the buyer and seller, the request is forwarded to a specific lawyer for verification. Once the lawyer approves the request and prepares the deed, a predetermined percentage of the negotiated lawyer fees is deducted from the buyer's account via PaymentSc. The request is subsequently transmitted to the Land Inspector to inspect the inspection request, and pre-defined fees are also paid. Finally, LRO receives the request along with the decided fee. Once LRO verifies the request, the land ownership is transferred to the buyer, and the transfer process is successful. At each payment stage, the PaymentSc maintains a record of all monetary transactions and notifies all parties involved.

3.5 *Government Monitoring*

In most countries, the Government monitors all the lands very closely. Our proposed approach entails government agencies or entities administering land monitoring. All land information and land transactions are accessible to administrators. They can view the land a specific citizen owns and trace its land history. They are permitted to review the buying and selling histories of all registered properties. So it helps government officials track any discrepancies in the pricing of the lands, in any of the land details entered, or any misalignment between the area of the land actually sold versus that entered in the system. It increases the system's reliability while minimizing the presence of any fraudulent activity. Additionally, the land search feature is enabled for the admins in case they are interested in buying some specific type of land in a specific region.

4 Security and Privacy Analysis

In this section, we explore the security and privacy implications of our proposed approach. We discuss the various types of attacks that could occur on our system and the countermeasures that could be taken.

- **Distributed Denial-of-Service Attacks**: Distributed Denial-of-Service attacks prevent the system's intended users from accessing services. It is accomplished by repeatedly making excessive calls to specified resources. There is a possibility of DDoS attacks in our proposed system, such as blocking payment channels through the use of false requests to the PaymentSc smart contract by an untrusted stakeholder. Because our proposal is built on the Ethereum blockchain technology, the inclusion of gas fees for transaction execution is a method of minimizing DDoS attacks. Apart from that, our proposed approach is strengthened against this attack by incorporating RegistrationSc which establishes a connection with a trusted issuer verification protocol and allows only legitimate stakeholders to participate in the system. To be more specific, government agencies have complete authority to verify the stakeholders' identity.
- **Sybil Attacks**: Sybil attacks are feasible when attackers hijack multiple nodes in the network and build a fictitious network to drive double-spending assaults. We simply prevent this threat in our proposed system by involving trusted government entities to verify stakeholders prior to their participation in the supply chain via the RegistrationSc smart contract.
- **Dictionary Attacks**: When an attacker attempts to get unauthorized access to a system by cracking the password using various combinations of words and phrases, this is known as a dictionary attack. Our proposed approach makes it impossible due to the use of the wallet. The wallet keeps the public key and private key like a secure cryptocurrency app. All the stakeholders of the system need to register

themselves into the system through a cryptocurrency wallet like MetaMask, and only then they can make transactions in the application.

– **Man-in-the-middle Attack**: An attacker who intercepts a connection between two parties might attempt to seize and tamper with the information they are exchanging. Such an attack is infeasible in our proposed framework because of the following factors: (1) the login method utilizes a cryptocurrency wallet, (2) the information we shared on the system is in encrypted form.

5 Proof of Concept

We developed a prototype with the following technologies: ReactJs[1] for the frontend, NextJs[2] for dynamic routing, and NodeJs[3] for the backend. The language Solidity is used to write smart contracts for performing various functionalities on the Ethereum blockchain. The Solc npm package compiles contracts into JSON-formatted ABI code. The ABI interface is then processed by the Web3 provider instance for contract deployment. We also use remix IDE to determine the deployed smart contract transaction and execution costs required for our system to operate. Infura,[4] which functions as a remote node, is utilized to connect to the Ethereum network.

In our system, a wallet called Metamask[5] has been used. Metamask is a browser extension that enables users to engage with any distributed application (dApps). Once a Metamask user account has been created, she can send Ether to her account. If there is sufficient Ether in the Metamask account, Metamask will inject a Web3 instance into the web browser, enabling interaction with the system.

Our system is linked to the Ethereum network through the Infura infrastructure. Instead of using Ethereum Mainnet, the identical Rinkeby testnet is utilized in our system. Our system has also utilized IPFS [3] to store documents (deeds, IR reports). The hyperledger caliper is used in our system to monitor our system's latency and throughput as the number of requestors and transmission rates change. The primary reason for using IPFS is because keeping large data such as PDFs and Images as Blockchain is not scalable to store a large file, and IPFS overcomes this issue. Figure 3 depicts the overall Implementation Architecture of our system.

[1] https://reactjs.org/.

[2] https://nextjs.org/.

[3] https://nodejs.org/.

[4] https://infura.io/.

[5] https://metamask.io/.

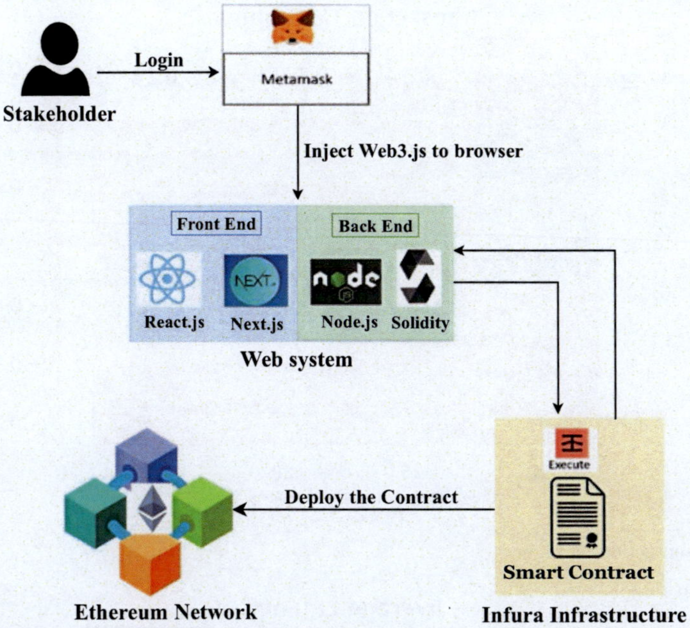

Fig. 3 Implementation architecture

6 Experimental Evaluation

We have done two phases of experimentation. In the first phase of experiments, we record transaction costs and the actual cost of various functions involved in our system. The gas costs are categorized into deployment gas costs for a smart contract and execution gas costs for various functions in the smart contract. Table 1 depicts the gas costs involved in the deployment of the RegistrationSc, LandSc, LISc, and RegistrySc smart contracts. It also depicts the gas costs for the execution of functions in these smart contracts. Note that, we set the gas price to Gwei, where Gwei = 1×10^{-9} Ether and 1 Ether = 3,030 US$ (On 17 Jan 2022).

Let us now discuss the evaluation of the performance of the smart contracts on the Ethereum blockchain platform using a blockchain benchmarking tool Hyperledger Caliper by executing the deployed smart contracts on the Rinkeby Test Network. We measure two parameters: send rate (total transaction requests sent per second) and concurrent users (total users at an instance). We varied the send rate from 10 transactions per second (TPS) to 30 TPS, with intervals of 5 TPS. For concurrent users, we varied the users from 10 to 25 users with an interval of 5 users. For evaluation metrics, we recorded the average latency and throughput of the system for a specific combination of send rate and concurrent users.

Figures 4 and 5 depict the transaction throughput and average latency we obtained. As we can see from Fig. 4, the throughput is increasing with an increase in the send

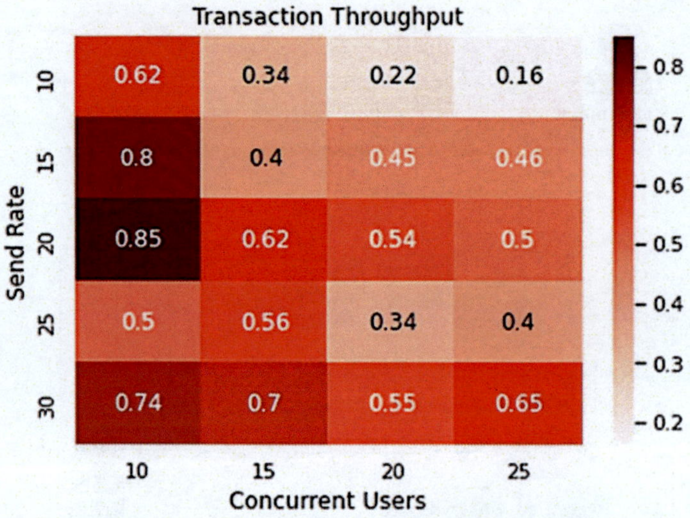

Fig. 4 Transaction throughput

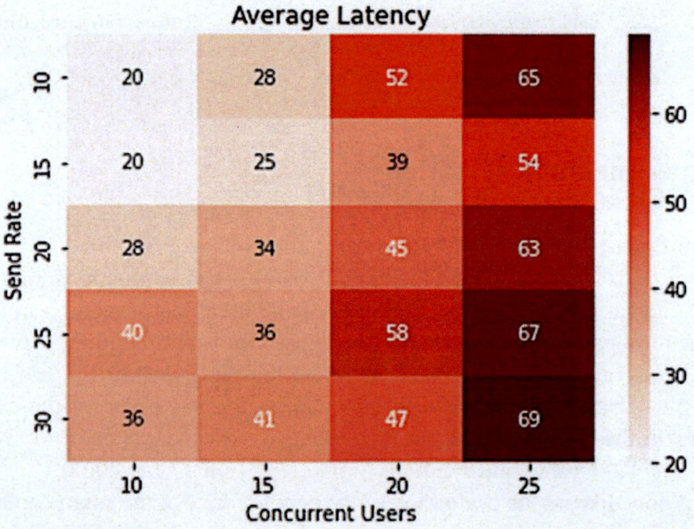

Fig. 5 Average latency

Table 1 Gas costs for deployment of RegistrationSc, LandSc, LISc and RegistrySc smart contracts and for execution of their respective functions

Function/Smart contract	Task	Gas (Gwei)	Cost dollars
RegistrationSc	Deployment	2116149	6.41
RegisterUser	Execution	1701279	5.16
CreateRequest	Execution	230993	0.70
RejectbyLawyer	Execution	32257	0.10
ApprovebyLawyer	Execution	71359	0.22
ApprovebyLI	Execution	63404	0.19
ChangeStatuLro	Execution	32614	0.10
RejectbyLI	Execution	33105	0.10

Function/Smart contract	Task	Gas (Gwei)	Cost dollars
LandSc	Deployment	2285746	6.92
RegisterLawyer	Execution	1810351	5.48
CreateRequest	Execution	280162	0.85
ApproveByLI	Execution	63510	0.19
RejectByLI	Execution	32698	0.10
ChangeStatus	Execution	33189	0.10
Approve	Execution	72821	0.22
Reject	Execution	33952	0.10

Function/Smart contract	Task	Gas (Gwei)	Cost dollars
LISc	Deployment	2047078	6.20
RegisterLI	Execution	1588044	4.81
Createrequest	Execution	281018	0.85
Approve	Execution	64302	0.19
Reject	Execution	31886	0.10
ChangeStatus	Execution	33083	0.10

Function/Smart contract	Task	Gas (Gwei)	Cost dollars
RegistrySc	Deployment	3794951	11.49
RegisteLro	Execution	1500092	4.54
RegisterLand	Execution	433251	1.31
RegisterLand Details	Execution	321682	0.97
TransferLand	Execution	81734	0.25
ChangeStatus Land	Execution	46115	0.14
CreateRequest	Execution	281580	0.85
Approve	Execution	31831	0.10
Reject	Execution	33853	0.10

rate. As for Fig. 5, the average latency increases with an increase in the number of requesters and with an increase in the send rate. According to this information, we can conclude that the latency is low and also that the throughput is high. Thus, the application will be fast. Note that, there were no failed requests during any of these transactions and we achieve a 100% success rate. All these factors ensure that none of the stakeholders will find it inconvenient to use this application.

7 Conclusion

In this paper, we propose a blockchain-based solution to overcome some of the major problems with today's land management systems. It may be possible to solve this issue by digitizing land records; however, even if the records are maintained in a database, they can be altered due to the database's centralized design and lack of proper security. Blockchain technology has been adopted in our system to address all gaps in the traditional system by enhancing the trust, security, transparency, and traceability of data shared across a network. Intriguingly, our system is resistant to a number of typical attacks. The findings of the experiments are encouraging, demonstrating gas's cost-effective performance in various phases. A secure communication route between stakeholders and using Drones and other IoT devices can be some of the future upgrades.

References

1. Aggarwal S, Chaudhary R, Aujla GS, Kumar N, Choo KKR, Zomaya AY (2019) Blockchain for smart communities: applications, challenges and opportunities. J Netw Comput Appl 13–48
2. Ashiquzzaman RIS, Pipash HK, Elme KM, Mahmud DM, Ethereum based land registry system for bangladesh
3. Benet J (2014) Ipfs-content addressed, versioned, p2p file system. arXiv:1407.3561
4. Buterin V et al (2013) Ethereum white paper. GitHub Repos 1:22–23
5. Deininger K, Goyal A (2012) Going digital: Credit effects of land registry computerization in india. J Dev Econ 99(2):236–243. https://doi.org/10.1016/j.jdeveco.2012.02.007, www.sciencedirect.com/science/article/pii/S0304387812000181
6. Eder GJ, Eder GJ (2019) Digital transformation: blockchain and land titles
7. Hu Y, Liyanage M, Mansoor A, Thilakarathna K, Jourjon G, Seneviratne A (2018) Blockchain-based smart contracts-applications and challenges. arXiv:1810.04699
8. Jung S, Dyngeland C, Rausch L, Rasmussen L (2021) Brazilian land registry impacts on land use conversion. Am J Agric Econ. https://doi.org/10.1111/ajae.12217
9. Khan R, Ansari S, Sachdeva S, Jain S (2020) Blockchain based land registry system using ethereum blockchain. Xi'an Jianzhu Keji Daxue Xuebao/J Xi'an Univ Architect Technol 12:3640–3648
10. Nandi M, Bhattacharjee RK, Jha A, Barbhuiya FA (2020) A secured land registration framework on blockchain. In: 2020 Third ISEA conference on security and privacy (ISEA-ISAP). IEEE, pp 130–138
11. Sahai A, Pandey R (2020) Smart contract definition for land registry in blockchain. In: 2020 IEEE 9th international conference on communication systems and network technologies (CSNT). IEEE, pp 230–235
12. Satoshi N (2008) Bitcoin: a peer-to-peer electronic cash system. Consulted 1(2012):28
13. Sharma R, Galphat Y, Kithani E, Tanwani J, Mangnani B, Achhra N (2021) Digital land registry system using blockchain. Available at SSRN 3866088
14. Shuaib M, Daud SM, Alam S, Khan WZ (2020) Blockchain-based framework for secure and reliable land registry system. Telkomnika 18(5):2560–2571
15. Shuaib M, Hassan NH, Usman S, Alam S, Bhatia S, Agarwal P, Idrees SM (2022) Land registry framework based on self-sovereign identity (ssi) for environmental sustainability. Sustainability 14(9). https://www.mdpi.com/2071-1050/14/9/5400
16. Underwood S (2016) Blockchain beyond bitcoin. Commun ACM 59(11):15–17
17. Xie J, Tang H, Huang T, Yu FR, Xie R, Liu J, Liu Y (2019) A survey of blockchain technology applied to smart cities: Research issues and challenges. IEEE Commun Surv Tutor 3:2794–2830
18. Zheng Z, Xie S, Dai H, Chen X, Wang H (2017) An overview of blockchain technology: architecture, consensus, and future trends. In: 2017 IEEE international congress on big data (BigData congress). IEEE, pp 557–564

Machine Learning

Classification of Kathakali Asamyuktha Hasta Mudras Using Naive Bayes Classifier and Convolutional Neural Networks

Pallavi Malavath and Nagaraju Devarakonda

1 Introduction

Kathakali dance form is a traditional Indian style dance show originated in the Indian territory of Kerala (south) started in the seventeenth century. The narrative related to 'Kathakali' dance execution is imparted to the crowd through single and double hand motions and looks including music. Kathakali dance is generally carryout by male artists in theaters and courts of Hindu locales. Kathakali hasta mudras are entertained as a total language without help from anyone else with essential linguistic components and kathakali language structures related with it. One can convey any message to others by totally utilizing hands and taking assistance from the 24 hasta mudras (Fig. 1).

It is extremely challenging for a common person to comprehend the significance of Kathakali dance dramatization because of its complicated hand motion language design and dance developments. Except if you realize all the hasta mudras and the sentences and various words you can express utilizing these hasta mudras. It is hard for a common person to know the significance conveyed by the Kathakali hand signals depending on an antiquated text lakshana Deepika, hasta [1]. There is an aggregate of 24 hasta mudras. Utilizing samyuktha and asamyuktha hastas, mudras are framed. Blend of these gestures pass specific significance on to them. Mudras are represented in Fig. 2 dataset which we have taken as mentioned in Fig. 3, and the hasta mudra denotations are summed up in Table 1. Various mixtures of hasta mudras in some ways address different importance and some are even contingent on the unique circumstance. Except that one is knowledgeable with these mudras, their blends and implications, it is troublesome for one to comprehend and like this workmanship. In this work, we have used Kathakali hand mudras dataset which is

P. Malavath (✉) · N. Devarakonda
School of Computer Science and Engineering, VIT-AP University, Amaravathi 522237, India
e-mail: pallavimalavath.20phd7126@vitap.ac.in

R. Chaki et al. (eds.), *Applied Computing for Software and Smart Systems*,
Lecture Notes in Networks and Systems 555,
https://doi.org/10.1007/978-981-19-6791-7_8

available in "Kathakali Mudras", the Mendeley Data that is available in [2]. Later we have applied preprocessing steps to the hasta mudra images, and eventually, we compare ML and DL approaches for asamyuktha hasta mudras classification.

The key contributions of this work are:

- Preprocessing, feature extraction, and classification of Kathakali hasta mudras.
- The naive classifier will absorb features as an input parameter and categorize them into various groups.
- Construct CNN for kathakali hasta mudras classification.
- Compare the results of Machine Learning classifier with CNN classification.

Fig. 1 Kathakali dance gestures

Fig. 2 1–12 Asamyuktha Hastha Mudras

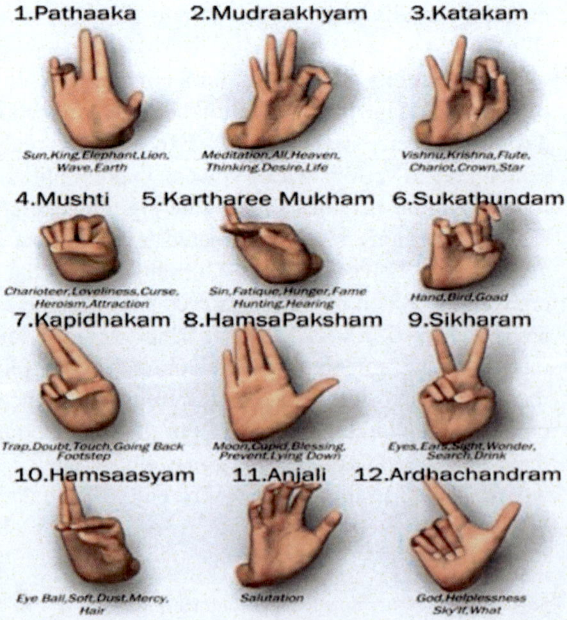

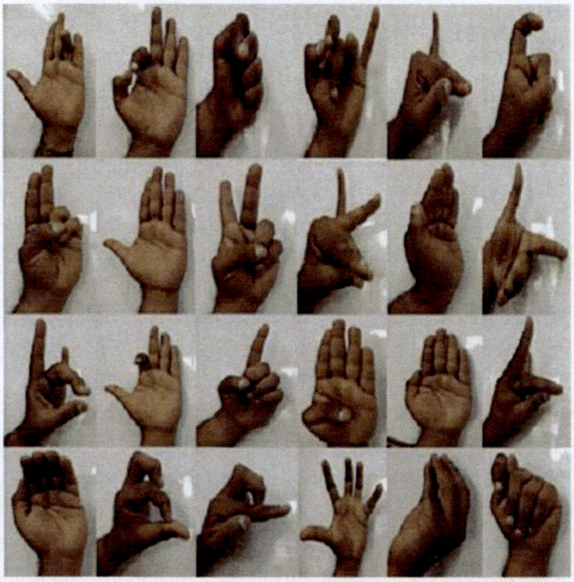

Fig. 3 24 Kathakali Hasta Mudras

Table 1 List for Kathakali Hasta Mudras

S.NO	Name of the mudra	S.NO	Name of the mudra
1	Pathaka	13	Bhramaram
2	Katakam	14	Mukuram
3	Mudraakhyam	15	Soochimukham
4	Musti	16	Thripathaaka
5	Sukathundam	17	Pallavam
6	Kartharee Mukham	18	Mrigaseersham
7	Kapidhakam	19	Vardhamanakam
8	HamsaPaksham	20	Sarpasirassu
9	Hamsaasyam	21	Araalam
10	Sikharam	22	Mukulam
11	Ardhachandram	23	Oornanabham
12	Anjaly	24	Katakaamukham

Remaining paper is coordinated as follows. Segment II momentarily makes sense of writing audit, area III portrays the proposed grouping strategies momentarily, area IV depicts the outcomes exhaustively as well as the impact of preprocessing strategies and arrangement results, lastly area V finishes up the paper.

2 Literature Survey

According to the consideration of existing phenomena, this is the second major project devoted to Kathakali hasta mudras. Kuchipudi, Bharatanatyam, Odissi, Khathak, Sattriya, Manipuri, Aceh (Indonesian) and Korean pop dance have all been featured in previous works. In this part, we primarily focused on providing current advances in human motion recognition and how it is used in arranging dance hand gestures in the dance structures discussed previously.

Anami et al. [3] proposed a three-step technique that includes extracting mudra contours from input photographs, feature extraction utilizing Eigen values, Hu moments, and crossing points, and finally categorizing hasta mudras using AI. The Bharatanatyam dataset has 2800 photos (100 pictures for each mudra). There are two types of mudras: clashing and non-clashing mudras. The claimed precision for complete hand mudras, clashing hast-mudras, and non-clashing hasta-mudras, respectively, is 97.1, 99.5, and 96.03%, respectively. For planning Indian traditional dance events, Kisore et al. [4] suggested a deep neural network design. This collection includes recordings from both the internet (YouTube, live performances) and albums. The overall accuracy of the hasta mudra categorization was 93.3%. Using the Histogram of oriented Gradients (HOG) for feature extraction and the Support vector machine (SVM) as the classifier, Kumar et al. [5] suggested a method for classifying kuchipudi dancing mudras into instant messaging (meaning of mudra).

The XBOX Kinetic sensors were used by Anbarsanthi et al. [6] to capture distinct poses of Aceh Traditional dance form. Matlab used Simulink Programming to create the identification system. For single hand mudra, they have a 94.86 percent accuracy rate. Tongpeang et al. [7] have presented a technology that detects errors, analyzes them, and provides feedback to the dancer so that they can improve their dancing skills. To achieve accuracy, the fundamental method collates the poses with a Thai dance expert and a real dancer. Kim et al. [8] originated an ELMC (Extreme Learning Machine Classifier) based on ReLU (Rectified Linear Unit) with 800 dance poses of 200 types of distinct dances. When compared to SVM and KNN, they have a high level of accuracy. Chen et al. [9] suggested a framework for classifying and predicting hand mudra labels. This structure will extract 1300 hand region images using the background subtraction approach, segmenting fingers and hand palm to identify the fingers, and using rule classifer to identify the labels. Using deep learning and machine learning approaches, Bhavanam et al. [10] produced Kathakali hasta mudras and discovered new ways to distinguish various mudras. There are 24 various types of hasta mudras that the Kathakali artist might use to portray the story. They gathered 645 photographs of Kathakali hasta mudras for this project, with each hand move having 27 images for classification. Literature survey is given in the Table 2.

Table 2 Literature Survey

S.NO	Dance form	Main objective	Learning	Dataset	Technique used
[1]	Kathakali	Classification of asamyuktha hastamudras	ML, CNN	645 images	Contour extraction, HOG, Haar Wavelet Features, Canny Edge Detection
[2]	Bharatanatyam	Classification of single hasta mudras	ANN	2800 images	Contour extraction, Hu-moments, Canny Edge Detector, eigen values
[3]	Kuchupudi	Classification of different Indian classical dance forms	SLIC	Images collected online	Watershed Algorithm
[4]	Kuchipudi	Classification of hasta mudras as text messages	SVM	Images collected from online	HOG, SIFT, SURF, HAAR and LBP
[5]	Indian Classical Dance	Classification of different Indian Traditional dance forms	CNN	2800 images	Not Required
[6]	Bharatanatym, Odissi and Kathak	Classification of different Indian Classical dance forms	SVM	30 Videos	(HOOF) Histogram of oriented optical flow
[7]	Sattriya	Classification of hasta mudras	SVM and Decision tree	1015 pictures collected offline	Background elimination utilizing GMM Smoothening the images by using Gaussian Filter
[8]	Kuchipudi and Bharatanatyam	Identifying complex Human Movements	Adaboost multi-class classifier	Images collected from offline and online	Local Binary Pattern (LBP) features, discreate wavelet transform
[9]	Aech (Indonesian)	Classifying the dance poses	HMM	2169 videos	Not Specified

(continued)

Table 2 (continued)

S.NO	Dance form	Main objective	Learning	Dataset	Technique used
[10]	Khathak, Kuchipudi, Manipuri, Bhratanatyam, Mohiniyattam, Sattriya, Odissi	Classifying different Indian dance forms	SVM	626 videos	SVM Motion Capturing: Optical Flow Algorithm
[11]	Korean pop Dance	Classifying korean dance movements	ELMC (extreme learning machine classifer)	800 images	Dimensionality Reduction: FLDA and PCA
[12]	Thai Dance form	Comparison of Thai dance between experts and real time dancers	Kinetic motion sensor	Real time dance movements	Not specified

3 Methodology

In this part, we describe about the model. We explain about the preprocessing steps which are applied to the input image and explains the classification using Naïve bayes classifier and deep learning (DL) technique for classification of images using CNN.

3.1 Data Preprocessing

We have taken Kathakali Dataset from "Kathakali Mudras", Mendeley Data 9 which is available in [2]. There are total 654 Kathakali hasta mudras where each gesture has 27 different images. Multiple factors like different background colors, different positions, different people are considered in creating the dataset. We employed four feature extraction approaches for Kathakali hand gesture classification: HOG, Haar wavelet features, canny edge detection, and Contour Extraction.

3.2 Naive Bayes Classification

We utilized a naive Bayes classification model to separate Kathakali dance hand motions to see how machine learning distinguishes our data. The naive Bayes classifier will absorb features as an input parameter and categorize them into various groups. Preprocessing, feature extraction, and classification of Kathakali hasta mudras are done in this work. Bayesian decision theory includes the concept of naive Bayes. Because it is mostly employed in text classification, it entails high-dimensional dataset training, which aids in the development of a quick machine learning model that can efficiently categorize images. It is called naïve because it speculates that features occurrence is not dependent on another feature occurrence. It uses the principle of Bayes theorem hence it is called naive bayes. The advantage of using naive Bayes is, it is suitable for multi class prediction if the interpretation of the feature occurrence is not dependent on another feature occurrence, then it performs more effectively than other models. Another advantage is it is highly scalable.

The naïve Bayes classifier correlates with Bayesian network, which is denoted in Eq. (1). Here 'm' attribute variables Xi and 'C' single class variable. Suppose 'C' resembles a class label and xi resembles a value of an attribute Xi. it can be formulated as:

$$p_r(c, x_{1,....,}x_m) = p_r(c).\prod_{i=1}^{m} pr(x_i|c) \tag{1}$$

So now we have conditional distributions *Pr(Xi|c)*. From the labelled data we can take these parameters based on MAP estimation. Once it is derived from the naïve-based classifier the new instances are labeled by class label i.e.; c* which has the utmost posterior probability of *sx*

$$c^* = argmaxp_r(c|x_1, ..., x_m) \tag{2}$$

3.3 Convolutional Neural Network Classification

CNN is one of the most eminently used for classification in deep learning. It is a part of neural networks which is used to improvize the image visualization. Generally the neural network uses matrix multiplication, whereas convolution neural network has a special approach called convolution. The workflow of CNN is that it initially extracts features for the given image, then the pixels data is used as an input, and it is returned with the inference of the pixels. CNN is prominently used in analyzing the image, the three main functionalities of CNN are the recognition of the image, object detection and segmenting the image. The image which is given as an input goes through many layers that includes convolutional layer in which there are set of kernels when each kernel convolves with the image and an activation map is generated, then the image will go through the pooling layer, which reduces the parameters. Next comes the Rectified linear unit (ReLU) where the activation map is implemented and finally moves to the fully connected layers where classification is done. We structured a mutilayer CNN technique with two Conv layers, 2 ReLU layers and 2 Maxpool layers. ReLU is used to increase the nonlinearity in the network, because

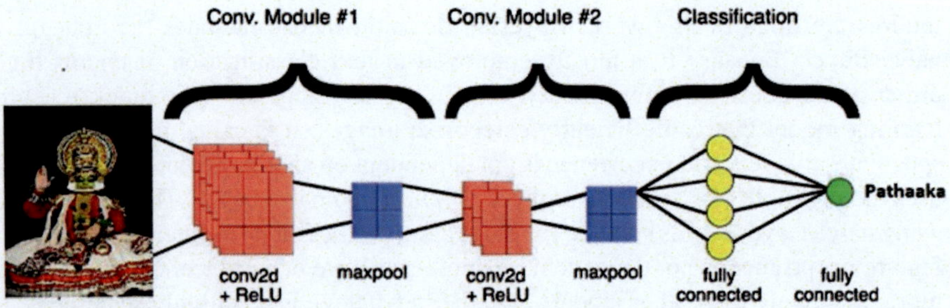

Fig. 4 Proposed CNN architecture

generally images are highly nonlinear and it eliminates all negative values from the activation map by mounting them to zero. Sigmoid activation function is used in the output layer for classification of kathakali mudras. We have implemented a batch normalization procedure for determining standard deviation and mean. The proposed CNN structure is depicted in Fig. 4. Initially the images go through preprocessing procedure, where images are resized and augmented. Resizing the input images will enhance the computational capacity. Later the preprocessed images are converted into gray scale images. In the next step these preprocessed images are fed into the convolutional neural network.

The Convolutional neural network model will learn from the extracted features. There are a total of 654 images, which are split-up into training and test images. In these 70% of images are considered as the training set and 30% images are considered as the testing set and again the training set is slit-up into the training set and validation set.

4 Outcomes

4.1 Naive Bayes Results

There are two steps in the Naive Bayes categorization. Feature extraction and preprocessing are performed in the first stage. The second stage is to classify the images using the Naive Bayes Classifier, which is based on feature extraction. Figure 5 shows the outcomes of Haar wavelet feature extraction. The outcomes show that, for example, in the Hamsa Paksham hastamudra, features are extracted precisely, however in other mudras such as Vardhamanakam, Mukulam, and Araalam, there may be a collision in feature extraction. Figure 6 shows the histogram of Oriented Gradients. The feature is extracted exactly from hamsa paksham as obtained from haar wavelet, although there is some misperception in the group set when compared to other mudras such as mukulam, Araalam, and Vardhamanakam. We've observed cany edge detection and contour extractions are used for extracting edge feature using

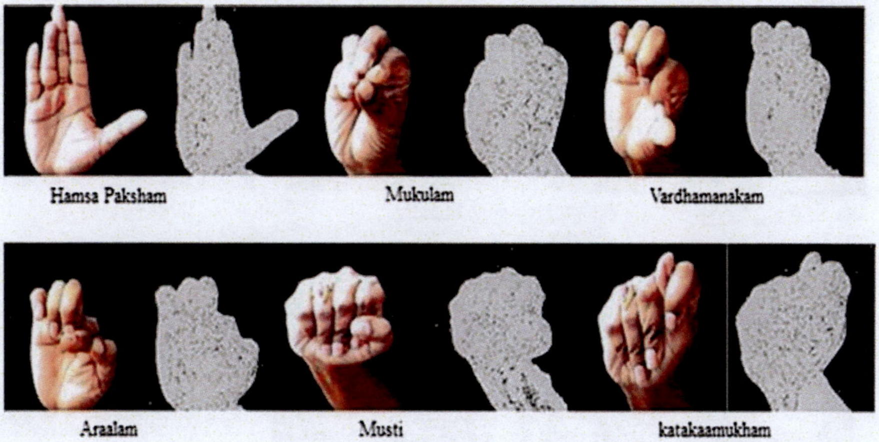

Fig. 5 Haar wavelet features

the edge detection method. The outcomes of the features data set using edge detection are explained in Fig. 7. The outcomes conclude that using canny edge and contour extraction the edges are not clearly determined and features are differing. Hence classification using the above feature does not obtain premier results. Precision, FP rate, recall, and accuracy are just a few of the parameters to consider when examining the Naive Bayes classifier. As a result, the performance metrics produced by Naive Bayes is shown Table 3 below. We can see from the table that there are 6 mudras (Musti, Kartharee Mukham, Bhramaram, Soochimukham, Pallavam, Mukulam) with a precision of 60%, which is not very good, and the outcomes for Recall and FP rate are depicted in Fig. 7. Lowest precision is observed for dance types (Katakam, Sukathundam), and the outcomes for recall and FP rate are compared in Fig. 8. Remaining dance types have a 40 percent accuracy rate and have an average precision. Based on the metrics, we may conclude that the Naive Bayes Classifier fails to classify the Kathakali Hasta Mudra.

4.2 Convolutional Neural Network (CNN) Results

We need a lot of manual processes for preprocessing and data cleansing, if we use Machine Learning techniques like Naive Bayes. As a result, for the classification of Kathakali Hasta Mudras, we chose a Deep Learning architecture. Initially, training and validation images are separated from the input photos. A validation set is made up of 20% of the training images. Later, we used kathakali photos to train our model. Figure 8 shows the recall and FP rate data for High Precision mudras. Figure 9 shows the recall and FP rate figures for Low Precision mudras. The outcomes of training this model numerous times with different numbers of epochs are shown in Fig. 10. The accuracy for training and testing images is 83 percent and 84 percent, respectively,

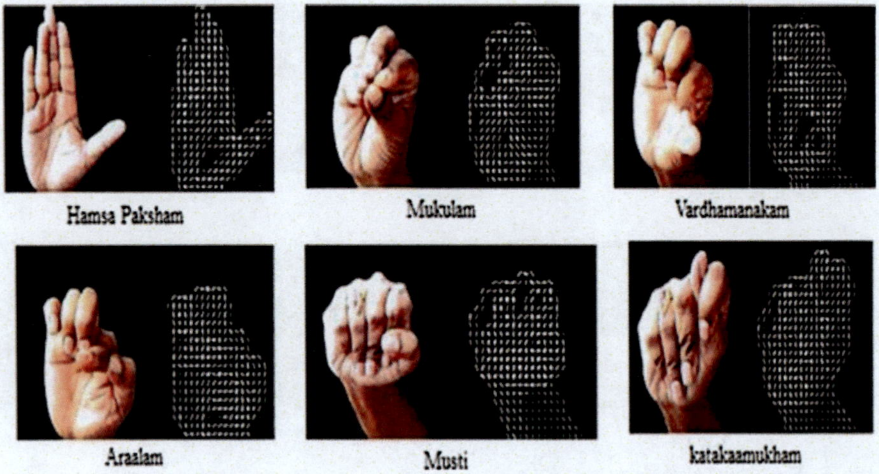

Fig. 6 Histogram of oriented features

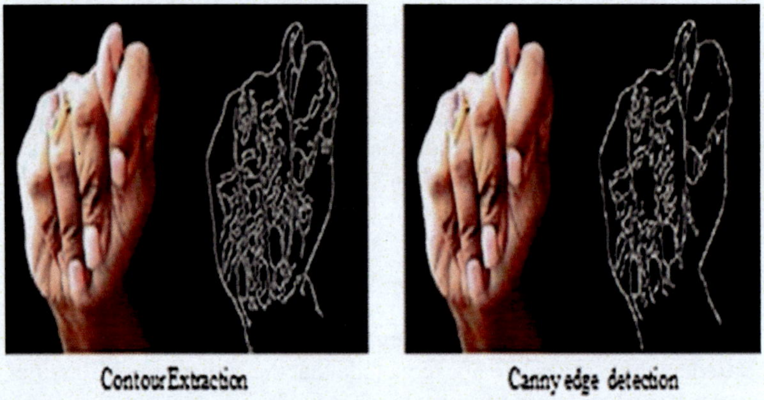

Fig. 7 Canny edge detection contour extraction

after 50 epochs. The accuracy for training and testing photos is 88 percent and 78 percent for 100 epochs, respectively. As the number of epochs increases, we can observe that the training loss decreases as each epoch is trained, and we get a new feature set from which classification is performed. Validation loss and training loss is depicted in Fig. 11. When the number of epochs taught by the CNN Model grows, plot validation grows as well. Classification results of Kathakali Asamyuktha Hasta Mudras are depicted in Fig. 8 (Fig. 12).

Table 3 Performance metrics

No	Mudra	Precision	Recall	FP rate
1	Pathaaka	1.00	0.90	1.00
2	Katakam	0.26	1.00	0.42
3	Mudraakhyam	1.00	0.65	0.60
4	Musti	1.00	0.30	0.30
5	Sukathundam	0.25	1.00	0.36
6	Kartharee Mukham	1.00	0.15	0.21
7	Kapidhakam	0.43	0.28	0.33
8	Hamsapaksham	0.68	0.50	0.56
9	Hamsaasyam	0.40	0.60	0.58
10	Sikharam	0.80	0.80	0.80
11	Ardhachandram	0.78	0.88	0.98
12	Anjaly	1.00	0.80	0.90
13	Bhramaram	1.00	0.80	0.90
14	Mukuram	0.80	0.57	0.89
15	Soochimukham	1.00	0.56	0.40
16	Thripathaaka	0.30	0.70	0.40
17	Pallavam	1.00	0.35	0.60
18	Mrigaseersham	0.50	0.56	0.73
19	Vardhamanakam	0.77	1.00	0.60
20	Sarpasirassu	0.70	0.87	0.65
21	Araalam	0.67	0.87	0.68
22	Mukulam	1.00	0.70	0.80
23	Oornanabham	0.80	0.79	0.78
24	Katakaamukham	0.90	0.70	0.79

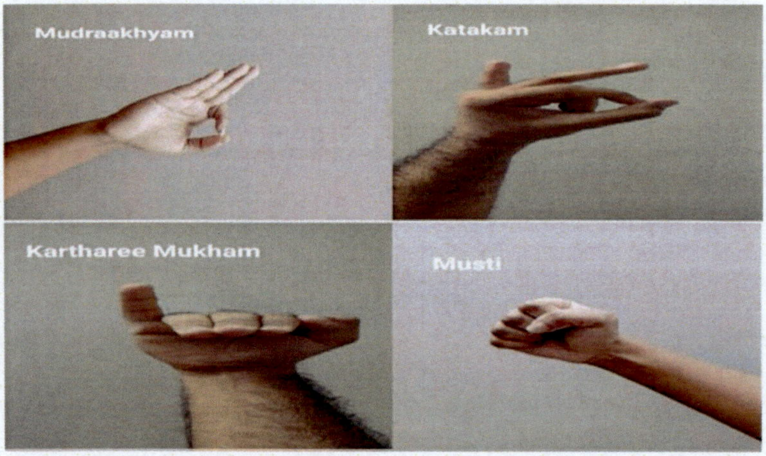

Fig. 8 Classification of Kathakali Asamyuktha Hasta Mudras

Fig. 9 Recall and FP rate scores for high precision mudras

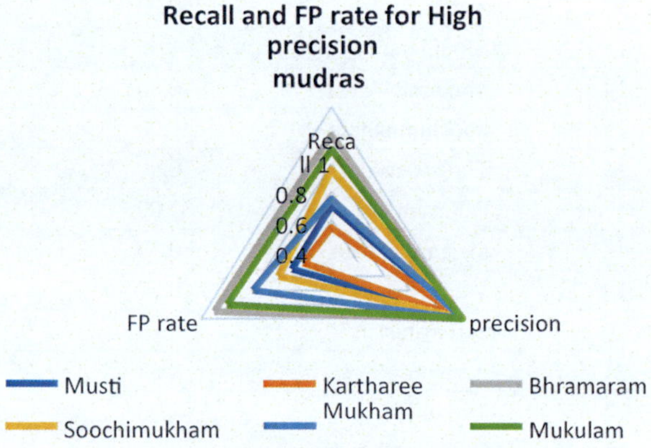

Fig. 10 Recall and FP rate values for low precision mudras

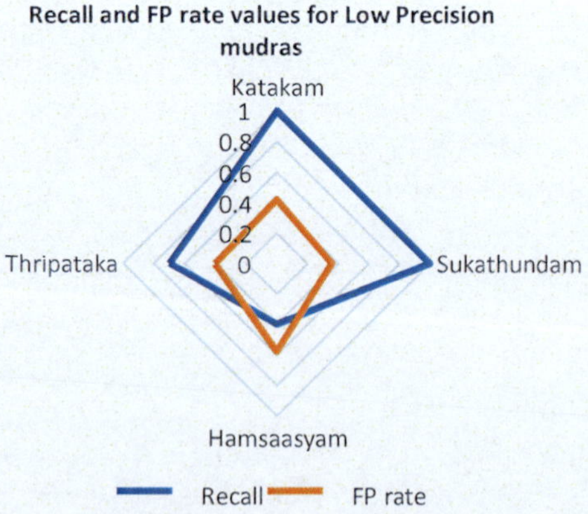

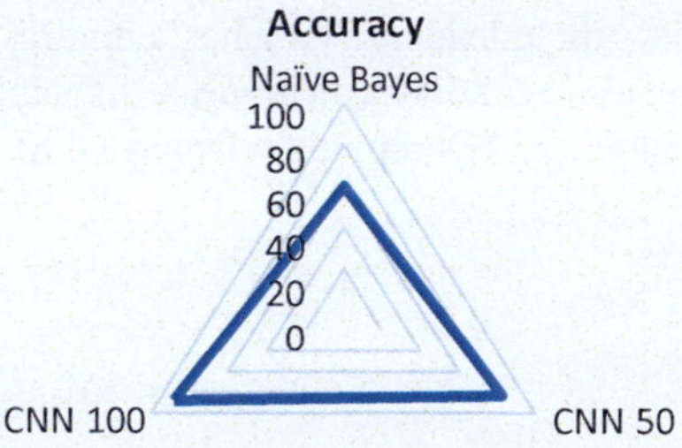

Fig. 11 Comparison between Naive Bayes classification, CNN 50 and CNN 100

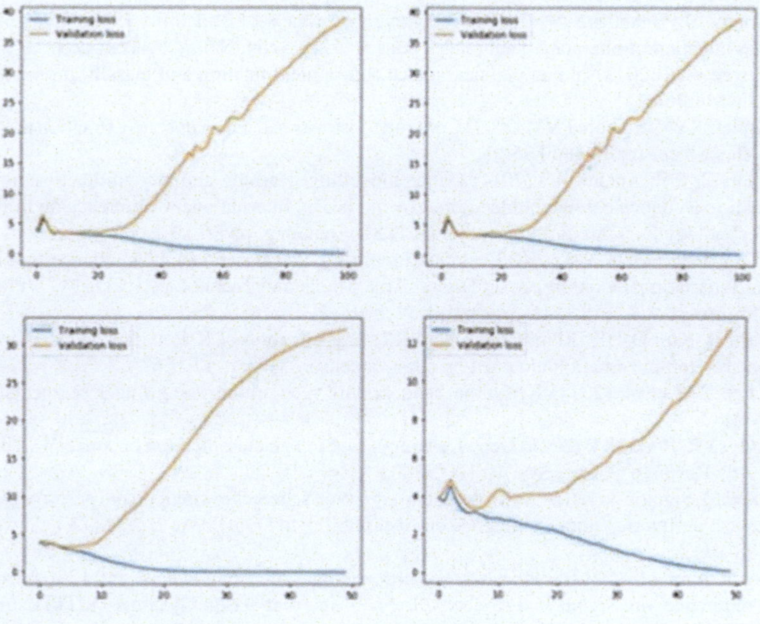

Fig. 12 Training and validation loss

5 Conclusion

In this paper, we have experimented the data set available in [2] and trained it using the CNN Model. Features are extracted and compared with Naive Bayes and CNN Model. Plot validation is increased when there is increase in the epochs that are trained by the CNN Model. The features derived are classified by the CNN Model and Naive Bayes and the outcomes consist of a total of 684 images in which 70% are of training images and 30% are of testing images. The extracted features are given as input to the Naive Bayes classifier. It classifies the images into various varieties of mudras and an accuracy of 6.79% was obtained. We got a low accuracy due to collision in hand mudras in feature extraction. In CNN scenario, we got an accuracy of upto 88%. Future works include identifying hasta mudras, generating large dataset including samyuktha and asamyuktha hasta mudras, displaying meaning of the mudra based on video and recognizing hasta mudras from real time videos etc.

References

1. Kadathanattu Udyavarma Thampuran, Hasta Lakshana Deepika, Janran-jinee Achukoodam (Printers), Nadapuram, 1892
2. Iyer N, Ganesh B, Tulasi L (2019), "Kathakali Mudras", Mendeley Data, V1. https://doi.org/10.17632/wdbm9srwn7.1
3. Anami B, Bhandage V (2018) A comparative study of suitability of certain features in classification of bharatanatyam mudra images using ANN. Neural Process Lett
4. Kisore et al (2018) Indian classical dance action identification and classification with CNN. Adv Multimed
5. Kumar KVV, Kishore PVV (2017) Computer vision based dance posture extraction using SLIC. J Theor Appl Inf Technol 95(9)
6. Santi H, Prihatmanto AS (2014) Dance modelling, learning and recognition system of aceh traditional dance based on hidden Markov model. In: International Conference on Information Technology Systems and Innovation (ICITSI). Bandung, pp 86-89
7. Tongpaeng NY et al. (2018) Evaluating real-time thai dance using Thai dance training tool. In: International Conference on Digital Arts, Media and Technology (ICDAMT). Phayao, pp 185-189
8. Kim D, Kim D-HK, Keun-Chang K (2017) Classification of K-Pop dance movements based on skeleton information obtained by a kinect sensor. Sensors 17:1261
9. Chen Z-H et al. (2014) Real-time hand gesture recognition using finger segmentation. Sci World J
10. Suresh R, Prakash P (2016) Deep learning-based image classification on amazon web service. J Adv Res Dyn Control Syst 10:1000-1003
11. Devi M, Saharia S (2016) A two-level classification scheme for single- hand gestures of Sattriya dance. In: International Conference on Accessibility to Digital World (ICADW). Guwahati, pp 193-196
12. Bisht A et al. (2017) Indian dance form recognition from videos. In: 13th International Conference on Signal-Image Technology & Internet-Based Systems (SITIS). Jaipur, pp 123-128
13. Internet-Based Systems (SITIS) (2017) Jaipur, pp 123–128
14. Prakash RM et al, Gesture recognition and fingertip detection for human computer interaction. In: International conference on innovations in information, embedded and communication
15. Gu D (2015) Fingertip tracking and hand gesture recognition by 3D vision. Int J Comput Sci Inf Technol 6(6): pp. 1–4. Systems (ICIIECS), Coimbatore (2017)

Multi-objective Fuzzy Reliability Redundancy Allocation for x_j-out-of-m_j System Using Fuzzy Rank-Based Multi-objective PSO

Satyajit De, Pratik Roy, and Anil Bikash Chowdhury

Abstract The main goal of this paper is to solve a fuzzy multi-objective reliability redundancy allocation problem (MORRAP) for x_j-out-of-m_j series-parallel system. We consider that system reliability and system cost are two conflicting objectives. Due to the incompleteness and uncertainty of input information, we formulate the objectives by considering the reliability and cost of each component as a triangular fuzzy number (TFN). Here, the fuzzy multi-objective optimization problem of system reliability and cost is analyzed simultaneously using our proposed fuzzy rank-based multi-objective particle swarm optimization (FRMOPSO) algorithm. Comparing the results of FRMOPSO with standard particle swarm optimization (PSO), we see that better optimum reliability and cost have been achieved in the FRMOPSO technique. To illustrate the effectiveness of our proposed technique, we consider the problem of the over-speed protection system of gas turbines containing two mutually conflicting reliability and cost objectives with entropy and several other constraints. We present graphically the effect of optimum system reliability and cost with respect to the percentage change of different parameters. We also compare the convergence rate of FRMOPSO with PSO. Our proposed algorithm shows better results.

Keywords Fuzzy-based reliability redundancy model · Triangular fuzzy number · MOPSO algorithm · Non-dominated sorting · Fuzzy ranking method · Averaged Hausdorff distance

S. De (✉)
Department of Computer Science, Maheshtala College, Maheshtala 700141, WB, India
e-mail: satyajitde973@gmail.com

P. Roy
Department of Computer Engineering and Applications, GLA University, Mathura 281406, UP, India
e-mail: pratik.roy@gla.ac.in

A. B. Chowdhury
Department of Computer Applications, Techno India University, Kolkata 700091, WB, India

R. Chaki et al. (eds.), *Applied Computing for Software and Smart Systems*, Lecture Notes in Networks and Systems 555,
https://doi.org/10.1007/978-981-19-6791-7_9

1 Introduction

The main goal of MORRAP of a multi-stage series-parallel system is to optimize the system reliability with other conflicting objectives by optimizing the number of redundant components in each stage by properly utilizing the other available resources. Multi-objective particle swarm optimization (MOPSO) introduced by Coello et al. [1] is a popular meta-heuristic technique used to optimize each objective of a MORRAP without being affected by any other solutions.

In this paper, we consider a fuzzy-based MORRAP of x_j-out-of-m_j series-parallel system based on the k-out-of-n system [3] to maximize the system reliability and minimize the system cost by maintaining maximum entropy constraint with a limited number of redundant components. Entropy can be used to measure the dispersal of allocation between stages in a redundancy allocation problem [4]. Therefore, in our fuzzy-based MORRAP we use entropy constraint. We propose the FRMOPSO algorithm where we update a personal best set of every particle in every iteration using a non-dominated sorting technique [5]. We fetch personal and global best solution using the fuzzy ranking method simultaneously. Using the over-speed protection system [6], we illustrate our proposed MORRAP. We compare the convergence rate of FRMOPSO with PSO using Average Hausdorff Distance (AHD) technique [7]. Sensitivity analysis with the change of different parameters is shown graphically.

2 Related Work

In recent years, various multi-objective optimization problems on series-parallel system or k-out-of-n system have been constructed with or without entropy constraints and solved using different meta-heuristic or evolutionary algorithms. Roy et al. [4] evaluated optimum system reliability and cost from a MORRAP with entropy constraint. Huang et al. [8] developed a model of fuzzy constraints for a series-parallel redundancy allocation problem and solved it using an improved swarm optimization. Kumar and Yadav [6] formulate a fuzzy MORRAP for over-speed protection system and solved it using a hybrid non-dominated sorting genetic algorithm (NSGA)-II. Sharifi et al. [9] considered a k-out-of-n load-sharing system with identical components. A consecutive k-out-of-n system is used by Dui et al. [10] to measure the joint importance of its optimal component sequence. Farhadi et al. [11] designed a k-out-of-n redundant system in their study. Wang [12] introduced a method for estimating the time-dependent reliability of both ordinary and weighted k-out-of-n systems.

Currently, various multi-objective optimization problems have been solved by using the MOPSO algorithm. Davoudi et al. [13] implemented a MOPSO algorithm to solve their developed optimization problem. Mahapatra et al. [2] solved a MORRAP under a hesitant fuzzy environment using the MOPSO technique. MahMond et al. [14] proposed the topsis fuzzy MOPSO technique to solve the Labyrinth Weir optimization problem. For the constrained MOPSO algorithm, Wang et al. [15]

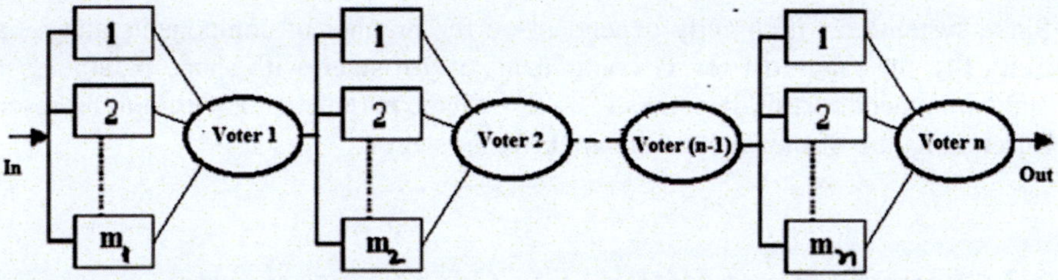

Fig. 1 x_j-out-of-m_j series-parallel system

proposed a constraint handling technique based on the Lebesgue measure. By effectively avoiding the local optimum problem, Yuan et al. [16] proposed an algorithm based on gray relationship analysis and MOPSO. Liang et al. [17] used Pareto-based MOPSO to optimize the supply air temperatures and velocities in their work. Ershadi et al. [18] calculated the optimal solutions using a parameter-tuned meta-heuristic algorithm based on MOPSO. The effect of inertia, cognitive and social parameters of the MOPSO algorithm is analyzed by Rajani et al. [19]. Khazaei et al. [20] used the MOPSO technique to optimize the stock portfolios in the stock price forecasting model.

3 MORRAP for x_j-out-of-m_j Series-Parallel System

3.1 x_j-out-of-m_j Series-Parallel System

Figure 1 based on k-out-of-n system shows a system which has n stages connected in series, and at jth stage m_j $(j = 1, 2...n)$ components are connected in parallel out of which $(m_j - 1)$ components are redundant. It is required that the system gets in working condition if at least x_j components out of total m_j must be operational at jth stage in order. For that, we call the system as x_j-out-of-m_j reliability redundancy series-parallel system. Generally, a perfect voter should be needed at every stage. The objective is to find out the optimum number of redundant components x_j at jth $(j = 1, 2...n)$ stage such that the system reliability will be maximum and system cost will be minimum, subject to the system entropy constraint with a total limited number of components. If all the components have the reliability R_j at jth stage, then the total reliability of that stage is obtained via binomial distribution defined in [3] as

$$\sum_{i=x_j}^{m_j} \frac{m_j!}{i!(m_j - i)!} R_j^i (1 - R_j)^{(m_j - i)}$$

Since system cost is directly dependent on the number of components and reliability [4], the total cost for x_j components at jth stage with shape parameter a_j and component cost C_j becomes $C_j x_j R_j^{a_j}$. Shape parameters are used to represent different nature of the components in different stages.

3.2 Formulation of MORRAP in Crisp Environment

If we have the maximum system reliability with minimum system cost as well as a limited number of redundant components with system entropy, then it will be highly accepted. Therefore, the optimization problem for the above system is defined in a crisp environment as follows:

$$Maximize\ R(x_1, x_2, x_3....x_n) = \prod_{j=1}^{n} \left\{ \sum_{i=x_j}^{m_j} \frac{m_j!}{i!(m_j - i)!} R_j^i (1 - R_j)^{(m_j - i)} \right\} \tag{1}$$

$$Minimize\ C(x_1, x_2, x_3....x_n) = \sum_{j=1}^{n} C_j x_j R_j^{a_j}$$

$$Subject\ to\quad - \sum_{j=1}^{n} \left\{ \frac{x_j}{\sum_i x_i} \ln \left(\frac{x_j}{\sum_i x_i} \right) \right\} \le E$$

$$S_l \le \sum_{i=1}^{n} x_i \le S_u$$

$$1 < x_j \le m_j \quad for\ j = 1, 2, 3, ...n$$

where E is the maximum entropy value, and S_l and S_u are the lower and upper limits of total components of the system, respectively.

3.3 Fuzzy MORRAP for Above System

In reality, component reliability and cost of a redundancy model may not be fixed and users can change it as per their requirements. So to make the x_j-out-of-m_j series-parallel redundant model more flexible and user- acceptable, we have to represent the reliability and cost by TFN [21]. We consider $\widetilde{R_j}$ and $\widetilde{C_j}$ as TFNs to represent the reliability and cost of every component at jth stage of the system, respectively. If $\widetilde{R}$ and $\widetilde{C}$ represent the system reliability and system cost as TFNs respectively, then the optimization problem (1) can be represented as

$$Maximize\ \widetilde{R}(x_1, x_2, x_3....x_n) = \prod_{j=1}^{n}\left\{\sum_{i=x_j}^{m_j}\frac{m_j!}{i!(m_j - i)!}\widetilde{R_j}^{i}(1 - \widetilde{R_j})^{(m_j - i)}\right\} \quad (2)$$

$$Minimize\ \widetilde{C}(x_1, x_2, x_3....x_n) = \sum_{j=1}^{n}\widetilde{C}_j x_j \widetilde{R_j}^{a_j}$$

$$Subject\ to\ same\ constraints\ as\ (1)$$

4 FRMOPSO Technique to Solve Fuzzy MORRAP

4.1 Standard PSO

Every particle of PSO is a random solution to the optimization problem in the search space. Suppose a search space contains N swarm particles of dimension n for an optimization problem. We consider the position and velocity of the ith particle at iteration t are $X_i(t) = (x_{i1}(t), x_{i2}(t), ..., x_{in}(t))$ and $V_i(t) = (v_{i1}(t), v_{i2}(t), ..., v_{in}(t))$, respectively. Each ith particle has its own best performance, and it is represented as $Pbest_i(t)$. $Gbest(t)$ represents the global best performance among the members of swarm particles at iteration t. Here, the updated velocity for next iteration $(t + 1)$ is defined as

$$V_i(t + 1) = \omega * V_i(t) + rnd_1 * cac * (Pbest_i(t) - X_i(t)) + rnd_2 * gac * (Gbest(t) - X_i(t))$$

where ω is the inertia coefficient, cac is the cognitive acceleration coefficient, gac is the social/global acceleration coefficient and rnd_1, rnd_2 are the uniformly distributed random values between 0 and 1. The modified location of the ith particle is calculated as

$$X_i(t + 1) = X_i(t) + V_i(t + 1)$$

4.2 FRMOPSO Algorithm

Step 1: Calculation of minimum and maximum value of each objective function:

To calculate the extreme values of individual objective functions, we solve them one by one considering as a single objective optimization problem. Suppose X^* and Y^* be the optimal solutions of equation reliability and cost objective functions, respectively. Then the lower bound and upper bound of system reliability are $R_{\min} = R(Y^*)$ and $R_{\max} = R(X^*)$ respectively. Similarly, the lower bound and upper bound of system cost are $C_{\min} = C(Y^*)$ and $C_{\max} = C(X^*)$ respectively.

Step 2: Construction of membership functions:

We define the membership functions for the objectives in Eq. (2). TFN for system reliability is considered as an ordered triplet $\widetilde{R_T} = (R_{\min}, R_{center}, R_{\max})$ such that $R_{\min} \le R_{center} \le R_{\max}$ and its membership function is defined as

$$\mu_{\widetilde{R_T}}(x) = \begin{cases} \frac{x - R_{\min}}{R_{center} - R_{\min}} \; if \; R_{\min} \le x \le R_{center} \\ 1 \; if \; x = R_{center} \\ \frac{R_{\max} - x}{R_{\max} - R_{center}} \; if \; R_{center} \le x \le R_{\max} \\ 0 \; otherwise \end{cases} \tag{3}$$

TFN for system cost is considered as an ordered triplet $\widetilde{C_T} = (C_{\min}, C_{center}, C_{\max})$ and its membership function is defined as

$$\mu_{\widetilde{C_T}}(x) = \begin{cases} \frac{x - C_{\min}}{C_{center} - C_{\min}} \; if \; C_{\min} \le x \le C_{center} \\ 1 \; if \; x = C_{center} \\ \frac{C_{\max} - x}{C_{\max} - C_{center}} if \; C_{center} \le x \le C_{\max} \\ 0 \; otherwise \end{cases} \tag{4}$$

TFN is chosen because decision-makers may change the modal value of system reliability (R_{center}) and system cost (C_{center}) in the uncertainty interval $[R_{\min}, R_{\max}]$ and $[C_{\min}, C_{\max}]$ respectively to meet their required level of satisfaction. Here, to get the high level of satisfaction for corresponding objective functions of the optimization problem (2), decision-maker may set the modal value R_{center} toward $R_{\max}$ and the modal value C_{center} toward $C_{\min}$.

Step 3: Defuzzification using center of gravity (COG):

The crisp value of system reliability $\widetilde{R} = (r_1, r_2, r_3)$ and system cost $\widetilde{C} = (c_1, c_2, c_3)$ in triangular fuzzy form can be calculated using COG [22] as $R^* = \frac{\sum_{i=1}^{3} r_i \mu_{\widetilde{R_T}}(r_i)}{\sum_{i=1}^{3} \mu_{\widetilde{R_T}}(r_i)}$ and $C^* = \frac{\sum_{i=1}^{3} c_i \mu_{\widetilde{C_T}}(c_i)}{\sum_{i=1}^{3} \mu_{\widetilde{C_T}}(c_i)}$ respectively where $\mu_{\widetilde{R_T}}$ and $\mu_{\widetilde{C_T}}$ are the membership functions for system reliability and system cost, respectively.

Step 4: Fuzzy multi-objective optimization problem construction from TFN-based MOOP (2)

Maximum value of the membership function of an objective function gives the maximum satisfaction level [6]. So our aim is to construct a fuzzy rank-based MOOP from Eq. (2) to maximize all membership functions of the corresponding objective functions as follows:

$$\begin{gathered} Maximize\; \mu(f) = [\mu_{\widetilde{R_T}}(R^*),\; \mu_{\widetilde{C_T}}(C^*)] \\ Subject\; to\; same\; constraints\; as\; (2) \end{gathered} \tag{5}$$

where R^* and C^* are the crisp values for the fuzzy objective functions $\widetilde{R}(x_1, x_2, x_3 \ldots . x_n)$ and $\widetilde{C}(x_1, x_2, x_3 \ldots . x_n)$ respectively.

Step 5: Non-dominated sorting technique:

non_dominated_sorting(z, A)

We consider z is a solution of dimension n and A is a set containing p number of non-dominated solutions of dimension n. Using the following algorithm, we check that the solution z is a non-dominated solution or not with the solutions of the set A. The algorithm returns 1 if the solution is non-dominated and set A is modified, else return 0.

Step 6: Representation of fuzzy MOOP of Eq. (5) into crisp format as follows [6]:

$$Maximize \left\{\left(1 \wedge \frac{\mu_{\widetilde{R_T}}(R^*)}{w_1}\right) \wedge \left(1 \wedge \frac{\mu_{\widetilde{C_T}}(C^*)}{w_2}\right)\right\} \tag{6}$$
$$Subject\ to\ same\ constraints\ as\ (2)$$

Here, the weights $w_k \in (0, 1]$, for $k = 1, 2$ are used to set preference for system reliability and system cost objective functions, respectively. These weights may be chosen by the decision-maker to set the priority into different objective functions such that $w_1 + w_2 = 1$. $\wedge$ indicates the aggregate operator used as an intersection. This optimization problem is solved by using standard PSO.

Step 7: Description of Fuzzy ranking method (FRM):

In FRM, rank will be provided to every solution of the solution set according to their satisfaction level. The highest satisfaction leveled solution will be the best compromise solution. Best membership value or satisfaction level can be calculated as [6]

$$\mu_{best} = \max_{X \in P}\left[\min\left\{\mu_{\widetilde{R_T}}(R^*),\ \mu_{\widetilde{C_T}}(C^*)\right\}\right]$$

Here, P is the set of Pareto optimal solutions and R^*, C^* are the defuzzified fitness value for system reliability and cost respectively following Step 4. Solution with maximum rank is the best solution.

Step 8: Initialize the number of stages n and the array of stage-wise available components as $[m_1, m_2, ...m_n]$ and shape array $[a_1, a_2, ..., a_n]$. Initialize the population size N, max_ite (maximum iteration), max_trial (maximum independent trials), E, S_l, S_u, cac, gac, and set the trial count variable $tc = 1$, $best\_sol = empty$. Consider the initial set of triangular fuzzy shape values for component reliability and cost of dimension n as $(\widetilde{R_1}, \widetilde{R_2}, ..., \widetilde{R_n})$ and $(\widetilde{S_1}, \widetilde{S_2}, ..., \widetilde{S_n})$, respectively.

Step 9: While ($tc \leq max\_trial$)

a. Set the value of inertia coefficient ω and its dumping ratio $\omega damp$.
b. Create a population set $Popset$ containing N particles (solutions) of dimension n with the following criteria.

  (i) Every particle is of the form $(x_1, x_2, x_3, ...x_n)$, where x_j is a randomly generated integer within $(1, m_j]$ representing the number of components at jth stage $(j = 1, 2, ..., n)$ and they satisfy the constraints of Eq. (1).
  (ii) Initialize the velocity vector V_i of dimension n of ith particle $(i = 1, 2, ..., N)$ such that its every component is a randomly generated number within [0, 0.1].
  (iii) Initially, we set the personal best solution set $Pbest_i$ of ith particle by the particle itself. Global best set $Gbest$ is initialized by the solutions of $Pbest_i$

$(i = 1, 2...N)$ which have the highest degree of satisfaction for the objective function of system reliability and cost. The highest degree of satisfaction can be chosen by applying the non-dominated sorting technique defined in Step 5 on the membership values of system reliability and cost.

c. Set the iteration counting variable $i\_count = 1$
d. while $(i\_count \leq max\_ite)$

d1. Set $i = 1$
d2. while $(i \leq N)$

(I) $rnd_1 =$ random value between 0 and 1 , $rnd_2 =$ random value between 0 and 1.
(II) $[pb, \ gb] =$ Fetch the best solution from the set $Pbest_i$ and $Gbest$ respectively using fuzzy ranking technique defined in Step 7.
(III) $V_i = \omega * V_i + rnd_1 * cac * (Pbest_i - X_i) + rnd_2 * gac * (Gbest - X_i)$.
(IV) $X_i = X_i + V_i$.
(V) if X_i satisfies all the constraints defined by Eq. (5) of Step 4, then

(i) Update $Pbest_i$ by applying the non-dominated sorting technique described in Step 5 on the membership values of X_i and the previous existing solutions of $Pbest_i$ following the objective function of Eq. (5).
(ii) if X_i included in $Pbest_i$ then
Update $Gbest$ by applying the non-dominated sorting technique described in Step 5 on the membership values of X_i and the previous existing solutions of $Gbest$ following the objective function (5).

(VI) $i = i + 1$

d3. $\omega = \omega * \omega damp$
d4. $i\_count = i\_count + 1$

e. Append all solutions of $Gbest$ to $best\_sol$ set
f. $tc = tc + 1$

Step 10. $best\_sol = unique(best\_sol)$ [Choose the unique solutions from best_sol].

Step 11. Display the solutions from $best\_sol$ whose membership values are non-zero.

Step 12. Display the best solution from the set $best\_sol$ using fuzzy ranking technique defined as in Step 7.

Step 13. Stop.

5 An Illustrative Example of a Series-Parallel Reliability Redundancy Model

Here, we solve a fuzzy reliability redundancy optimization problem of a four- stage over-speed protection gas turbine system [6] using FRMOPSO and PSO techniques and compare their efficiency and applicability.

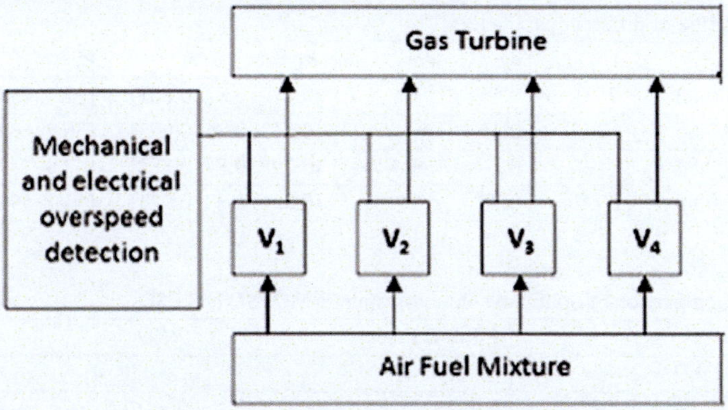

Fig. 2 Over-speed protection system of gas turbine

5.1 *Over-Speed Protection Gas Turbine System*

Over-speed protection gas turbine has a mechanical and electrical system to detect the speed continuously and the fuel supply to it is cut off by 4 control valves v_1 to v_4 if an over-speeding is found as shown in Fig. 2. We may consider the over-speed protection system as a x_j-out-of-m_j series-parallel reliability redundancy system as in Fig. 1 in which jth stage contains extra $(m_j - 1)$ identical valves (same reliability and cost) in parallel to increase the reliability of the system.

In general, the reliability and cost of the components may change as user requirements. So to make the over-speed protection system more flexible, the inputs for reliability and cost of every valve of the model are considered as TFN. Our aim is to optimize the number of active valves x_j $(1 < x_j \leq m_j)$ at jth $(j = 1, 2, 3, 4)$ stage such that we achieve the maximum system reliability and minimum system cost subject to satisfying the constraints of equation (2).

5.2 *Fuzzy Reliability Redundancy Optimization Problem for Over-Speed Protection System*

Suppose $\widetilde{R}_i$ and $\widetilde{C}_i$ are reliability and cost of every component at ith $(i = 1 \text{ to } 4)$ stage for the x_j-out-of-m_j over-speed protection series-parallel reliability redundancy system. If $\widetilde{R}$ and $\widetilde{S}$ represent the system reliability and system cost respectively for the over-speed protection system, then the optimization problem of Eq. (2) can be represented considering $n = 4$. For our proposed approach, we construct the fuzzy MOOP for the over-speed protection system from Eq. (5) by Step 4. For the PSO technique, we represent the fuzzy MOOP of Eq. (6) for the over-speed protection system in a crisp format using Step 6.

Table 1 Value of different parameters

j	R_j	C_j	a_j	m_j	$\widetilde{R}_i$	$\widetilde{C}_i$
1	0.82	75.0	0.65	7	[0.779, 0.82, 0.861]	[67.5, 75.0, 82.5]
2	0.86	63.0	0.75	8	[0.817, 0.86, 0.903]	[56.7, 63.0, 69.3]
3	0.84	71.0	0.7	9	[0.798, 0.84, 0.882]	[63.9, 71.0, 78.1]
4	0.88	59.0	0.78	6	[0.836, 0.88, 0.924]	[53.1, 59.0, 64.9]

Table 2 Comparison of optimum solutions using FRMOPSO and PSO

Solutions	$FRMOPSO$	PSO
$[x_1, x_2, x_3, x_4]$	[2, 2, 4, 4]	[2, 3, 3, 4]
R^*	0.99222	0.99177
C^*	603.501	612.162
$\mu(R^*)$	0.99037	0.96786
$\mu(C^*)$	0.98048	0.82695

6 Numerical Presentation

In this section, we discuss the result obtained by the optimizing system using Eqs. (2) and (5) for FRMOPSO and Eqs. (2) and (6) for PSO. The value of the parameters may be changed by the decision- maker/system executor as per their requirements depending on the nature of the problems. In Table 1, different chosen parameter values are shown.

We consider the parameters value as follows: $S_l = 10, S_u = 30,\ N = 60,$ $max\_ite = 200$, maximum number of independent trials $max\_trial = 20$, $E = 1.6$, $pac = 0.05$, $gac = 0.05$, $w = 0.99$ and $wdam = 0.99$. The optimization problem (6) is solved by using PSO considering the weighted values $w_1 = 0.5$ and $w_2 = 0.5$.

Using PSO, first we calculate the minimum and maximum values of system reliability and system cost following Step 1 using above-defined parameter values, and we get $R_{\min} = 0.9737,\ R_{\max} = 0.99859,\ C_{\min} = 583.6$ and $C_{\max} = 658.81$. We consider a value toward $R_{\max}$ as $R_{center} = 0.9924$ and a value toward $C_{\min}$ as $C_{center} = 602.4$.

From Table 2, we observe that the high system reliability and low system cost are found in our proposed FRMOPSO algorithm. Also, the high membership values are shown in the FRMOPSO algorithm than the PSO algorithm. Therefore, the performance of the FRMOPSO algorithm is better than the PSO algorithm.

Figure 3 shows the comparisons of optimum system reliability and cost with Pareto front solutions using FRMOPSO and PSO algorithms. Here, we see that the maximum optimized solutions of the MOPSO algorithm lie in the high reliability and low system cost region. Also, the high membership values of system reliability and system cost are shown for the FRMOPSO technique.

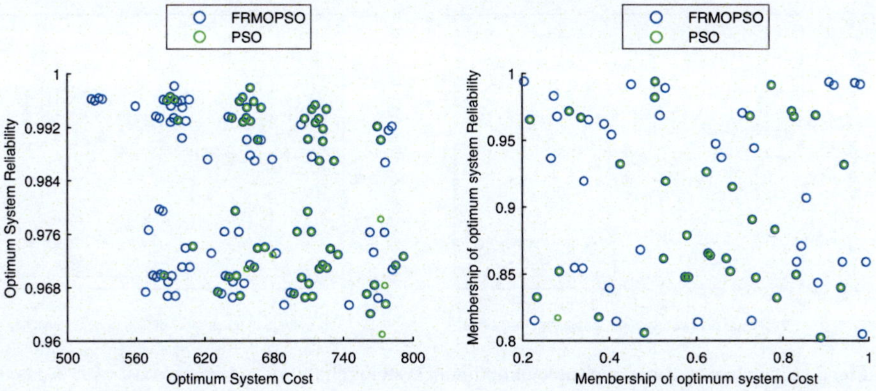

Fig. 3 Comparison of Pareto front solutions and their membership values

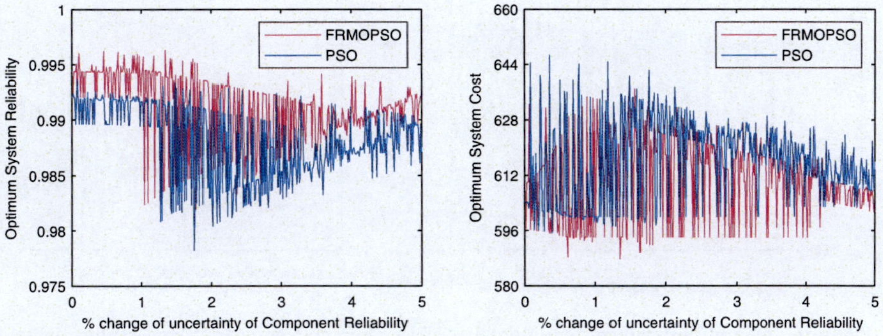

Fig. 4 Optimum system reliability and system cost with % change of uncertainty of component reliability

6.1 Sensitivity Analysis

In the optimization problem (2) of over-speed protection system, we consider the 10% uncertainty of the center value of component cost represented as TFN. Figure 4 shows the effect of optimum system reliability and system cost with respect to the change of uncertainty of every component reliability from 0 to 5%. Here, our proposed FRMOPSO solutions show better optimum system reliability and system cost than PSO solutions. Figure 5 shows the effect of membership value of the optimum system reliability and membership value of optimum system cost with respect to the change of uncertainty of every component reliability from 0 to 5%. In Fig. 5, we see that our proposed FRMOPSO algorithm shows maximum satisfaction level in both the cases than PSO technique. Similarly in the optimization problem (2) of the over-speed protection system, we consider 5% uncertainty of the center value of component reliability represented as TFN. Figure 6 shows graphically the effect of optimum system reliability and system cost with respect to the change of uncertainty of every

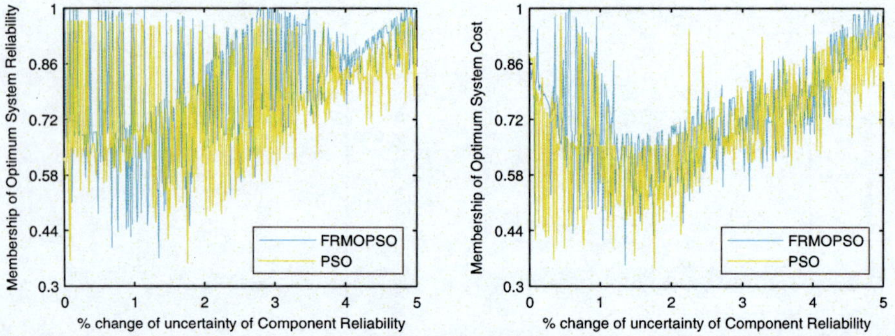

Fig. 5 Change of membership value of optimum system reliability and system cost with % change of uncertainty of reliability

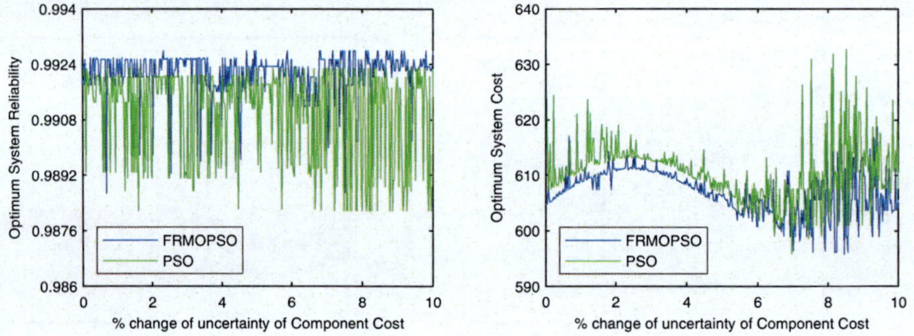

Fig. 6 Change of optimum system reliability and system cost with % change of uncertainty of component cost

component cost from 0 to 10%. Here, our proposed FRMOPSO solutions show better optimum system reliability and system cost than PSO solutions. Figure 7 shows the effect of the membership value of optimum system reliability and membership value of optimum system cost with respect to the change of uncertainty of every component cost from 0 to 10%. In Fig. 7, we see that our proposed FRMOPSO algorithm shows a maximum satisfaction level in both the cases than PSO technique.

6.2 Performance Measurement of FRMOPSO and PSO

Convergency testing of MOPSO and PSO

If X is a solution of an optimization problem and the set Y is an approximation of true Pareto front solutions, then the Average Hausdorff Distance (AHD) is defined from [7] as $\Delta_p(A, X) = \max(IGD_p(Y, X), IGD_p(X, Y))$ where IGD is the Inverted Generational Distance. Lower value of Δ_p shows a better convergence rate of an

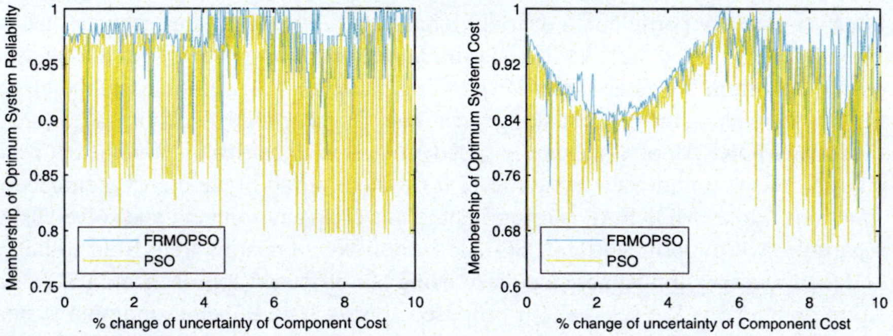

Fig. 7 Membership value of optimum system reliability and system cost with % change of uncertainty of component cost

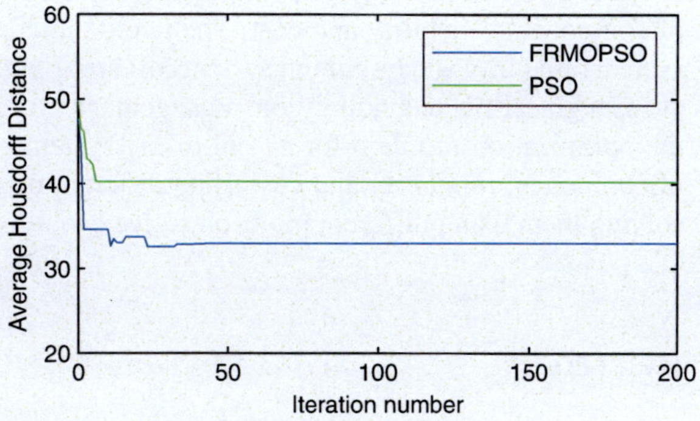

Fig. 8 Comparison of convergence rate to true Pareto front set

algorithm. AHD indicator Δ_p for $p = 3$ is determined up to 200 iterations using FRMOPSO and PSO algorithms with the value of the parameters defined in the above numerical example. Figure 8 shows a comparison graph of the best result of AHD indicator values from 30 individual trials with respect to the number of iterations. The graph shows that the convergence rate of MOPSO is higher than the PSO algorithm.

7 Conclusion

In this paper, we propose a fuzzy multi-objective reliability redundancy allocation problem (MORRAP) on x_j-out-of-m_j series-parallel components system. To make the system more flexible, the uncertainty of component reliability and cost are represented by triangular fuzzy numbers. Our aim is to find out the optimum number of redundant components in each stage of the model such that the degree of satisfaction level of system reliability and cost become maximum. We propose a fuzzy ranking multi-objective particle swarm optimization algorithm where the personal best solutions set of every particle and overall global best solutions set are reconstructed by using non-dominated sorted solutions among the previous existing solutions and the current solution. We fetch the corresponding particle's best solution and the global best solution using the fuzzy ranking technique. Our proposed algorithm is applied to the fuzzy MORRAP of 4-stage over-speed protection system to find out the optimum solution of maximum satisfaction level for system reliability and cost membership functions. Fuzzy MORRAP is reformulated in a crisp environment and solved using a particle swarm optimization (PSO) algorithm. We compare Pareto front solutions and their membership values calculated using our proposed approach and PSO algorithm graphically. We see that our proposed approach gives better optimum system reliability and cost with maximum membership value of the two objectives than PSO. The convergence graph shows that a high convergency rate occurs for our approach. So our proposed algorithm is stable to solve the fuzzy MORRAP. We analyze the effect of optimum system reliability and cost with respect to the uncertainty change of component reliability and cost. The two conflicting objectives that is maximum system reliability and minimum system cost are applicable in most of the fields of science, engineering, technology and management. Our future work will be to construct the optimization problem for x_j-out-of-m_j system by representing the uncertainty of component reliability and cost using different other types of fuzzy numbers and solving them using different multi-objective evolutionary algorithms.

References

1. Coello C, Pulido G, Lechuga M (2004) Handling multiple objective with particle swarm optimization. IEEE Trans Evol Comput 8(3)
2. Mahapatra GS, Maneckshaw B, Barker, K (2022) Multi-objective reliability redundancy allocation using MOPSO under hesitant fuzziness. Expert Syst Appl 116696. ISSN 0957-4174
3. Xie M, Dai Y-S, Poh K-L (2004) Computing systems reliability models and analysis. Kluwer Academic Publishers, New York
4. Roy P, Mahapatra BS, Mahapatra GS, Roy PK (2014) Entropy based region reducing genetic algorithm for reliability redundancy allocation in interval environment. Expert Syst Appl 41(14):6147–6160
5. Deb K, Pratap A, Agarwal S, Meyarivan T (2002) A fast and elitist multi-objective genetic algorithm: NSGA-II. IEEE Trans Evol Comput 6:2
6. Kumar H, Yadav SP (2019) Hybrid NSGA-II based decision-making in fuzzy multi-objective reliability optimization problem. SN Appl Sci 1:1496
7. Schutze O, Esquivel X, Lara A, Coello CCA (2012) Using the averaged housdorff distance as a performance measure in evolutionary multiobjective optimization. IEEE Trans Evol Comput 16:4
8. Huang C-L, Jiang Y, Yeh WC (2020) Developing model of fuzzy constraints based on redundancy allocation problem by an improved swarm algorithm. IEEE Access 8:155235–155247
9. Sharifi M, Taghipour S, Abhari A (2022) Condition-based optimization of non-identical inspection intervals for a k-out-of-n load sharing system with hybrid mixed redundancy strategy. Knowl-Based Syst 240:108153
10. Dui H, Tian T, Zhao J, Wu S (2022) Comparing with the joint importance under consideration of consecutive-k-out-of-n system structure changes. Reliab Eng Syst Saf 219:108255

11. Farhadi M, Shahrokhi M, Rahmati SHA (2022) Developing a supplier selection model based on Markov chain and probability tree for a k-out-of-N system with different quality of spare parts. Reliab Eng Syst Saf 222:108387
12. Wang C (2021) Time-dependent reliability of (weighted) k-out-of-n systems with identical component deterioration. J Infrastruct Preserv Resil 2:3
13. Davoudi M, Jooshaki M, Moeini-Aghtaie M, Barmayoon MH, Aien M (2022) Developing a multi-objective multi-layer model for optimal design of residential complex energy systems. Int J Electr Power Energy Syst 138:107889
14. Mahmoud A, Yuan X, Kheimi M, Almadani MA, Hajilounezhad T, Yuan Y (2021) An improved multi-objective particle swarm optimization with TOPSIS and fuzzy logic for optimizing trapezoidal labyrinth weir. IEEE Access 9:25458–25472
15. Wang H, Cai T, Li K, Pedrycz W (2021) Constraint handling technique based on Lebesgue measure for constrained multiobjective particle swarm optimization algorithm. Knowl-Based Syst 227:107131
16. Yuan X, Liu Y, Bucknall R (2021) Optimised MOPSO with the grey relationship analysis for the multi-criteria objective energy dispatch of a novel SOFC-solar hybrid CCHP residential system in the UK. Energy Convers Manag 243:114406
17. Liang S, Li B, Tian X, Cheng Y, Liao C, Zhang J, Liu D (2021) Determining optimal parameter ranges of warm supply air for stratum ventilation using Pareto-based MOPSO and cluster analysis. J Build Eng 37:102145
18. Ershadi MJ, Ershadi MM, Haghighi Naeini S et al (2021) An economic-statistical design of simple linear profiles with multiple assignable causes using a combination of MOPSO and RSM. Soft Comput 25:11087–11100
19. Rajani, Kumar D, Kumar V (2020) Impact of controlling parameters on the performance of MOPSO algorithm. Procedia Comput Sci 167:2132–2139
20. Khazaei A, Karimi BH, Mozaffari MM (2021) Optimizing the prediction model of stock price in pharmaceutical companies using multiple objective particle swarm optimization algorithm (MOPSO). J Optim Ind Eng 14(2):73–81
21. Lee KH, Dai Y-S, Poh K-L (2005) First course on fuzzy theory and applications. Springer, Berlin
22. Kim D, Choi Y-S, Lee S-Y (2002) An accurate COG defuzzifier design using Lamarckian co-adaptation of learning and evolution. Fuzzy Sets Syst 130(2):207–225

Image Binarization with Hybrid Adaptive Thresholds

Yanglem Loijing Khomba Khuman, O. Imocha Singh, T. Romen Singh, and H. Mamata Devi

Abstract The process of image binarization divides pixel values into two groups, with the back-and-forth represented by white and black, respectively. A novel image binarization approach is presented in this research, "Image Binarization with Hybrid Adaptive Thresholds (IBHAT)" with two adaptive thresholds being proposed. It associates two threshold values which are based on global and local mean for each pixel that is to be transformed to either in background or foreground. One of the threshold values has to be selected for the concerned pixel. Hence the term hybrid adaptive is used. The local contrast of pixels inside a block of 3×3 is used to make the decision. This technique can switch to three different modes such as local, global, or hybrid adaptive thresholding technique simply by using a specific control parameter value. Global adaptation technique gives a result similar to global techniques, local technique gives a result similar to other local techniques, while the hybrid technique gives a result which is not similar to the previous two but is a very effective one. So it is convenient to apply in degraded document image binarization. This technique is compared with other global as well as local techniques at different conditions of the controlling factor based on the types of the image applied. Since it uses only 3×3 block size, its computational time is similar to the global one. In comparison to previous relevant strategies, the suggested methodology takes less time to compute and produces effective results, according to the results of the experiment.

Keywords Adaptive · Binarization · Global · Hybrid · Local · Thresholding

---

Y. L. K. Khuman (✉) · O. I. Singh · T. R. Singh · H. M. Devi
Department of Computer Science, Manipur University, Imphal, India
e-mail: loijingyanglem@gmail.com

R. Chaki et al. (eds.), *Applied Computing for Software and Smart Systems*,
Lecture Notes in Networks and Systems 555,
https://doi.org/10.1007/978-981-19-6791-7_10

1 Introduction

The technique of separating pixel values into two groups, white for the background and black for the foreground of an image, is known as image binarization. There will be a specific value that will partition the pixels into two groups called the threshold. The selection of appropriate threshold values is essential, and this process is called thresholding. The threshold can be classified as either a global or a local threshold depending on the application. For scanned documents with such a homogeneous background and foreground contrast distribution, global thresholding is preferable over local thresholding.

In degraded documents [1–4], with significant background noise or fluctuation in contrast and luminance, multiple pixels are difficult to classify as foreground or background. Binarization of documents must be done by integrating and considering the outcomes of a variety of binarization approaches, especially for documents with much ambiguity. In such a case, the local thresholding technique takes a significant role in binarization. In general, scanned papers are categorized as mixed-type documents because they contain line drawings, letters, and image regions. Many implementations necessitate the recognition or improvement of document text content.

In such instances, converting the documents to binary format is preferable. As a result, in most document analysis systems, document binarization is the initial step. Thresholding is a basic yet helpful strategy for identifying objects from their surroundings. Document image analysis is an example of a thresholding application. The purpose of image analysis is to retrieve textual content [5, 6], graphical information, logos, or musical compositions; map analysis, which requires the location of lines, legends, and characters [7]; analysis of a landscape in which a target must be found [8]; and inspection of the quality of the materials [9, 10], where faulty components must be restricted.

The thresholding technique creates a binary picture with one state indicating foreground things like a legend, a target, a faulty part, and textual content of a material, and so on, and the other state representing background objects like the background. Depending on the requirements, the front could be gray-level 0 (black for letters) and the background can be the maximum intensity for the document (255 in 8-bit graphics as white) or white for the forefront and black for the background. There are two sorts of binarization procedures for grayscale documents, depending on the thresholding technique used: global binarization and local binarization. Otsu and Nobuyuki [11] recommended using a global binarization strategy to get a single document-wide threshold value. Each pixel's gray value is then utilized to decide whether it belongs in the page's foreground or background. For ordinary scanned documents, global binarization algorithms are speedy and produce decent results. For many years, grayscale documents were binarized using global thresholding techniques [11–16], which are based on statistical methods. As clustering approaches, these methods may be sufficient for transforming any grayscale image to binary form.

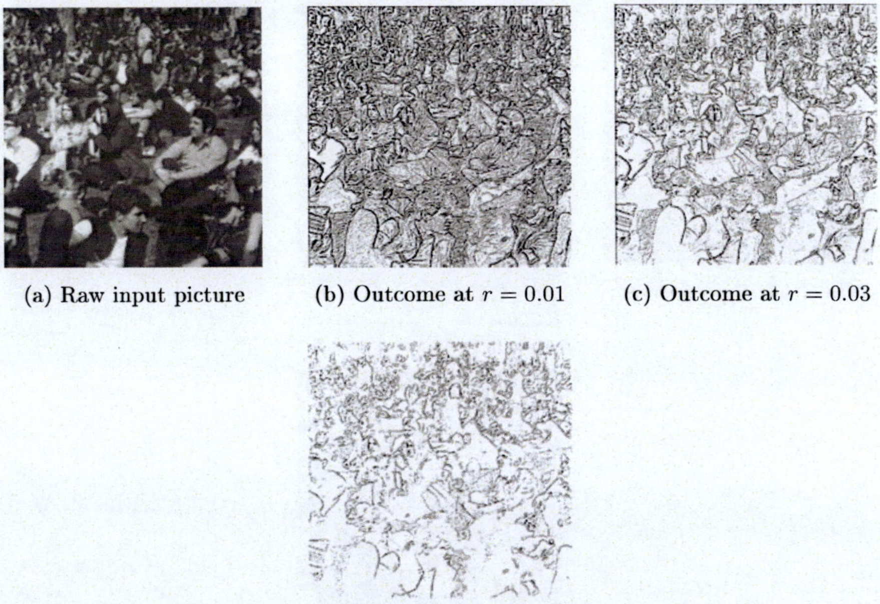

(a) Raw input picture (b) Outcome at $r = 0.01$ (c) Outcome at $r = 0.03$

(d) Outcome at $r = 0.06$

Fig. 1 Background removal in local technique ($r_l = 0$) at different values of r

They are, however, unsuitable for complicated documents and, much more so, for damaged materials.

When the brightness over the document isn't really equal, as is the case of digitized book pages or camera-captured documents, global binarization approaches produce extra distortion at the page boundaries. Local thresholding strategies for document binarization have been developed to resolve these complications. Depending on the grayscale characteristics of nearby pixels, these algorithms establish a unique threshold for every pixel. [17–24] approach all fall under this group. But these techniques still have problems for the complex illuminated images. Hybrid approaches, such as those proposed by [25–27] to overcome the shortcomings of separate (global/local) techniques, combined global and local threshold information.

In this study, we use a hybrid adaptive thresholding methodology to binarize grayscale document pictures. This technique involves two thresholds which are based on local and global means. It uses only a 3×3 block for local adaptation. As a result, unlike other local procedures, the computational time is unaffected by block size, the binarization's speed is equivalent to global binarization methods, and its performance is comparable to local binarization schemes. It can take 3 different roles of local, global, and hybrid techniques with the help of a control parameter. As a consequence, even on highly deteriorated documents, this procedure frequently produces satisfactory results.

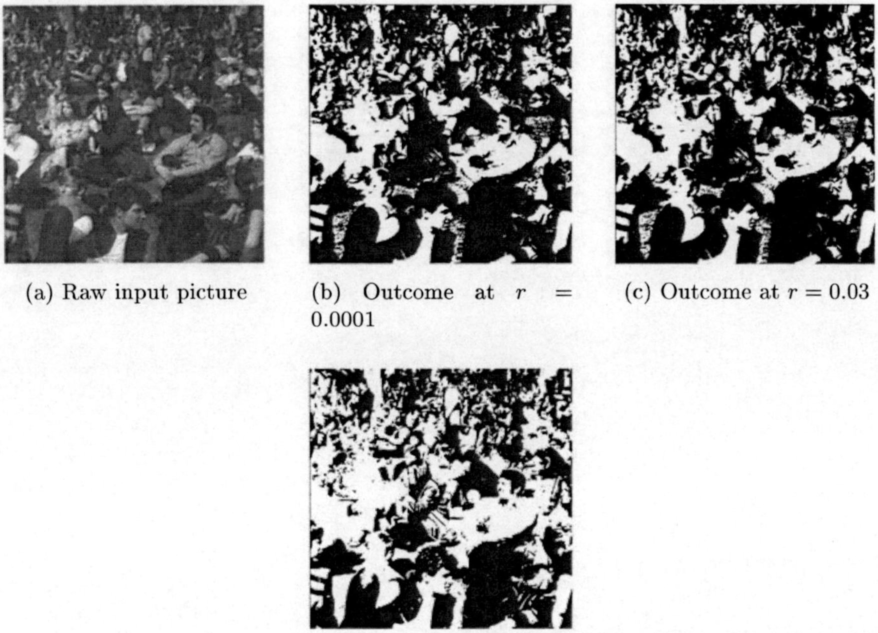

(a) Raw input picture (b) Outcome at $r = 0.0001$ (c) Outcome at $r = 0.03$

(d) Outcome at $r = 0.3$

Fig. 2 Background removal in global technique ($r_l = 0.5$) at different values of r

2 Methodology

This proposed system is based on two adaptive threshold values, T_l and T_g, calculated using the local and global mean of the pixels of an input picture I, respectively. T_l is calculated based on the local mean m_l of pixels inside a block of 3×3, while T_g is based on the global mean m_g of the pixel of the entire picture. When an input picture is binarized, both thresholds may or may not be used according to the switched-mode: local, global, or hybrid.

2.1 Binarization Techniques

Binarization of an input pixel $I(c, d)$ to $b(c, d)$ with a threshold value $T(c, d)$ based on local (within a local window $w \times w$) or global technique (within the entire image) can be formulated as

$$b(c, d) = \begin{cases} 0, & \text{if } I(c, d) \leq T(c, d) \\ 255, & \text{otherwise} \end{cases} \tag{1}$$

(a) Raw input picture

(b) At $r = 0.01$ and $r_l = 0.01$

(c) At $r = 0.03$ and $r_l = 0.025$

(d) At $r = 0.1$ and $r_l = 0.15$

Fig. 3 Background removal in hybrid technique (different values of r_l) at different values of r

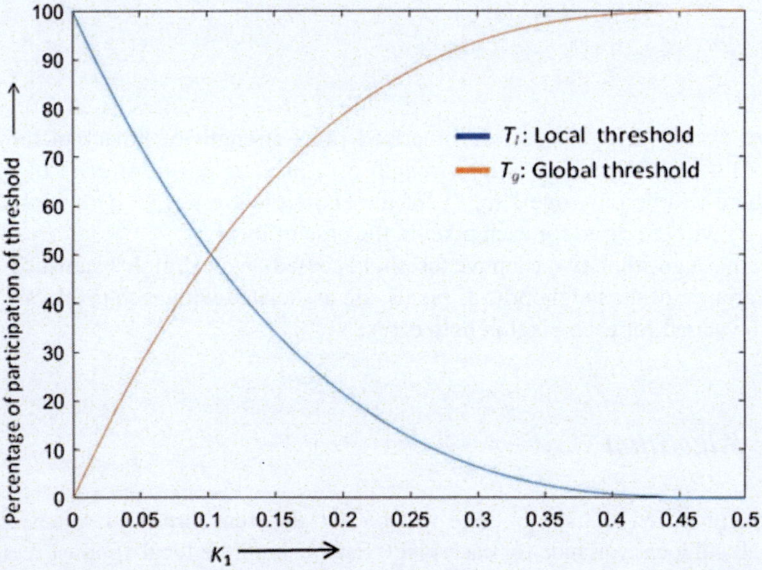

Fig. 4 Percentage of participation of local (T_l) and global (T_g) threshold in binarization

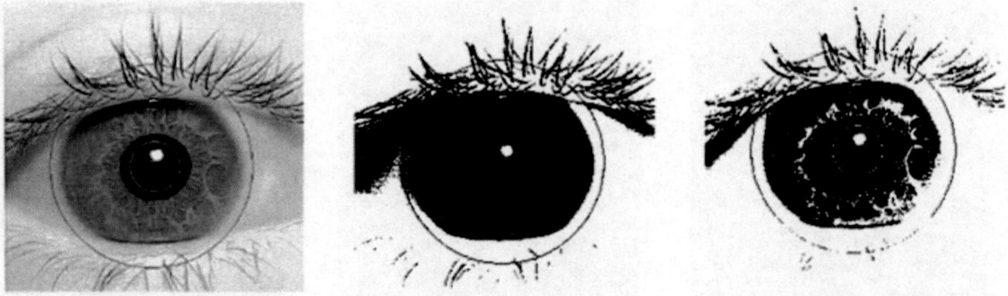

(a) Raw input picture

(b) Outcome of Otsu's

(c) Outcome of Global ($r_l = 0.5$) condition at $r = 0.2$

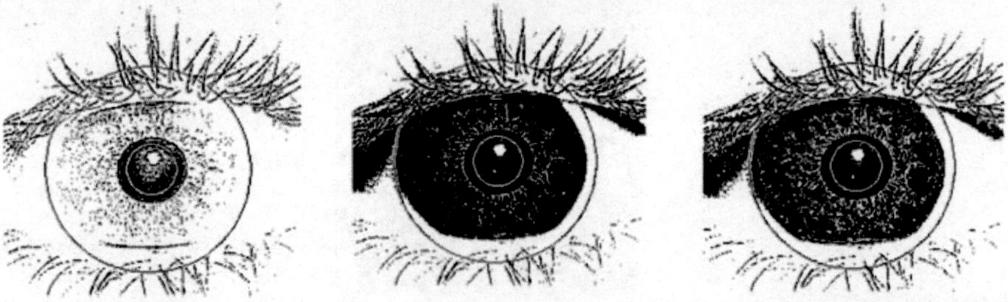

(d) Outcome of Local ($r_1 = 0$) condition at $r = 0.07$

(e) Outcome of Hybrid at $r = 0.07$ and $r_l = 0.03$

(f) Outcome of Hybrid at $r = 0.08$ and $r_l = 0.02$

Fig. 5 Outcome of Otsu and proposed techniques at different values of r and r_l applied on document image

where $b(c, d) \in [0, 255]$ is the binarized pixel strength of binary picture b and $I(c, d) \in [0, 255]$, which is the strength of a pixel at position (c, d) of a picture I. There will be only one $T(c, d)$ for the entire picture if $T(c, d)$ is global, while $T(c, d)$ will be varied for each pixel in the case of local.

In this algorithm, two adaptive thresholds based on local/global statistics, such as the variance of the neighborhood pixels, are associated with each pixel. So, $T(c, d)$ will be varied for each pixel in both cases.

2.2 Algorithm

In this proposed technique, there will be two different threshold values based on global and local variance for each pixel. Establish on the local contrast region, one of the threshold values should be chosen. Local contrast adaptation is based on the

Fig. 6 Outcome of Otsu and proposed techniques at different values of r and r_l applied on document image

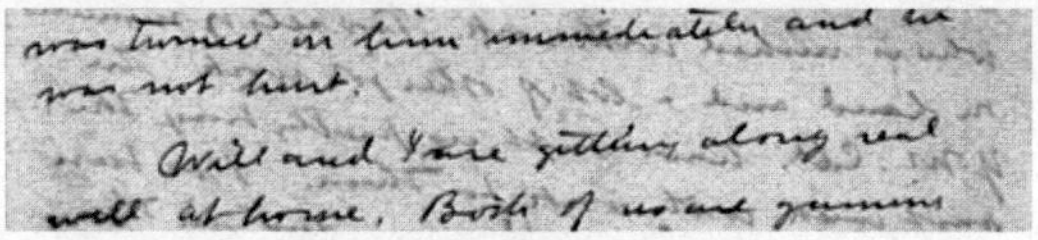

(a) Raw input picture

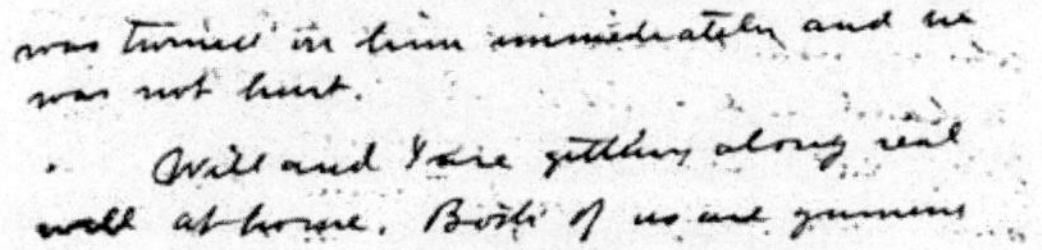

(b) Outcome of Otsu's technique

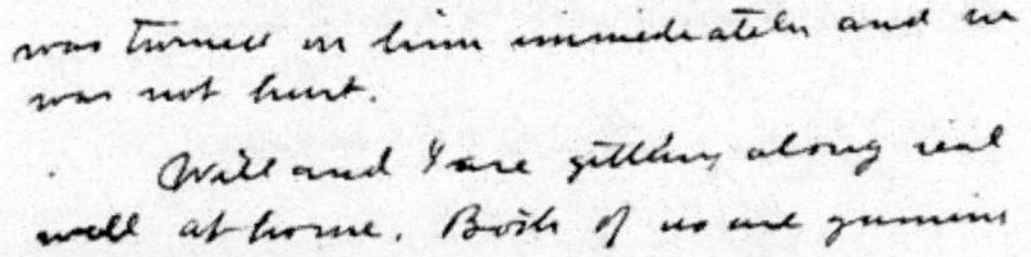

(c) Outcome of Global ($r_l = 0.5$) condition at $r = 0.35$

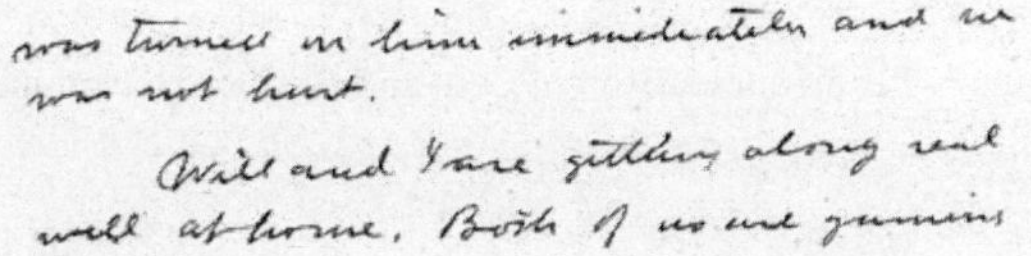

(d) Outcome of Local ($r_l = 0.0$) condition at $r = 0.2$

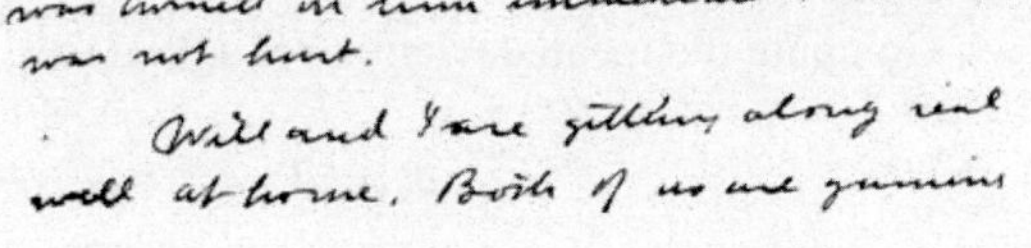

(e) Outcome of Hybrid at $r = 0.35$ and $r_l = 0.3$

local standard deviation and variances of pixels within a 3×3 block. The following is the designed algorithm for this technique.

(i) Take an input image $I_{m \times n} \in [0, 1]$ (in normalized form).

(ii) Compute the global mean.

$$m_g = \frac{1}{m \times n} \sum_{i=1}^{m} \sum_{j=1}^{n} I(c, d) \tag{2}$$

(iii) Compute the local mean of the neighboring pixels of $I(c, d)$ within a 3×3 block $I(c, d)$ being at the center.

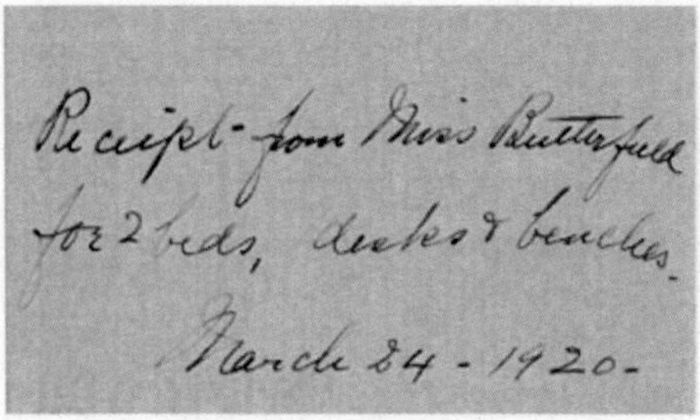

(a) Raw input picture

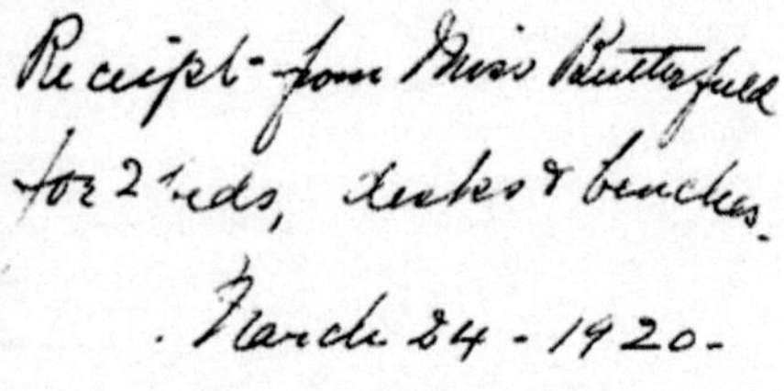

(b) Outcome of Otsu's technique

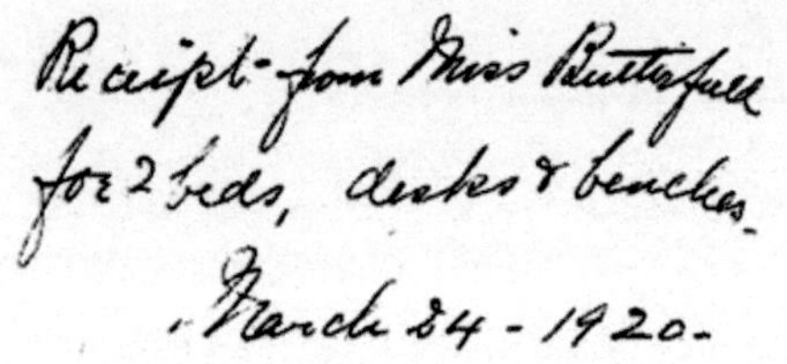

(c) Outcome of proposed technique at $r = 0.1$ and $r_l = 0.2$

Fig. 7 Result comparison of Otsu and proposed techniques

$$m_l = \frac{1}{9} \sum_{p=c-1}^{c+1} \sum_{q=d-1}^{d+1} I(p, q) \tag{3}$$

(iv) Compute the mean deviations of $I(c, d)$ as

$$\partial_g(c, d) = l(c, d) - m_g(c, d) \tag{4}$$

$$\partial_l(c, d) = l(c, d) - m_l(c, d) \tag{5}$$

(v) Compute the local standard deviation of pixels within the 3×3 block.

$$\delta(c, d) = \frac{1}{9} \sqrt{\sum_{p=c-1}^{c+1} \sum_{q=d-1}^{d+1} \{I(p, q) - m_l(c, d)\}^2} \tag{6}$$

(vi) Compute two threshold values.

$$T_g = m_g(c, d)\{1 + r(\partial_g(c, d) - 1)\} \tag{7}$$

$$T_l = m_l(c, d)\{1 + r(\partial_l(c, d) - 1)\} \tag{8}$$

where $r \in [0, 0.5]$ is a bias to control background removal.

(a) Raw input picture

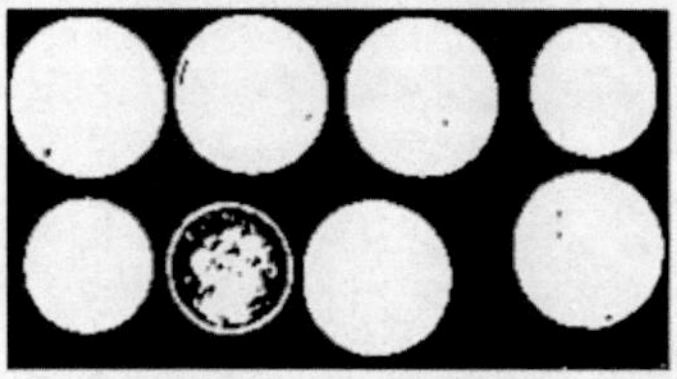

(b) Outcome of Otsu's

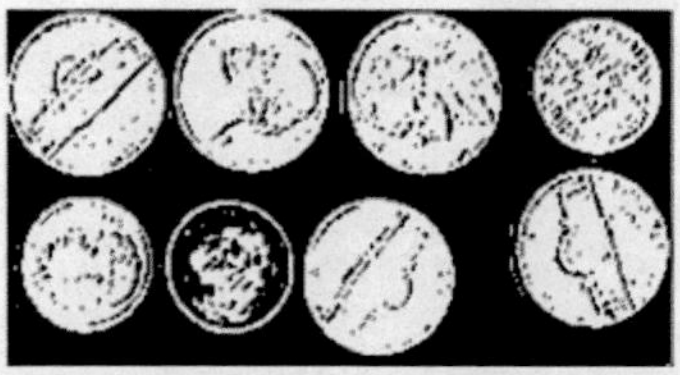

(c) Outcome of proposed technique at $r = 0.001$ and $r_l = 0.03$

Fig. 8 Outcome comparison of Otsu and proposed techniques

(a) Raw input picture (b) Outcome of Otsu's technique (c) Hybrid at $r = 0.028$ and $r_l = 0.015$

Fig. 9 Outcome comparison of Otsu and proposed techniques

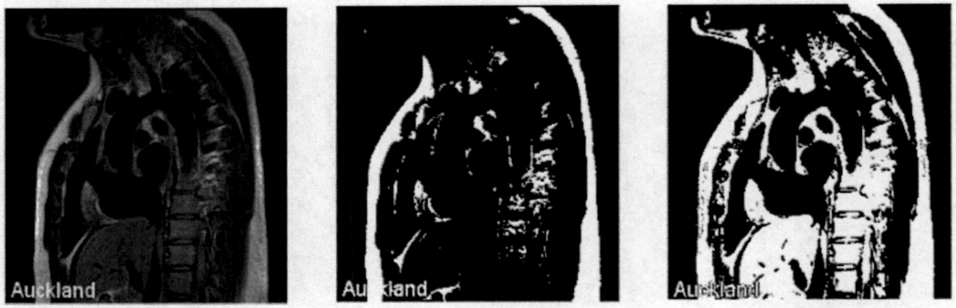

(a) Raw input picture (b) Outcome of Otsu's (c) Outcome of proposed technique at $r = 0.01$ and $r_l = 0.1$

Fig. 10 Outcome comparison of Otsu and proposed techniques

(a) Raw input picture (b) Outcome of Otsu's technique (c) Hybrid at $r = 0.04$ and $r_l = 0.06$

Fig. 11 Outcome comparison of Otsu and proposed techniques

(vii) Select a suitable threshold value as

$$T(c,d) = \begin{cases} T_g(c,d), \text{ if } |\partial_g(c,d)| < r_1 and \delta(c,d) < r_1 \\ T_l(c,d), \ otherwise \end{cases} \tag{9}$$

where $r_1 \in [0, 0.5]$ is a constant that controls the selection of the threshold.

(viii) Transform the pixel $I(c, d)$ to binarized picture based on the selected threshold values as

$$b(c,d) = \begin{cases} 1, \text{ if } T(c,d) < I(c,d) \\ 0, \text{ otherwise} \end{cases} \tag{10}$$

(ix) Repeat steps (iii) to (viii) until the last pixel of the image.

3 Discussion

In this algorithm, the two threshold values T_g and T_l take adaptive global and local threshold roles individually. The bias r is used to control background removal for the two thresholds. The higher the r results in more background (white) in the binarized image, which is illustrated in Figs. 1, 2, and 3. Either of these two thresholds has to apply for selecting the group in which a pixel belongs in the binarized image. As in Eq. 9, selection is dependent on the local variance and standard deviation (SD) of neighboring inner pixels of 3×3 block. Here, r_l controls the percentage of participation of the two thresholds to binaries the image as in Fig. 4.

The lower the value of r_l results, the more the participation of T_l in binarization. The percentage of participation of the two thresholds is inversely proportional. If $r_l = 0$, only T_l participate, and the result is similar to the local technique, and if $r_l = 0.5$, only T_g participates in the binarization whose result is similar to global techniques. But if $0< r_l <0.5$, the result will be a hybrid one. Different results at different values of r_l are shown in Fig. 5. Thus, we can achieve a good result of the binarized image, which is different from others' techniques, by adjusting the two parameters r and r_l to adapt the nature of the applied input image. This feature is not found in the other local/global techniques so far. Unlike the other local/global techniques, it is straightforward to calculate with minimum operations and local window size 3×3, while other local techniques used complex calculation with at least 31×31 block size. Since it uses a local technique, it can detect the low contrast region, which is not found in global techniques.

4 Experimental Result

The algorithm is tested on different categories of images for result analysis and comparison. The result is compared with only Otsu's technique as in Figs. 5, 6, 7, 8, 9, 10, and 11. No local techniques are associated for comparison because they use a larger local block size, while this technique uses only 3×3. Since it uses only 3×3 block size, the order in which it takes to compute is $O(n^2)$ for a picture $I_{n \times n}$. According to the results of the experiments, this strategy outperforms.

5 Conclusion

This technique IBHAT is a hybrid type adaptive binarization technique which can adapt to local environment so as to get a proper threshold value of a pixel. It entails each pixel having two dynamic threshold values. One is chosen based on the type of the pixel's neighbor. As a result, it may be used as a local, global, and hybrid thresholding strategy. This technique is suitable for document image binarization. Its computational time complexity is only $O(n^2)$ since it uses only 3×3 local block sizes. Hence, the result is compared with only the standard global technique of Otsu. Comparatively we found this technique outperforms.

Acknowledgements We appreciate the editors and anonymous reviewers' thorough perusal of our paper as well as their numerous insightful remarks and recommendations.

References

1. Gatos B, Pratikakis I, Perantonis SJ (2006) Adaptive degraded document image binarization. Pattern Recogn. 39(3):317–327. https://doi.org/10.1016/j.patcog.2005.09.010
2. Le T, Ngan H, Bui TD, Suen CY (2011) Ternary entropy-based binarization of degraded document images using morphological operators. In: 2011 international conference on document analysis and recognition, pp 114–118. https://doi.org/10.1109/ICDAR.2011.32

3. Lu S, Su B, Tan CL (2010) Document image binarization using background estimation and stroke edges. Int J Doc Anal Recognit (IJDAR) 13(4):303–314. https://doi.org/10.1007/s10032-010-0130-8
4. Oh H-H, Lim K-T, Chien S-I (2005) An improved binarization algorithm based on a water flow model for document image with inhomogeneous backgrounds. Pattern Recognit 38(12):2612–2625. https://doi.org/10.1016/j.patcog.2004.11.025
5. Kamel M, Zhao A (1993) Extraction of binary character/graphics images from grayscale document images. CVGIP: Graph Model Image Process 55(3):203–217. https://doi.org/10.1006/cgip.1993.1015
6. Abak AT, Baris U, Sankur B, The performance evaluation of thresholding algorithms for optical character recognition. In: Proceedings of the fourth international conference on document analysis and recognition, pp 697–700. https://doi.org/10.1109/ICDAR.1997.620597
7. Trier OD, Jain AK (1995) Goal-directed evaluation of binarization methods. IEEE Trans Pattern Anal Mach Intell 17(12):1191–1201. https://doi.org/10.1109/34.476511
8. Bhanu B (1986) Automatic target recognition: state of the art survey. IEEE Trans Aerosp Electron Syst AES-22(4):364–379. https://doi.org/10.1109/TAES.1986.310772
9. Sezgin M, Tasaltin R (2000) A new dichotomization technique to multilevel thresholding devoted to inspection applications. Pattern Recognit Lett 21(2):151–161. https://doi.org/10.1016/S0167-8655(99)00142-7
10. Sezgin M, Sankur B (2001) Selection of thresholding methods for nondestructive testing applications. In: Proceedings: 2001 international conference on image processing (Cat. No.01CH37205), pp 764–767. https://doi.org/10.1109/ICIP.2001.958231
11. Otsu, Nobuyuki (1979) A threshold selection method from gray-level histograms. In: IEEE Trans. Sys, Man, Cybern 9(1):62–66. https://doi.org/10.1109/TSMC.1979.4310076
12. Viola Paul, Jones Michael J (2004) Robust real-time face detection. Int J Comput Vis 57(2):137–154. https://doi.org/10.1023/B:VISI.0000013087.49260.fb
13. Cattoni R, Coianiz T, Messelodi S, Modena CM (1998) Geometric layout analysis techniques for document image understanding: a review. ITC-irst Tech Rep 9703(09)
14. Shafait F, Keysers D, Breuel TM, Performance comparison of six algorithms for page segmentation. In: Document analysis systems VII, pp 368–379
15. White JM, Rohrer GD (1983) Image thresholding for optical character recognition and other applications requiring character image extraction. IBM J Res Dev 27(4):400–411. https://doi.org/10.1147/rd.274.0400
16. Chi Z, Yan H, Pham T (1996) Fuzzy algorithms: with applications to image processing and pattern recognition. World Sci 10. https://doi.org/10.1142/3132
17. Bernsen J (1986) Dynamic thresholding of gray level image. In: ICPR' 86—proceedings of international conference on pattern recognition, pp 1251–1255
18. Chow CK, Kaneko T (1972) Automatic boundary detection of the left ventricle from cineangiograms. Comput Biomed Res 5(4):0010–4809. https://doi.org/10.1016/0010-4809(72)90070-5
19. Eikvil L, Taxt T, Moen K (1991) A fast adaptive method for binarization of document images. In: Proceedings of 1st international conference on document analysis and recognition (ICDAR). Saint-Malo, France, pp 435–443
20. Mardia KV, Hainsworth TJ (1988) A spatial thresholding method for image segmentation. IEEE Trans Pattern Anal Mach Intell 10(6):919–927. https://doi.org/10.1109/34.9113
21. Niblack W (1986) An introduction to digital image processing. Englewood Cliffs, Prentice-Hall
22. Taxt T, Flynn PJ, Jain AK (1989) Segmentation of document images. IEEE Trans Pattern Anal Mach Intell 11(12):1322–1329. https://doi.org/10.1109/34.41371
23. Yanowitz SD, Bruckstein AM (1989) A new method for image segmentation. Comput Vis, Graph, Image Process 46(1):82–95. https://doi.org/10.1016/S0734-189X(89)80017-9
24. Sauvola J, Seppanen T, Haapakoski S, Pietikainen M (1997) Adaptive document binarization. In: Proceedings of the fourth international conference on document analysis and recognition, vol 1, pp 147–152. https://doi.org/10.1109/ICDAR.1997.619831

25. O'Gorman L (1994) Binarization and multithresholding of document images using connectivity. CVGIP: Graph Model Image Process 56(6):494–506. https://doi.org/10.1006/cgip.1994.1044
26. Gatos B, Pratikakis I, Perantonis SJ, Improved document image binarization by using a combination of multiple binarization techniques and adapted edge information. In: 2008 19th international conference on pattern recognition. https://doi.org/10.1109/ICPR.2008.4761534
27. Sezgin M, Sankur B (2004) Survey over image thresholding techniques and quantitative performance evaluation. J Electron Imag 13(1):146–165. https://doi.org/10.1117/1.1631315

BEN-CNN-BiLSTM: A Model of Consequential Document Set Identification of Bengali Text

Taniya Seal, Debapratim Das Dawn, Abhinandan Khan, Sanjit Kumar Setua, and Rajat Kumar Pal

Abstract Document set identification assigns a text to its predefined text set. Therefore, the objective of any classification work is to create a model that can categorise various texts and objects into distinct classes. In this paper, three consequential deep learning based models, viz. BiLSTM, BiLSTM with attention layer, and CNN-BiLSTM have been used, which have the auto-learning capability in Bengali corpora. The CNN-BiLSTM model has been designated the classification model for its best performance in categorising Bengali text documents. This consequential model, named BEN-CNN-BiLSTM, was learned with Bengali text documents to determine the category of an unknown Bengali document. At first, more than four lac news articles from renowned Bengali newspapers are processed. After that, the training data is processed and entered into the proposed model. Finally, the model performance is assessed using the test dataset to calculate recall, precision, F-score, and accuracy. Compared to other standard classification algorithms in Bengali text classification, our proposed BEN-CNN-BiLSTM model achieved 93.94% accuracy. Thus, it can be said that the proposed BEN-CNN-BiLSTM model can be a new document set identification technique for Bengali datasets.

Keywords Text classification · Deep learning · CNN · Bidirectional LSTM · Attention mechanism

1 Introduction

Topic-wise, systematically conserving digital text data is necessary for future use or betterment of a task. Automatic document categorisation, a difficult task in natural language processing (NLP), is used for this purpose. It automatically allocates a text document or a succession of text documents to one or more predetermined categories. About 245 million people worldwide communicate in the Bengali language, considered the seventh most spoken language globally. It appears in practically

T. Seal (✉) · D. D. Dawn · A. Khan · S. K. Setua · R. K. Pal
Kolkata, India
e-mail: taniyaseal19922@gmail.com

R. Chaki et al. (eds.), *Applied Computing for Software and Smart Systems*,
Lecture Notes in Networks and Systems 555,
https://doi.org/10.1007/978-981-19-6791-7_11

every text document, including newspaper articles, social media posts, and blog posts. Therefore, the size and variety of Bengali digital text documents have gradually increased daily due to extensive Internet use. However, as text categorisation by humans involves considerable time and cost, it has some limitations. Therefore, an automatic document categorisation system will be developed to handle a large amount of text data so that documents can be organised and analysed easily and quickly. Bengali document set identification is a significant research issue that must be solved. Automatic document classification has been done in English and other European languages using a variety of linguistic and statistical methodologies. However, only a few have been proposed in Bengali till now. Yet, Bengali document set identification has several applications, viz.,

- Security agencies can surmise corrupted Bengali digital data or detect spam.
- The daily newspapers can structure their news category-wise.
- A library person can categorise papers or books the library has.
- Patients can be grouped according to their diagnosis reports in hospitals.
- Official data from government or non-government organisations can be archived.
- Better Bengali content can be searched, and soon.

Convolutional neural networks (CNNs) and recurrent neural networks (RNNs) have been extensively used for text categorisation in recent years [23–27]. CNN can learn local features from temporal or spatial data but not from sequential correlations. RNNs, on the other hand, are built for sequential modelling rather than concurrent extraction of features. Hence, text classification can be thought of as a sequential modelling process. RNNs are utilised more commonly in text categorisation because of their properties. For prolonged data sequences, typical RNNs, on the other hand, cause exploding and disappearing states against their gradient.

Long short-term memory (LSTM) [28] is a kind of RNN architecture that successfully solves the problem while simultaneously capturing long-term reliance. LSTM is essential in NLP because of its incredible capacity to extract high-level text information. Bidirectional long short-term memory (BiLSTM) [29] is a type of LSTM that combines the forward and backward hidden layers to give access to both previous and subsequent contexts. In contrast to BiLSTM, LSTM relies solely on historical data. As a result, BiLSTM outperforms LSTM in sequential modelling tests.

Therefore, the BEN-CNN-BiLSTM model has been formed by combining the features of CNN and BiLSTM. Although BiLSTM can extract contextual information from a text, it is not feasible to find the most meaningful information within that data. Hence, classification performance will improve if one focuses on critical information. Moreover, by varying the weights of the contextual information, the attention mechanism can focus on the crucial information. Therefore, using BiLSTM in conjunction with an attention mechanism can boost classification accuracy even more.

For Bengali report classification, different approaches have been proposed. This paper uses three systems to categorise documents: BiLSTM, BiLSTM with an attention mechanism (BiLSTM-AM), and BEN-CNN-BiLSTM. We have evaluated the models based on the achieved accuracy and the time needed for training. We have used more than four lac news articles from the renowned Bengali newspaper Prothom Alo as a dataset in this experiment. We have pre-processed the data with the one-hot encoder algorithm. This research aims to demonstrate how to classify Bengali texts using the lately popular deep, consequential models and compare them to find the best model. The term consequential refers to the self-learning properties of the deep learning based models.

The rest of the paper has been organised as follows. Section 2 describes the literature survey. Section 3 depicts the corpus and working methodology. The result analysis and comparison have been studied in Section 4. Finally, Section 5 discusses the conclusion, the proposed approach's limits, and future research scope.

2 Literature Survey

Text categorisation (TC) is one of the most demanding areas in Bengali NLP. Several remarkable research works have been published on Bengali TC in the past decade. The approaches for Bengali TC can be categorised mainly into three classes: conventional-based, fuzzy logic-based, and deep learning based.

2.1 *Text Categorisation Based on Conventional Approaches*

Various researchers in Bengali TC use different conventional feature identification and extraction approaches and classification algorithms, as detailed below. Mandal et al. [1] developed a Bengali Corpus from various websites and applied it in a system using four supervised learning methods. The support vector machine (SVM) technique achieved a maximum accuracy of 89.14%. Alam et al. [2] curated an enormous dataset with around 376,226 Bengali articles from several news portals. This large variety of data is trained using five supervised algorithms: logistic regression, neural networks, naïve Bayes (NB), AdaBoost, and random forest. In addition, several textual features, such as term frequency-inverse document frequency (TF-IDF) with 3000-word vectors, TF-IDF with 300-word vectors, and Word2Vec [2], are utilised. They achieved a maximum precision of 0.96 using the Word2Vec feature in Neural Network. By including inverse class frequency into the TF-IDF scheme, Dhar et al. [3] established the TF-IDF-ICF feature extraction approach. Pearson's correlation coefficient (PCC) minimised the feature collection. Using the multi-layer perceptron (MLP) on four thousand text documents from eight text domains, this experiment achieved an accuracy of 98.03%. Dhar et al. [4] also offered a novel approach to the problem of Bengali text classification. As baseline feature extraction methods, PART [4], a rule-based classification system, is formulated based on term association and aggregation.

2.2 Text Categorisation Based on Fuzzy Logic

A fuzzy logic-based strategy for classifying Bengali text documents was proposed by Dhar et al. [5]. They used TF-IDF-ICF to cluster 8600 text documents from 8 domains using a modified fuzzy C-means (MFCM) clustering technique and achieved 99.19% accuracy.

2.3 Text Categorisation Based on Deep Learning

Automatic Bengali document classification based on deep convolutional networks was done by Hossain et al. [6]. The accuracy of this method was 94.96%. Next, Rahman et al. [7] used a deep recurrent neural network with BiLSTM to classify Bengali documents. The accuracy of this method was 98.33%, with an F-score of 0.98. BiLSTM has been around for quite some time as a model and was first used in English at the turn of the twentieth century [29]. However, its use in the Bengali language is relatively recent [7]. Finally, Rahman et al. [8] reported the classification of a Bengali document utilising a transformer-based learning model for text classification.

Bidirectional encoder representations from transformers (BERT) and efficiently learning an encoder that classifies token replacements accurately (ELECTRA) models are used on three Bengali text datasets. The ELECTRA model outperformed the BERT model in terms of accuracy and F-score. Additionally, the recognition of Bengali sign language gesture images by a CNN was done in real-time [9].

Using an enlarged lexical data dictionary, Bhowmik et al. [10] developed deep learning models for sentiment analysis on Bengali text. The authors introduced the Bangla text sentiment score (BTSC) method, which extracts polarity from large texts using a rule-based approach. The extensive study of several specific deep learning models, such as hierarchical attention based LSTM, dynamic routing based capsule neural network with BiLSTM, and BERT-LSTM with some fine-tuning, is also described in this research. BERT-LSTM provided the maximum accuracy of 84.18% to the authors. Ahmed et al. [11] developed two deep learning models for Bengali document categorisation: current neural network attention layer and current neural network with BiLSTM. These two models achieved 97.72 and 86.56% accuracies, respectively.

3 Proposed Methodology

This section demonstrates the steps to furnish our Bengali document set identification research. In the first step, data is collected. Then, this data is pre-processed and cleaned. Next, this cleaned and pre-processed data is divided into training and test data. Subsequently, BiLSTM, BiLSTM with attention layer, and BEN-CNN-BiLSTM models are designed and applied to the training dataset. Finally, model performances are evaluated on the test dataset.

3.1 Corpus

The outcome of any experiment is influenced by both the quality and quantity of the data collected. In the digital realm, Bengali datasets are scarcely organised by category, and only a few documents are freely accessible. The corpus has been built with the data collected from Nabil [12], comprising 437,948 news articles having 32 different categories from the renowned Bengali newspaper Prothom Alo. Some research has been done with this dataset because of its availability. We have selected this as it is freely accessible, for its various categories, and for getting the scope to compare our results with similar research works. Each data possesses the following ten attributes:

- Author—Author of the news article
- Category—Category Name in English
- Category_bn—Category Name in Bengali
- Comment_count—Total number of comments for a particular news
- Content—A detailed description of the news
- Modification_date—Date and time of news modification
- Published_date—Date and time of news publication
- Tag—A single unique word for a news article.
- Title—Headline of a news article.
- URL—Uniform resource locator of the news article.

3.2 Data Pre-processing

Some data set categories occur for a very short time that would not contribute much to the experiment. Thus, nine (out of 32) categories have been selected with occurrences higher than 8000. The number of samples in each category is depicted in Fig. 1. Table 2 gives several examples of the category in its original (Bengali) and English versions. Some statistical information of this corpus, namely, the total number of characters, the total number of words, the total number of sentences, vocabulary size, the total number of news articles, five most occurred words, and the count of most occurred words of the corpus, have already been given in Table 1. Deep learning algorithms perform better with numerical inputs. Thus, a One-Hot-Encoder converts text or categorical data into numerical data. Here, categorical features are encoded as a one-hot numeric array. First, content data are cleaned by removing unnecessary punctuations, and then word-level tokenisation is performed. The average number of words in a sample is 274. Hence, all samples are padded with a 300-word sequence.

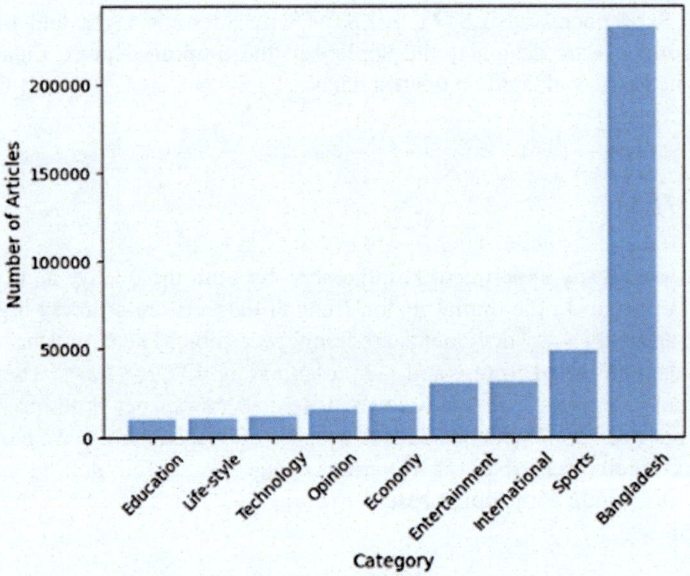

Fig. 1 The number of samples in each category

Table 1 Statistical information about the corpus

Total characters	731,896,821
Total words	114,647,003
Total sentences	8,750,868
Vocabulary	996,496
Total number of articles	408,471
Five most occurred word	গাজীপুরের কালিয়াকৈর উপজেলার তেলিরচালা এলাকায় আজ বৃহস্পতিবার রাতের টিফিন খেয়ে একটি পোশাক কারখানার ৫০০ শ্রমিক অসুস্থ হয়ে পড়েছেন। এঘটনায় বিক্ষোভ করেছেন ওই কারখানার শ্রমিকেরা।............
Count of most occurred word	1,443,792

3.3 Classification Model

After pre-processing, it is now time to train the Bengali news corpora. Three different experiments are conducted on these corpora. These three experiments deal with consequential models, viz. BiLSTM, BiLSTM-AM [14], and the proposed BEN-CNN-BiLSTM for the classification of Bengali news articles. Here, the term consequential alludes to the auto-learning capability of our proposed BEN-CNN-BiLSTM

Table 2 News sample with category

Text (Bengali)	Text (English)	Category
দেশের সরকারি মেডিকেল কলেজ গুলোতে এমবিবিএস কোর্সে ভর্তিতে ৫০০ আসন বাড়ানোর সিদ্ধান্ত নিয়েছে স্বাস্থ্যও পরিবার কল্যাণ মন্ত্রণালয়। এই সিদ্ধান্ত ২০১৮- ১৯ শিক্ষাবর্ষ থেকেই কার্যকর হবে।…………	Five hundred garment factory workers fell ill after eating tiffin in the Telirchala area of Kaliakair Upazilla of Gazipur on Thursday night. The workers of the factory protested against this incident…………	Bangladesh
খুক খুক শুকনো কাশি যখন-তখন অনেককে ভোগায়, বিব্রতও করে। কারও সারা বছর খুসখুসে কাশি লেগেই থাকে। সবসময় কাশি সারাতে অ্যান্টিবায়োটিক বা কাশির ওষুধের দরকার নেই।…………	The Ministry of Health and Family Welfare has decided to increase the number of seats for MBBS courses in government medical colleges by 500. This decision will be effective from the 2018–19 academic year…………	Education
সহজ পাসওয়ার্ডের কারণে অনলাইন অ্যাকাউন্ট সহজেই হ্যাক হয়ে যেতে পারে। জটিল পাসওয়ার্ড মনে রাখা কষ্টকর। তবে জটিল পাসওয়ার্ড মনে রাখার কষ্টের দিন হয়তো ফুরাল।…………	Whooping dry cough sometimes afflicts and embarrasses many. Someone coughs all year round. You do not need antibiotics or cough medicine to cure cough all the time…………	Lifestyle
প্রথম আলো ট্রাস্ট মাদক বিরোধী আন্দোলনের উদ্যোগে গত ১৫ জুন বিকেল সাড়ে চারটায় পরামর্শ সহায়তা-৩৯-এর আসরটি অনুষ্ঠিত হয় ধান মন্ডির ডব্লিউভিএ মিলনায়তনে। এ আয়োজনে মাদকাসক্ত ব্যক্তি ও তাদের পরিবারের সদস্যরা উপস্থিত ছিলেন।…………	Online accounts can be easily hacked due to simple passwords. Unfortunately, complex passwords are difficult to remember. However, the difficult days of remembering complex passwords may be over…………	Technology
সপ্তাহের শেষ কার্য দিবসে আজ বৃহস্পতিবার প্রধান শেয়ার বাজার ঢাকা স্টক এক্সচেঞ্জ (ডিএসই) লেনদেন ৯০০ কোটি টাকা ছাড়িয়েছে। আর চট্টগ্রাম স্টক এক্সচেঞ্জের (সিএসই) লেনদেনও গতকালের চেয়ে বেড়েছে।…………	The Prothom Alo Trust organised the 39th counselling support session, as a part of their anti-drug movement, at the WVA auditorium in Dhanmondi on 15th June at 4:30 pm. Drug addicts and their family members were present at the event…………	Opinion
একক-মিশ্র অ্যালবামে অনেক গানই করেছেন বাপ্পা মজুমদার ও মমতাজ। পাশাপাশি গেয়েছেন চলচ্চিত্রের গানও।…………	The Dhaka Stock Exchange (DSE) traded above 900 crores today, Thursday, the last working day of the week, and the Chittagong Stock Exchange (CSE) turnover has also increased compared to yesterday…………	Economy

(continued)

Table 2 (continued)

Text (Bengali)	Text (English)	Category
একক-মিশ্র অ্যালবামে অনেক গানই করেছেন বাপ্পা মজুমদার ও মমতাজ। পাশাপাশি গেয়েছেন চলচ্চিত্রের গানও।...........	Bappa Majumder and Mumtaz have done many songs in solo-mixed albums. In addition, they have also sung movie songs…………	Entertainment
দক্ষিণ আফ্রিকার বর্ণবাদ বিরোধী নেতা নেলসন ম্যান্ডেলাকে কৃত্রিম ব্যবস্থায় বাঁচিয়ে রাখা হয়েছে। আজ বৃহস্পতিবার আলজাজিরা অনলাইনের প্রতিবেদনে জানানো হয়, মৃত্যুর পর ম্যান্ডেলাকে সমাহিত করা নিয়ে পরিবারের পক্ষ থেকে আদালতে দেওয়া নথিপত্র সূত্রে ম্যান্ডেলার এভাবে বেঁচে থাকার বিষয়টি জানা গেছে।...........	South Africa's anti-apartheid leader Nelson Mandela was kept alive in an artificial system. According to an Al Jazeera online report on Thursday, documents provided by the family to the court regarding the burial of Mandela after his death have revealed that Mandela survived in this way…………	International
এবারের উইম্বলডনটা স্মরণীয় করে রাখার মিশনেই যেন নেমেছেন সাবিনা লিসিকি। চতুর্থ রাউন্ডের লড়াইয়ে সেরেনা উইলিয়ামসকে হারিয়ে শুরু করেছিলেন স্বপ্নযাত্রা।...........	Sabina Lisicki seems to be on a mission to make this year's Wimbledon memorable. She started her dream run by defeating Serena Williams in the fourth round …………	Sports

model in Bengali corpora. This model produced the most efficient result with respect to accuracy over the other two models. We use the power of the CNN model, which extracts local features, and the BiLSTM model for capturing long-distance dependencies in Bengali text sequences and blend these features into one single BEN-CNN-BiLSTM model. Lesser training time is another reason for choosing the BEN-CNN-BiLSTM model as the classification model. The total training time of the BiLSTM and BEN-CNN-BiLSTM models for ten epochs are 4109 and 2306 s, respectively.

In comparison, BiLSTM with the attention model takes 5860 s in one epoch. Thus, BiLSTM with an attention model cannot be associated with any real-life application because it takes a huge amount of time. It could be because it increases the number of weight components in the model, which increases the training time, especially if the input data is in long sequences. In a sequence-to-sequence model, the context vector for time t is created by combining the encoder output at each time interval with the decoder output at time t. (e.g., an RNN-based encoder-decoder). However, we have employed trainable weights for each encoder time interval to calculate the context vector. If the encoder input data is long, as it is for our task, this can result in a vast number of weight parameters. As a result, the larger a model becomes (in terms of parameter count), the longer it takes to train.

Furthermore, overfitting can be caused by the model's training time. If the model is trained for too long on the training data, it learns the noise or irrelevant data in the dataset. As a result, the performance of the BiLSTM-AM model degraded. Due to space limits, the detailed architecture of the BiLSTM, CNN, and BiLSTM-AM network could not be discussed in detail. As a result, we have only discussed the characteristics of these models that are relevant to our study, along with appropriate citations. Our proposed BEN-CNN-BiLSTM model uses BiLSTM, consisting of two LSTMs–one process input in a forward direction and the other in a backward direction. This LSTM is a special kind of Recurrent Neural Network (RNN). The neurons of RNN are connected in the shape of a directed cycle. The information is processed sequentially in the RNN model because it uses internal memory to process its inputs. The output of RNN is dependent on the inputs of all the previous nodes and remembered information for further processing. The LSTM model maintains long-term dependencies of input words with the help of three different gates efficiently. The architecture of LSTM with a detailed explanation is given in [20].

Another part of our proposed model is CNN. Feature maps are generated by applying feature detectors to the input text or the feature map output of the preceding layers. The pooling layer reduces the computational overhead preserving vital information. We have used max-pooling, where the maximum value element is selected within the pooling window. The architecture of CNN with a detailed explanation is given in [21]. The architectures of the three models are described below.

The Architecture of BiLSTM

This section describes the layers with their hyperparameters value used in the BiLSTM model. Figure 2 shows the architecture of BiLSTM. In the embedding layer, the input vectors of size 408471×300 are embedded with a lower-dimensional vector of size 100. This layer starts with random weights and learns an embedding for every word in the training dataset. Two BiLSTMs with 128 memory units each are used to extract features from the embedding layer. The output of the BiLSTM layer is fed into a fully connected neural network layer with a Softmax activation function [21], with the number of categories as the output dimension. As a loss function, we employed categorical cross-entropy and Adam [15] as the model's optimiser.

The Architecture of BiLSTM with Attention Mechanism

This section describes the layers used in the BiLSTM with the attention mechanism and its hyperparameters value. Figure 3 shows the architecture of BiLSTM with the attention mechanism. In the embedding layer, the input vectors of size 408471×300 are embedded with a lower-dimensional vector of size 100. One BiLSTM with 128 memory units, 0.2 dropout rates, and recurrent dropout is employed to extract features

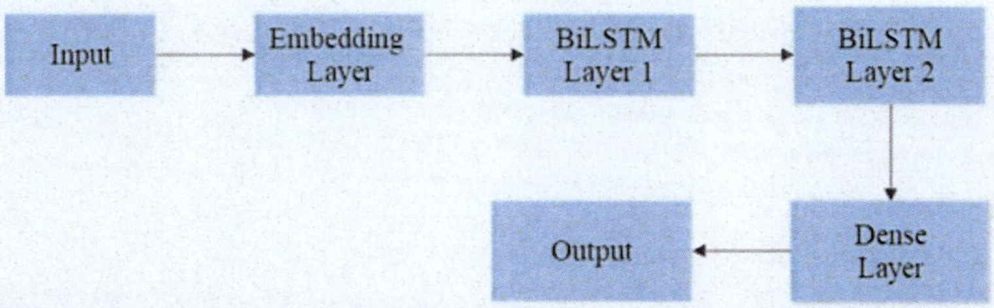

Fig. 2 BiLSTM architecture

from the embedding layer. The subsequent attention layer generates the weight vector. By multiplying these weight vectors, word-level features produced in every time step are combined into a sentence-level feature vector. This layer employs the sigmoid activation function. The attention layer's features are then transferred into a one-dimensional vector. Finally, the output of the flattening layer is fed into a single fully connected neural network layer with a Softmax activation function, with the number of categories as the output dimension. As a loss function, we employed categorical cross-entropy, with Adam as the model's optimiser.

The Architecture of BEN-CNN-BiLSTM

This section describes the layers with their hyperparameters value used in the CNN-BiLSTM model. Figure 4 shows the architecture of BEN-CNN-BiLSTM. In the Embedding layer, the input vectors of a size of 408471 × 300 are embedded with a lower-dimensional vector of size 500. The Conv1D layer produces a convolution kernel convolved with the output from the embedding layer. We used 128 filters, a kernel size of five, and ReLU as an activation function in this layer. MaxPooling 1D layer down samples its input by selecting the maximum value in the window of size five. Stride is used to move the window. Two bidirectional LSTMs with 64 memory units each and a 0.2 rate of dropout are utilised to extract features from the max-pooling layer. The output from the BiLSTM layer is fed into two fully connected neural network layers with a Softmax activation function, resulting in a 28-dimensional output. Next, the Dense layer's features are put into a one-dimensional vector. Finally, the previous layer's output is fed into a single fully connected neural network layer with a Softmax activation function, with the number of categories as the output dimension. As the loss function, we employed categorical cross-entropy and Adam as the model's optimiser.

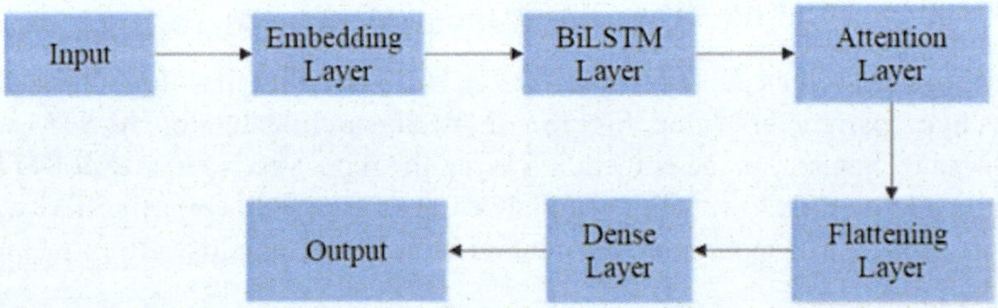

Fig. 3 BiLSTM with attention mechanism architecture

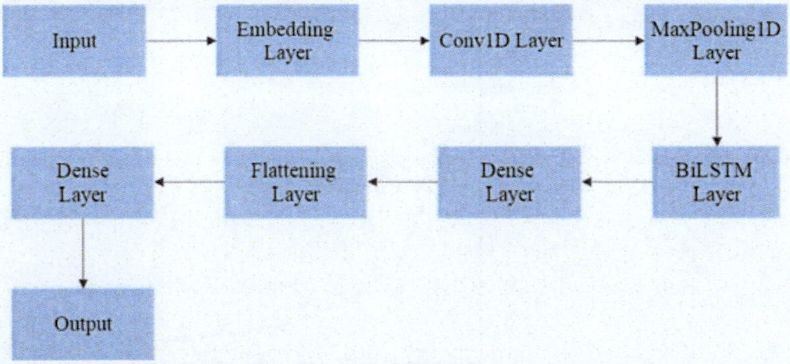

Fig. 4 BEN-CNN-BiLSTM architecture

4 Result Analysis and Comparison

4.1 Results Analysis

In a 70:30 ratio, the dataset is separated into two parts: training and testing. The number of training and testing samples are 285,929 and 122,542, respectively. Further training data is divided into training and validation datasets of sizes 228,742 and 57,186, respectively. Figure 5 shows the model training and validation accuracy for ten epochs.

Here, it can be seen that the training accuracy sharply increases from zero to four epochs, then slowly increases between four and eight epochs, and remains the same for eight to ten epochs. On the other hand, the validation accuracy remained the same, more or less within a fixed limit. Figure 6 shows the model training and validity loss throughout ten epochs. We can see that the training loss declines significantly between zero and two epochs, gradually decreases between two and six epochs, and

Fig. 5 The training and validation accuracy of CNN-BiLSTM

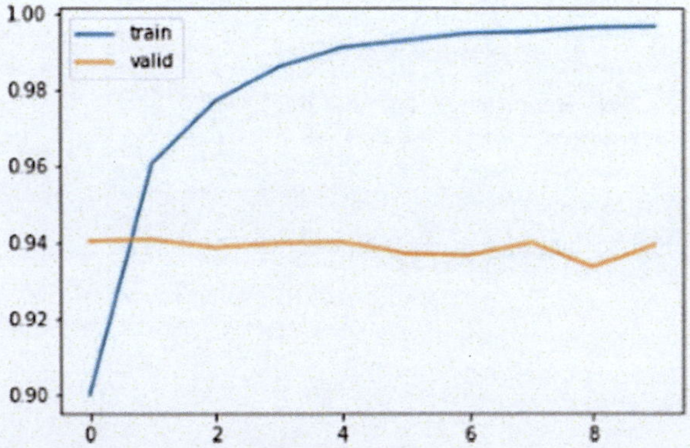

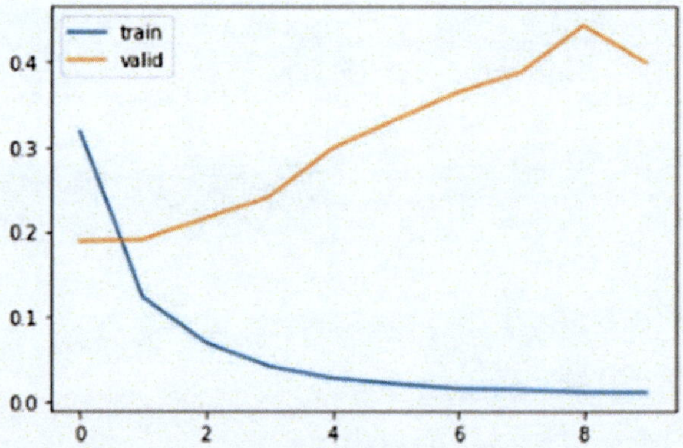

Fig. 6 The training and validation loss of CNN-BiLSTM

remains constant between six and ten epochs. The validation loss, on the other hand, grew gradually from zero to seven epochs and then dropped.

The performance of this consequential model having self-learning characteristics is evaluated using four evaluation matrices: precision, recall, F-score, and support [22]. A confusion matrix is shown in Fig. 7 to measure the performance of the classification algorithm of the BEN-CNN-BiLSTM model. The classification report is presented in Table 3. Figure 8 shows the different experimental results in terms of testing accuracy. We have conducted some statistical analysis. However, we could not provide all of these data due to space limits and a lack of comparable test results in the current literature. In Bengali text classification, our proposed consequential model BEN-CNN-BiLSTM achieved the best accuracy of 93.94% compared to the other two algorithms.

	Opinion	Education	Economy	Sports	Entertainment	International	Technology	Bangladesh	Life-Style
Opinion	**67540**	698	183	360	437	132	238	46	117
Education	1047	**3828**	10	25	108	12	47	9	87
Economy	322	9	**2439**	33	1	41	35	6	30
Sports	386	4	5	**8589**	95	19	9	25	8
Entertainment	347	22	2	95	**8631**	14	56	11	79
International	221	29	15	131	81	**2604**	50	26	99
Technology	350	35	30	51	86	29	**4090**	26	13
Bangladesh	117	6	2	78	95	10	8	**14383**	5
Life-Style	183	95	21	40	201	58	14	8	**3015**

Fig. 7 Confusion matrix with BEN-CNN-BiLSTM

Table 3 Classification report of BEN-CNN-BiLSTM

Category	Precision (%)	Recall (%)	F-score (%)	Support
Opinion	95.78	96.83	96.30	69,751
Education	81.00	74.00	77.34	5173
Economy	90.10	83.64	86.75	2916
Sports	91.35	93.97	92.64	9140
Entertainment	88.66	93.24	90.89	925
International	89.21	79.98	84.34	3256
Technology	89.95	86.84	88.37	4710
Bangladesh	98.92	97.82	98.37	14,704
Lifestyle	87.32	82.94	85.07	3635

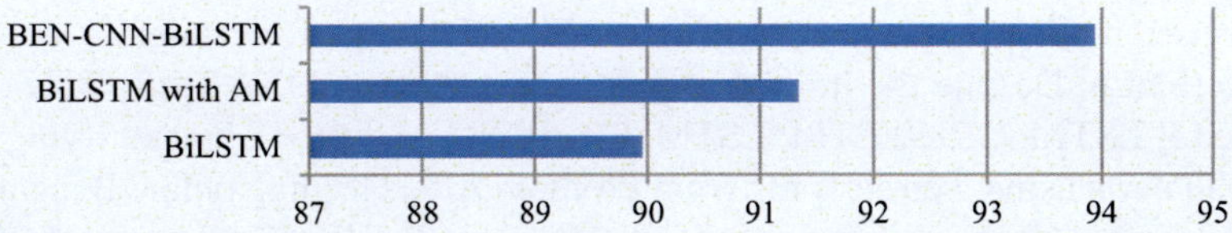

Fig. 8 Test result with three models

4.2 Result Comparison

Now it is important to compare our model performance with others. Unfortunately, in the case of the Bengali language, benchmark datasets are not openly available. Furthermore, existing research works only offer the overall accuracy of their proposed Bengali text classification algorithms. Thus, we could only compare the results of our proposed methodology based on overall accuracy. Methods whose results are compared are described here.

Method 1: Multi-layer dense neural network with TF-IDF [16] exhibits accuracy (84.58%) and F-score (0.84) compared to the other popular classification algorithms (SVM, Multinomial NB, and soon) for Bengali text documents classification.

Method 2: The authors demonstrated 85.21% accuracy in determining the polarity of Facebook Bengali comments representing people's perceptions of Bangladesh [17].

Method 3: In [1], the authors developed a Bengali Corpus from various websites and applied it to a system using four supervised learning methods. They found maximum accuracy of 89.4% with the Support Vector Machine (SVM) algorithm.

Method 4: The authors of [19] used bidirectional LSTM with an attention mechanism to categorise text based on news captions and achieved 91.37% accuracy. Even though the work [19] employed the same dataset and BiLSTM-AM as us, we were unable to compare the complexity of the BiLSTM-AM network, such as the number

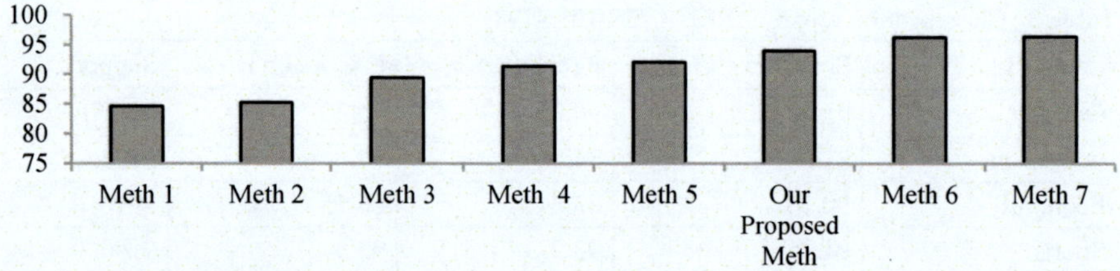

Fig. 9 Result comparison (accuracy) with recent works in Bengali text classification

of trainable parameters, training duration, and so on, of our proposed methodology to that of the article [19].

Method 5: In [18], authors have shown 92% accuracy using SGD with TF-IDF for Bengali document categorisation. SVM classifier [1] and NB Classifier are two supervised models that respond well, but our model outperforms them.

Method 6: Despite the limited sample size, [30] found that Text- GCN outperformed BiLSTM, GRU-LSTM, LSTM, Char-CNN, and BERT in identifying online Bengali news using web scraping from Prothom Alo, a leading online Bengali news source. Text-GCN had a precision of 96.25%, recall of 96.2%, F-score of 96.25%, and accuracy of 96.25%.

Method 7: The authors of this paper [10] used the most recent transformer-based models to classify Bengali text data. They applied the BERT and ELECTRA models for Bengali text classification to three different Bengali text datasets: BARD [2], OSBC [31], and Prothom Alo [32]. ELECTRA had the highest Recall value of 94.36% for the Prothom Alo dataset. When comparing Recall values, ELECTRA surpassed BERT in every situation. When analysing the F-score and accuracy, the ELECTRA model achieved higher values in most cases (94.18 and 96.39%, respectively).

Proposed Method: Our model's performance has surpassed the version of the model proposed in [19]. Result comparison in terms of accuracy with recent works in Bengali text classification is shown graphically in Fig. 9.

5 Conclusion

This paper presents a Bengali text classification scheme based on deep learning techniques having auto-learning features. These three consequential models, namely BiLSTM, BiLSTM-AM, and BEN-CNN-BiLSTM, are used for this task. The accuracy of BiLSTM is 89.96%, BiLSTM-AM is 91.34%, and BEN-CNN-BiLSTM is 93.94%. This work has shown that our proposed BEN-CNN-BiLSTM model, ensemble deep neural network (DNN) architecture achieves higher accuracy than DNN with attention layer and other well-known supervised learning models like this

k-nearest neighbour, SVM, NB, Decision tree, etc. for classifying Bengali text documents. However, the dataset of 400 K + news articles used here is insufficient for our purpose. We shall try with significant and other news articles in the future, and our results might improve. Compared to previous research, BERT and ELECTRA, two deep learning models, perform well in classifying Bengali text. As a result, these models will be used in the future to produce more efficient results.

References

1. Mandal AK, Sen R (2014) Supervised learning methods for Bangla web document categorisation. arXiv:1410.2045
2. Alam MT, Islam MM (2018) BARD: Bangla article classification using a new comprehensive dataset. In 2018 international conference on bangla speech and language processing (ICBSLP). IEEE, pp 1–5
3. Dhar A, Dash NS, Roy K (2018) Categorisation of Bangla web text documents based on TF-IDF-ICF text analysis scheme. In: Annual convention of the computer society of India. Springer, Singapore, pp 477–484
4. Dhar A, Dash NS, Roy K (2018) An innovative method of feature extraction for text classification using part classifier. In: International conference on information, communication, and computing technology. Springer, Singapore, pp 131–138
5. Dhar A, Dash NS, Roy K (2018) A fuzzy logic-based Bangla text classification for web text documents. J Adv Linguist Stud 7:159–187
6. Hossain M, Hoque MM (2019) Automatic Bengali document categorisation based on deep convolution nets. In: Emerging research in computing, information, communication and applications. Springer, Singapore, pp 513–525
7. Rahman S, Chakraborty P (2021) Bangla document classification using a deep recurrent neural network with BiLSTM. In: Proceedings of the international conference on machine intelligence and data science applications. Springer, Singapore, pp 507–519
8. Rahman MM, Pramanik MA, Sadik R, Roy M, Chakraborty P (2020) Bangla documents classification using transformer based deep learning models. In: 2020 2nd international conference on sustainable technologies for industry 4.0 (STI). IEEE, pp 1–5
9. Ahammad K, Shawon JAB, Chakraborty P, Islam MJ, Islam S (2021) Recognising Bengali sign language gestures for digits in real-time using convolutional neural network. Int J Comput Sci Inf Secur (IJCSIS) 19(1)
10. Bhowmik NR, Arifuzzaman M, Mondal MRH (2022) Sentiment analysis on Bangla text using extended lexicon dictionary and deep learning algorithms. Array 100123
11. Ahmed M, Chakraborty P, Choudhury T (2022) Bangla document categorisation using deep RNN model with attention mechanism. In: Cyber intelligence and information retrieval. Springer, Singapore, pp 137–147
12. Zan N, 400K+ Bengali newspaper Article. https://www.kaggle.com/furcifer/bangla-newspaper-dataset
13. Rehman AU, Malik AK, Raza B, Ali W (2019) A hybrid CNN-LSTM model for improving accuracy of movie reviews sentiment analysis. Multimed Tools Appl 78(18):26597–26613
14. Bahdanau D, Cho K, Bengio Y (2014) Neural machine translation by jointly learning to align and translate. arXiv:1409.0473
15. Kingma DP, Ba J (2014) Adam: a method for stochastic optimisation. arXiv:1412.6980
16. Chakraborty M, Huda MN (2019) Bangla document categorisation using multi-layer dense neural network with TF-IDF. In: 2019 1st international conference on advances in science, engineering, and robotics technology (ICASERT). IEEE, pp 1–4

17. Faruque MA, Rahman S, Chakraborty P, Choudhury T, Um J-S, Singh TP (2022) Ascertaining polarity of public opinions on Bangladesh cricket using machine learning techniques. Spat Inf Res 30(1):1–8
18. Islam M, Jubayer FEM, Ahmed SI (2017) A comparative study on different types of approaches to Bengali document categorisation. arXiv:1701.08694
19. Shopon M (2020) Bidirectional LSTM with attention mechanism for automatic Bangla news categorisation in terms of news captions. In: Electronic systems and intelligent computing. Springer, Singapore, pp 763–773
20. van Houdt G, Mosquera C, Nápoles G (2020) Are view on the long short-term memory model. Artif Intell Rev 53(8):5929–5955
21. Mercioni MA, Holban S (2020) The most used activation functions: Classic versus current. In: 2020 international conference on development and application systems (DAS). IEEE, pp 141–145
22. Dalianis H (2018) Evaluation Metrics and Evaluation. Clin Text Min 45–53
23. Hossain MR, Hoque MM, Siddique N, Sarker HI (2021) Bengali text document categorisation based on very deep convolution neural network. Expert Syst Appl 184:115394
24. Rahman MM, Sadik R, Biswas AA (2020) Bangla document classification using character level deep learning. In: 2020 4th international symposium on multidisciplinary studies and innovative technologies (ISMSIT). IEEE, pp 1–6
25. Hossain M, Hoque MM, Sarker IH (2020) Text classification using convolution neural networks with FastText embedding. In: International conference on hybrid intelligent systems. Springer, Cham, pp 103–113
26. Rahman S, Chakraborty P (2021) Bangla document classification using deep recurrent neural network with BiLSTM. In: Proceedings of the international conference on machine intelligence and data science applications. Springer, Singapore, pp 507–519
27. Dhar A, Mukherjee H, Obaidullah S, Santosh KC, Dash NS, Roy K (2019) Web text categorisation: A LSTM-RNN approach. In: International conference on intelligent computing and communication. Springer, Singapore, pp 281–290
28. Hochreiter S, Schmidhuber J (1997) Long short-term memory. Neural Comput 9(8):1735–1780
29. Graves A, Schmidhuber J (2005) Framewise phoneme classification with bidirectional LSTM and other neural network architectures. Neural Netw 18(5–6):602–610
30. Rahman MM, Khan MAZ, Biswas AA (2021) Bangla news classification using graph convolutional networks. In: 2021 international conference on computer communication and informatics (ICCCI). IEEE, pp 1–5
31. https://scdnlab.com/corpus/
32. https://www.kaggle.com/twintyone/prothomalo

Smart Healthcare

A Machine Learning Model for Automatic Sleep Staging Based on Single-Channel EEG Signals

Santosh Kumar Satapathy, Hari Kishan Kondaveeti, and A. S. Venkata Praneel

1 Introduction

Sleep is a transitory mental and physical condition denoted by elevated levels of consciousness and visual separation from the external environment. It differs from wakefulness because it cannot respond to environmental stimuli. Sleep is vital for holistic well-being. Insufficient sleep can profoundly influence our mental health and impair our cognitive abilities. Maintaining good health for humans is a crucial activity. Sleep deprivation and poor sleep quality interfere with regular exercise and cause physical and mental problems [2, 3].

The ability to continuously monitor sleep quality helps identify sleep disorders quickly. Polysomnography (PSG) is a common technique used in sleep research, especially in diagnosing sleep disorders[4]. PSG, or sleep test, is an official medical procedure to capture important biological signals for detecting sleep disorders. PSG is made of an electroencephalogram (EEG), electrocardiogram (EOG), electromyogram (EMG), and an electrocardiogram (ECG) [5, 6]. This is a slightly subjective procedure, where a consensus may not be achieved between the assessments conducted by two different sleep experts. Therefore, accurate automatic sleep scoring

S. K. Satapathy (✉)
Department of Information and Communication Technology, Pandit Deendayal Energy University (PDEU), Gandhinagar, Gujarat, India
e-mail: Santosh.Satapathy@sot.pdpu.ac.in

H. K. Kondaveeti (✉)
School of Computer Science Engineering, VIT-AP University, Amaravati, Andhra Pradesh, India
e-mail: kishan.kondaveeti@vitap.ac.in

A. S. V. Praneel
Department of Computer Science and Engineering, GITAM (Deemed to Be University), Visakhapatnam, Andhra Pradesh, India

R. Chaki et al. (eds.), *Applied Computing for Software and Smart Systems*,
Lecture Notes in Networks and Systems 555,
https://doi.org/10.1007/978-981-19-6791-7_12

can be less time-taking, more effective, and provide an actual assessment of sleep that is free from inter-scorer variations.

Each 30 s epoch of nocturnal sleep, as per R&K [7] standards, corresponds to either of the seven distinct stages: Wake, S1 (drowsiness), S2 (light sleep), S3 (deep sleep), S4 (deep or slow-wave sleep), REM (rapid eye movement), and movement time (MT). The AASM [8] rules were updated in 2007 to address certain new flaws in the R&K norms[9–11].

The rest of the paper is organized as follows: the brief description of the related work is described in Sect. 2. A brief overview of the dataset is described in Sect. 3. Section 4 presents the proposed methodology. The experimental settings and result analysis are briefly described in Sect. 5. Section 6 presents the concluding remarks and the future scope of this proposed work.

2 Related Works

Yan [12] worked on a model that involved different combinations of the four channels of polysomnography signals, namely, EEG, EOG, EMG, and ECG. Eight such combinations were made and supplied to the system as input. A Cyclic Alternating Pattern (CAP) Physio-Net was considered the source for these signal recordings. Comprising 232 characteristics, including entropy characters, fractal characters, non-linear characters, frequency characters, statistical characters, and time–frequency characters, this database was used for training through machine learning-based techniques. Out of the multiple algorithms tested, Random Forest performed the best with an accuracy of 86.24%. Hence, this study concluded that the automatic sleep scoring method was economical and would significantly reduce the time taken and effort required. Apart from that, the limitation of this method was that Stage S1 is often erroneously detected as wakefulness and REM by automatic sleep scoring.

Shen [13] used three freely available datasets: Sleep-EDF, Dreams Subjects, and ISRUC. Firstly, every epoch of EEG is broken down into low-level (LSB) and high-level sub-bands. Then the DSSM is estimated, and the LSB causes the LE calculation to be performed in the HSB. Finally, the model reported 79.90 and 81.65%, respectively, with the Dreams and ISRUC-Sleep databases. But the drawback of the proposed method was that the misclassified ratio was more for the S2 stage. Huang et al. [14] proposed a novel feature screening and signaled pre-processing method. They employed multi-channel signal superposition to enhance the information present in the original signal. Sixty-two features were first selected for feature screening. To pick the relevant features out of 62 of them, a ReliefF algorithm was used. Fourteen significant features were extracted and considered essential determiners of the varied sleep stages. Ghimatgar et al. [15] presented a method of selecting features based on relevance and redundancy analysis. A collection of convenient features was initially chosen from a vast group of features obtained from each epoch of sleep data containing only EEG recordings. A random forest classifier was then used to classify EEG segments. Finally, in the wake of a temporal pattern of transitions between sleep stages observed by the study, a Hidden Markov Model (HMM) was deployed to discard the false positives. The suggested technique was tested on four public datasets based on EEG signals according to the famous scoring standards, i.e., R & K and AASM norms. They obtained an accuracy of 79.4–87.4% using a leave-one-out validation strategy. The only limitation of the presented work is that a reliable interpretation couldn't be achieved for the wake stage, which is often the most difficult to detect. Cooray et al. [16] provide a wholly automated RBD detection approach that includes automatic sleep staging and RBD identification. A restricted polysomnography montage was used to analyze 53 patients with RBD and 53 age-matched healthy controls. A Random Forest (RF) classifier was trained on 156 characteristics derived from electroencephalogram (EEG), electrooculogram (EOG), and electromyogram (EMG) channels were used to classify sleep stages. An RF classifier was designed to detect RBD by integrating established techniques for measuring muscle atonia with extra variables such as sleep architecture and the EMG fractal exponent. This study surpasses existing measurements and shows that combining sleep architecture and sleep stage transitions might help detect RBD. But the study gave more misclassification in the N2 stage. Mohammed et al. [17] aimed to create a novel automatic approach for classifying EEG-sleep stages using a statistical model and weighted brain networks. According to the sliding window approach, each EEG segment is divided into several blocks. Each block has a certain number of statistical features taken from it, resulting in a vector representation for each EEG segment. The features vector is then transformed into a weighted undirected network. A least square support vector machine (LS-SVM) classifier extracts the networks' different structural characteristic spectral features. Simultaneously, the attributes of the network are comprehensively studied. The network's properties are discovered to fluctuate depending on their sleep stages. The critical elements of each sleep stage's networks are best portrayed. The drawback of this method is imbalanced EEG data.

Sriraam et al. [18] used a multi-channel EEG signal and multilayer perceptron feedforward neural network. The model achieved a general accuracy of 92.9% with 20 hidden units. Memar et al. [19] used the Kruskal–Wallis test and a Random Forest classifier to attain an accuracy of 95.31% for fivefold cross-validation and 86.64% for subject-wise cross-validation. Silveira et al. [20] obtained an RF classifier and the model gave 90% accuracy. In [21] considered RF classifier. The model classified the sleep accurately for about 85.95% of the sleep data. Zhu et al. [22] selected nine features from the raw data sent to an SVM classifier. For two-stage sleep classification, the proposed method attained an accuracy of 87.50%. In [23] put forward a portable system for classifying sleep duration into stages. The authors made use of varied combinations of features derived from EEG signals. The presented model reported 97.1% accuracy.

Despite improved performance in the existing state-of-the-art works, still, it is a challenging task due for three main reasons, (i) the traditional classification models are less flexible and only accept fewer training parameters, which causes the model not to be properly adjusted with the input data during the training stage. Hence there should be a lack of a generalized model which can able to classify two-class to five-class sleep states classification problems, (ii)less available labeled data and (iii) low sleep staging classification accuracy. Therefore it's highly essential for a generalized automated sleep staging system that gives good classification accuracy performance even with fewer numbers of training samples. Considering all these challenges, in the present study, we consider two different categories of subjects: (1) Individuals with clear sleep disorders, from whom we obtained recordings from two sleep sessions; and (2) Healthy control subjects without any sleep-related symptoms. This work is novel in that it analyses the sleep behavior of subjects suffering from sleep disorders during two different recording sessions, which were recorded on two different dates. This analysis may make it easier for sleep clinicians to achieve a proper diagnosis because many studies have shown that sleep disturbances are affected by professional and social pressure. For this reason, if we cannot assume that a subject has a sleep disorder, we should analyze the subject's sleep behavior on more than one occasion, as doing so will help to reach the correct diagnosis. This research aims to devise an effective mechanism for identifying sleep pattern irregularities to improve sleep stage classification efficiency with different subjects' health conditions. Develop an enhanced machine learning framework for categorizing data into sleep stages with accurate identification of the changes in sleep behavior. The uniqueness of this experimental study is the assessment of the sleeping patterns of subjects. This analysis sometimes makes it easier for sleep clinicians to go for a proper diagnosis because it has been found from so much literature that sleep disturbances occur due to some professional and social pressure on the subjects. So, in that scenario, we cannot assume that the issue has a sleep disorder problem; in that case, we must analyze the sleep behavior of the subject in more than one session of recordings, which will help to reach one correct conclusion related to diagnosis. The model works on EEG epochs and is designed to classify two-class sleep state classification problems accurately. Hence the substantial motivation of this study is to improve the accuracy of sleep staging and organize sleep stages by monitoring their behavior using an enhanced machine learning model.

Our work differs from similar studies in that we obtained recordings during two different sleep sessions for subjects suffering from a sleep disorder. The current work assesses the classification performance of statistical models using from single-channel EEG signals from subjects of different genders with various symptoms of sleep problems. The results of the proposed system suggest that extracted time and frequency domain features are more relevant to classification than other features.

3 Experimental Data

The required data has been acquired at the Hospital of Coimbra University (CHUC) [25]. This proposed study evaluates three different medical condition subjects' details from the ISRUC-Sleep dataset. This dataset contains three different categories of subjects. In subgroup-I, 100 adult recordings with symptoms of mild sleep problems are analyzed using a data acquisition session. In subgroup-II, recordings of eight adults with symptoms of mild sleep problems were recorded during two data acquisition sessions. Finally, in sub-group-III, eight healthy controlled subjects were recorded. The authors have considered two subjects from each section of the ISRUC-Sleep data repository. One issue is male, and the other is female. The subjects for this study are Subject-16 and Subject-23 from Subgroup-I, Subject-03 and Subject-04 are selected from Subgroup-II, and Subject-02 and Subject-05 from Subgroup-III (Table 1).

Figure 1 presents the sample of the EEG signal of the subjects having sleep disorders. Figure 2 shows the sample behavior of the healthy controlled matter.

This study includes three categories of subjects. One is affected by sleep problems with one session recording sleep data. Two recording sessions extracted the second data category from subjects with sleep disorders. Finally, the third category includes healthy subjects. The authors have obtained the EEG channel of C3-A2 signals of six subjects. The stage annotations are also provided in the data repository according to AASM rules. The epochs are labeled as W, N1, N2, N3, and R for the Wake, N-REM1, N-REM2, N-REM3, and REM stages. The unscored epochs are not considered for further analysis in the experimental work. Each epoch is regarded as a 30 s-time length in this present work. To research sleep behavior, we have presented samples of all sleep stages of EEG signals extracted from different categories of subjects. The sleep EEG signals behavior changes according to the sleep cycle covered by the subject. Every step of sleep is characterized by other behavior from sleeping brains, such as low amplitude, mixed frequency, sawtooth waveforms, low amplitude in muscle movements, and rapid eye movements. However, the EEG signal behaviors are complex since they are not periodic and the continuous amplitude, frequency, and phase range change according to the sleep stage. The characteristic changes verified in the different sleep stages during sleep are represented in Figs. 3 and 4, respectively.

4 Proposed Automatic Sleep Stage Detection Method

Figure 5 provides a complete structure diagram of our proposed research work.

The simplified execution process is shown as follows: (i) First of all, the required physiological data are collected from the public sleep repository for sleep staging, (ii) Then the acquired EEG is not directly processed for sleep staging because it is contaminated with the different irrelevant noise compositions, so, we have considered a pre-processing step to further eliminate the noisy data, (iii) It's quite difficult to view the recorded waveform and take interpret and identify the changes in sleep behavior over the individual sleep stage. So we have extracted different types of features to study its behavior over the changes in time and frequency domains. It helps to make a certain decision on discriminating the irregularities over the individual sleep stages, (iv) Afterward, the extracted features sometimes may not be suitable for applying to the subject cases, so we have obtained the feature selection step for screening the best suitable features, and (v) Finally the selected features are fed into the classification model for classifying the sleep patterns.

Table 1 Description of the percentage of epoch per individual sleep stages

Subject ID/Session	Wake (%)	N-REM1 (%)	N-REM2 (%)	N-REM3 (%)	REM (%)
Subject-1 Subgroup-II/One	13.87	16.67	24.40	24.13	20.93
Subject-1 Subgroup-II/Two	45.60	16.27	10.80	16.93	10.40
Subject-2 Subgroup-II/Two	40.53	18.80	26.80	7.33	6.53
Subject-2 Subgroup-II/Two	30.40	8.13	26.67	23.60	11.20
Subject-3 Subgroup-II/One	8.93	16.93	36.13	23.33	14.67
Subject-3 Subgroup-II/Two	10.53	14.80	37.07	19.47	18.13
Subject-4 Subgroup-II/One	9.60	21.73	43.87	10.13	14.67
Subject-4 Subgroup-II/Two	14.80	34.27	27.87	10.53	12.53
Subject-5 Subgroup-II/One	8.93	8.67	38.27	33.47	10.67
Subject-5 Subgroup-II/Two	10.40	9.47	48.53	26.53	5.07
Subject-6 Subgroup-II/One	11.60	18.40	45.20	14.00	10.80
Subject-6 Subgroup-II/Two	16.53	18.53	40.93	15.60	8.40
Subject-1 Subgroup-III/One	19.87	12.13	35.60	21.07	11.33
Subject-2 Subgroup-III/One	40.53	18.80	26.80	7.33	6.53
Subject-5 Subgroup-III/One	67.00	8.67	38.27	33.47	10.67
Subject-6 Subgroup-III/One	7.20	14.80	34.80	32.93	10.27
Subject-7 Subgroup-III/One	27.47	7.07	20.53	34.67	10.27
Subject-8 Subgroup-III/One	45.73	9.73	15.20	15.47	13.87
Subject-9 Subgroup-III/One	13.47	19.33	31.47	30.00	5.60
Subject-10 Subgroup-III/One	16.80	29.07	23.07	13.07	18.00

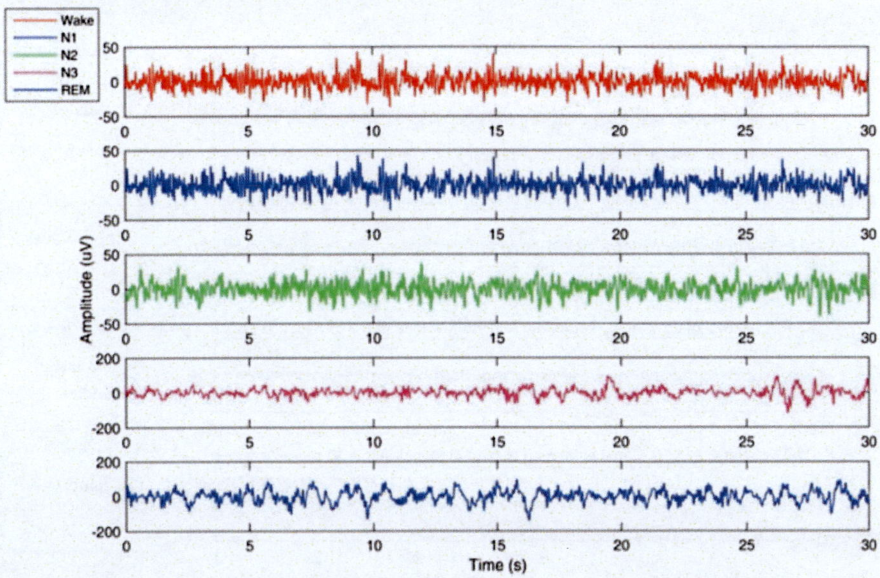

Fig. 1 Sleep-EEG signal sample of sleep syndrome problem, Subgroup-II subject-03

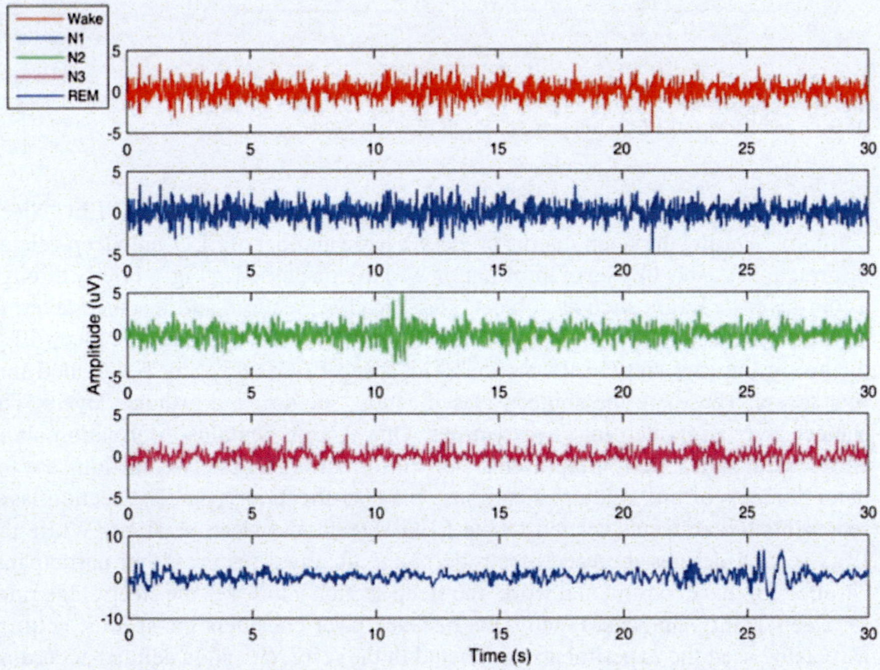

Fig. 2 Sample of a healthy subject, subgroup-III subject-05

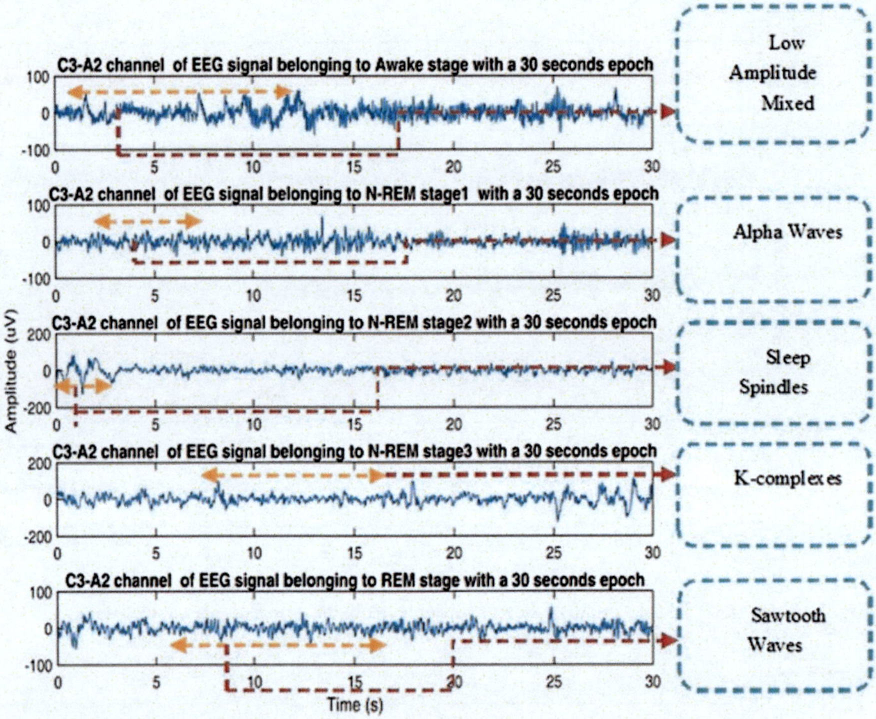

Fig. 3 C3-A2 channel of EEG signal for the different sleep stages: Awake, N-REM (stage1, stage2, stage3), and REM stage from subject-16 (Sleep Problem Subject)

Meanwhile, other sleep-related information is directly mapped to "EEG-Sleep ontology." During the sleep staging process, a large number of EEG and Sleep-related information is used, this set of information we have discussed in Fig. 6 briefly through a four-layer structure such as subject, classification, section, and instance layer. In this section, the authors have the purpose of information mapping for the Sleep-EEG method using raw signal information. The Sleep-EEG study can be divided into four layers. These are the subject, classification, section, and instance layers. The subject layer is divided into two sections. One section contains information about EEG, and another section represents sleep information. These two sections are the main domains of this study and compose the classification layer. The section layer represents the concepts related to the EEG signals and sleep sections. While the EEG section defines the used electrode name, obtained features, and participants enrolled for this experimental work, the sleep section addresses the sleep stage rules for sleep quality analysis. Finally, the instance layer considers the specific settings associated used the experiments conducted in the research and is defined according to the other layers. The block diagram in Fig. 6 presents a graphical representation of the executed activities during automatic sleep stage detection development diagnosis. As we have observed from the state-of-the-art works, all the extracted features may

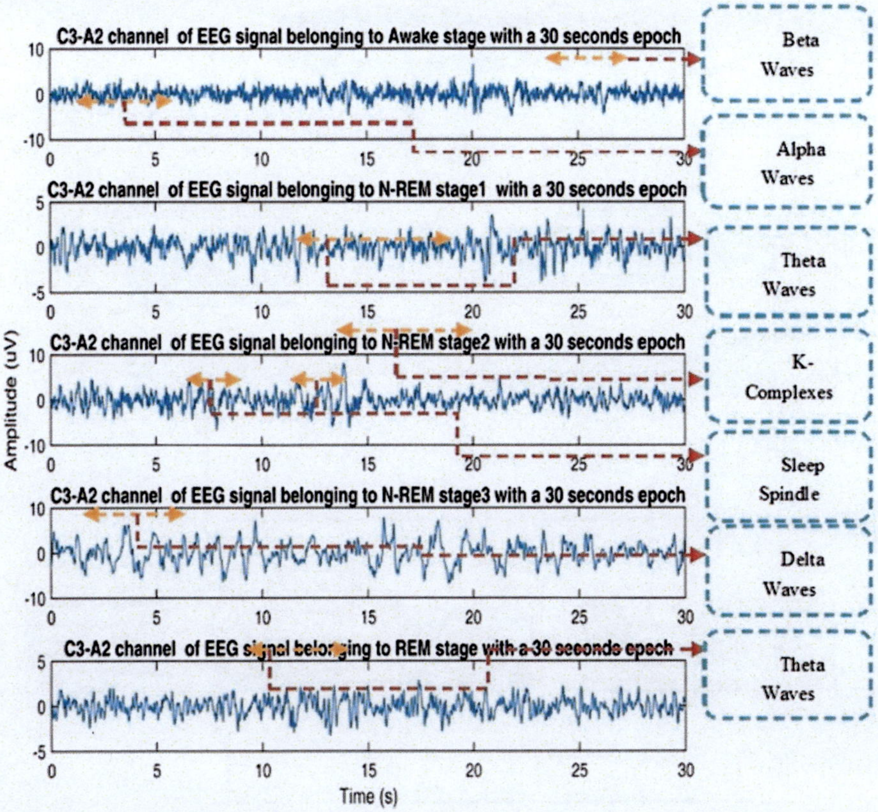

Fig. 4 C3-A2 channel of EEG signal for different stages behavior of sleep: Awake, N-REM (stage1, stage2, stage3), and REM stage from subject-05 (Healthy Subject)

sometime affect the classification performance due to contradictory behavior of the features. To find the most discriminative features, a feature screening step method Online Streaming Feature Selection (OSFS) technique is proposed and described in Algorithm 2, the fewer relevant feature types are eliminated.

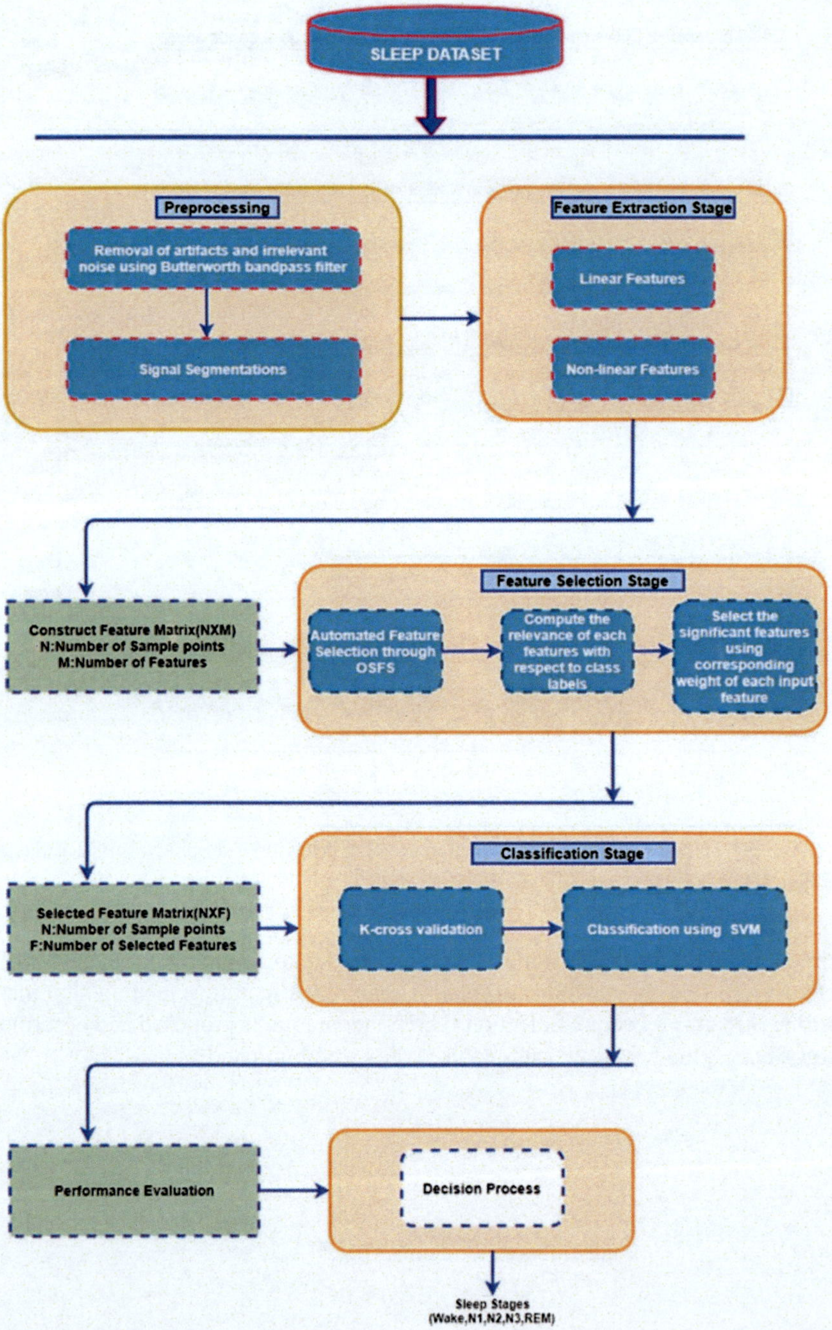

Fig. 5 Block diagram of the proposed method

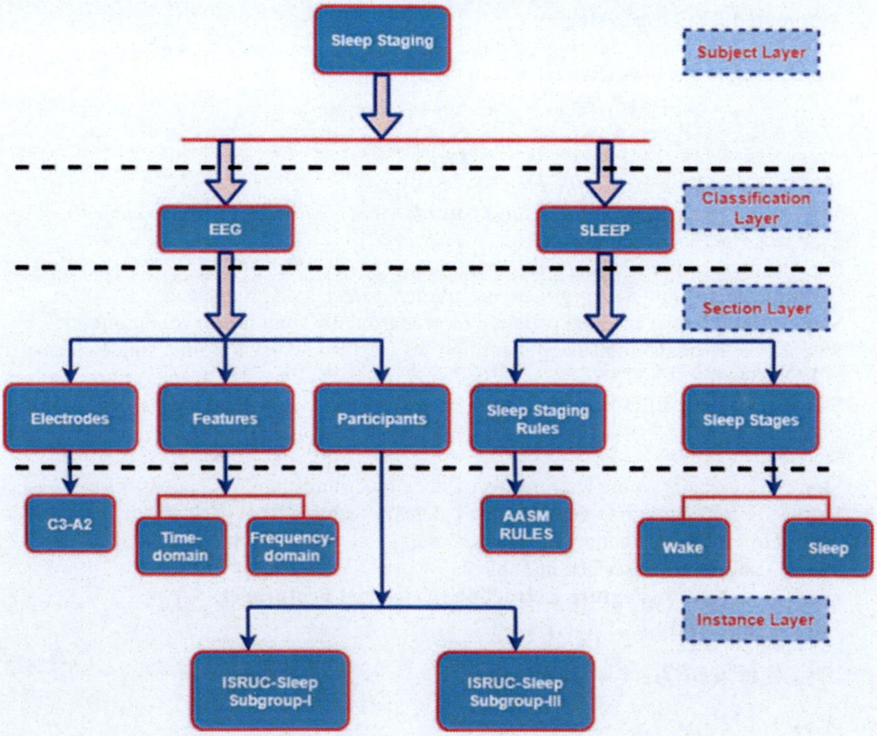

Fig. 6 The structural framework of the "Sleep-EEG ontology"

Algorithm 1

Automated EEG-Sleep staging
Procedure: *Sleepstage_Scoring = SleepStage_Classify (Feature List, Labels)*
Input: *Patient Records* [Extracted from Sleep_Dataset]

$$P_V = \{P_1, P_2, \dots\dots\dots, P_N\}$$
$$P_{SI} = \{S_1, S_2, \dots\dots\dots, S_M\}$$
$$P_{SE} = \{S_{I1}, S_{I2}, \dots\dots\dots, S_{IZ}\}$$

N=Number of Patients enrolled for EEG-Sleep test
M=Number of the Session recordings considered per individual enrolled subjects
Z=Number of Sleep stages epochs
Pre_Predict_Label=*Pre-predicted annotations of sleep stages*
Output: Sleepstage_Scoring=*Correct predicted sleep stage sequences*
Step-1: Let P be the enrolled patient's records from different health conditions
Step-2: Let K be the total sleep recordings of enrolled subjects from a single channel of EEG signal
Step-3: For each EEG recording from individual enrolled patients in the EEG-Sleep study, do
Step-4: P=P+1
Step-5: Divide the whole K segments of sleep recordings into 30s epochs-wise
Step-6: Extract both time and frequency domain features from each epoch and stored them into individual feature vectors per individual enrolled patient. The detailed extracted features are described in Table 2.

Feature Extraction (Extracted FeatureSet)

$$FV_{P1} = \{P1_{F1}, P1_{F2}, \dots\dots\dots, P1_{FE}\}$$
$$FV_{P2} = \{P2_{F1}, P2_{F2}, \dots\dots\dots, P2_{FE}\}$$
....
$$FV_{PN} = \{PN_{F1}, PN_{F2}, \dots\dots\dots, PN_{FE}\}$$
$$F_j = \{y_1, y_2, y_3, \dots\dots\dots\dots, y_T\}$$

E=number of feature types, T=number of feature elements
Step-7: Forwarded the extracted features to the proposed feature selection techniques and selected the suitable components and store them in the selected feature vector patients-wise. The detailed algorithm is discussed in Algorithm 2 and finally selected features are mentioned in Table 3.

Feature Selection (Selected FeatureSet)

$$FS_{P1} = \{P1_{y1}, P1_{y2}, P1_{y3}, \dots\dots\dots\dots, P1_{yT}\}$$
$$FS_{P2} = \{P2_{y1}, P2_{y2}, P2_{y3}, \dots\dots\dots\dots, P2_{yT}\}$$
....
$$FS_{PN} = \{PN_{y1}, PN_{y2}, PN_{y3}, \dots\dots\dots\dots, PN_{yT}\}$$

%Piyj: an element of feature selection set
Step-8: Forwarded selected feature vector list and predicted sleep stage segments into the proposed classification model
Sleepstage_Scoring=SleepStage_Classify (Feature_Selection_PN, Pre_Predict_Label)
Step-9: if the Sleep stage classification accuracy is suitable, then recommend the sleep stage scoring approach in the diagnosis of sleep-related problems; go to step 11
Step-10: else go to step 6 [For new feature extraction from sleep recordings of the enrolled subjects]
Step-11: end if
Output: *Automated Corrected Sleep Stage Sequences of the Subjects*

Algorithm 2

Pseudocode of the OSFS Feature Selection
Input: Target label Feature vector
Output: Best_Candidate_Featureset vector
Step-1: Initialization
Best_Candidate_Featureset= {}
Step-2: Initialize i=1
Step-3: for i to r
/*Relevance Analysis*/
Step-4: count: =0
/*Extracting new Features*/
Step-5: $X_i \leftarrow X_j \leftarrow$get_nextFeature ()
Step-6: if Dependent ($X_i == C$)
/*Add relevant features X_i to BCF*/
Step-7: Best_Candidate_Feature=Best_Candidate_Featureset $\cup X_i$
Step-8: count: =count+1
Step-9: end if
Online redundancy analysis/
Step-10: if (count: =True)
Step-11: if $\exists S \in BCF \backslash X_i$,tates that Ind ($X_i, C|S$)
* remove redundancy feature X_i*/
Step-12: BestCF=BCF-Y
Step-13: end if
Step-14: end for
Step-15: *end if*
Step-16: i=i+1
Step-17: Till the stopping condition satisfied
Step-18: Output BCF

5 Experimental Results and Discussion

In the study of this article, the developed automated sleep staging system is mainly composed of feature extraction, feature screening, and a binary class classification problem. The complete illustrations of the practical steps of this proposed work are shown in Fig. 7.

The entire experiments were executed using the MATLAB R2017a software. According to the different ways of dividing the datasets, one kind of test scheme is utilized and compared, named a subject-dependent test in this paper. Mainly the proposed methodology is conducted upon two different subgroups of sleep datasets with varying recordings of sessions. Epochs from all the six subjects from ISRUC-Sleep Subgroup-II and eight subjects from ISRUC-Sleep subgroup-III are independently together for tenfold cross-validation and the performance is evaluated on the level of epochs instead of subjects for both the subgroup categories. In this type of test, part of the testing data and the training data could come from the same subjects. It has been noticed that most of the sleep studies [24–35] preferably selected the C3-A2 channel for better monitoring the sleep behavior; therefore, in this work, we

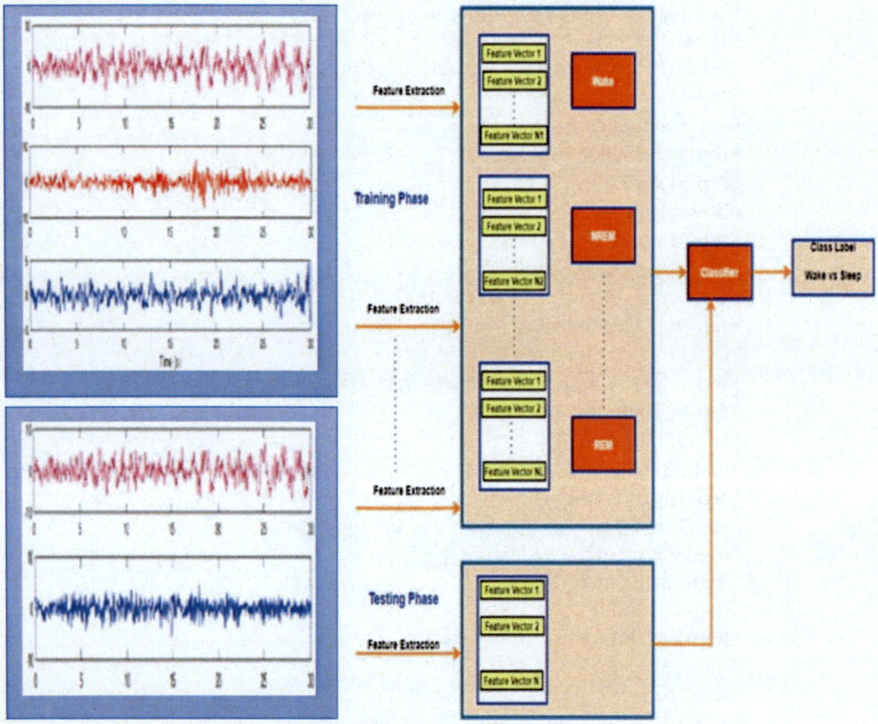

Fig. 7 An illustration of experimental procedures

have considered that in both the experiments, the input channel is regarded as the C3-A2 channel. Initially, we approached the signal pre-processing technique to eliminate the irrelevant noise compositions. The details on the extracted feature vector are presented in Table 2. We have selected the most discriminative features from the extracted feature vectors by implementing the Online Streaming Feature Selection technique. The set features for each subject are described in Table 3. Finally, the selected parts are classified through SVM classifiers.

5.1 Experiment 1: Sleep Staging with Sleep-Disordered Subjects

The complete result analysis with the subjects affected by the sleep-related diseases are shown in Table 4.

Table 2 Extracted features

Feature type	Feature name	Feature number
Time domain	Mean_Val	1
	Max_Val	2
	Min_Val	3
	Standard_Deviation_Val	4
	Median_Val	5
	Variance_Val	6
	Zero-Crossing Rate_Val	7
	75percentile_Val	8
	Skewness_Val	9
	Kurtosis_Val	10
	Hjorth parameters_Val	11–13
Frequency domain	Rel_Spec_Pow_Val	14–17
	Seven_Power Ratios_Val	18–24
	B_Power_Val	25–28

5.2 Classification Accuracy of Category-III Subject ISRUC-Sleep Database

The performances achieved for all the healthy controlled subjects are presented in Table 5.

It has observed that the proposed classification model achieved an overall accuracy of 92.5, 92.26, 91.73, 91.6, 96.53, 66.13, 95.06 and 93.06% through SVM classifier with ISRUC-Sleep subgroup-III data for subject-1, 2, 5, 6, 7, 8, 9 and 10 respectively. The highest precision was reported from subject-07 (97.26%). The specificity performance was the highest from subject-07 (92.71%). Furthermore, the F1-Score performance reports the highest value of 97.75% from subject-02. The kappa score for subject-02 and subject-07 is suitable with SVM classification models.

5.3 Summary of Results

The reported classification accuracy from different session recordings for all the three category subjects is presented in Table 7.

Table 3 Screened best candidate features

Subject	Selected features	Total
Subject1/Se1/Subgroup-II	$F1_{01\_1}, F5_{01\_1}, F6_{01\_1}, F7_{01\_1}, F8_{01\_1}, F11_{01\_1}, F12_{01\_1}, F15_{01\_1}, F16_{01\_1}$ $F1701_{\_1}, F18_{01\_1}$	11
Subject1/Se2/Subgroup-II	$F1_{01\_2}, F10_{01\_2}, F11_{01\_2}, F12_{01\_2}, F13_{01\_2}, F14_{01\_2}, F15_{01\_2}, F16_{01\_2}, F19_{01\_2}F22_{01\_2}$	10
Subject2/Se1 Subgroup-II	$F1_{02\_1}, F4_{02\_1}, F5_{02\_1}, F6_{02\_1}, F7_{02\_1}, F9_{02\_1}, F12_{02\_1}, F16_{02\_1}, F18_{02\_1}, F19_{02\_1}$	10
Subject2/Se2 Subgroup-II	$F1_{02\_2}, F3_{02\_2}, F4_{02\_2}, F5_{02\_2}, F10_{02\_2}, F13_{02\_2}, F17_{02\_2}, F18_{02\_2}, F28_{02\_2}$	9
Subject3/Se1 Subgroup-II	$F1_{03\_1}, F2_{03\_1}, F3_{03\_1}, F4_{03\_1}, F5_{03\_1}, F7_{03\_1}, F8_{03\_1}, F9_{03\_1}, F11_{03\_1}, F12_{03\_1}, F13_{03\_1}, F14_{03\_1}, F21_{03\_1}, F22_{03\_1}, F25_{03\_1}$	15
Subject-3/Se2 Subgroup-II	$F1_{03\_2}, F2_{03\_2}, F3_{03\_2}, F4_{03\_2}, F5_{03\_2}, F7_{03\_2}, F8_{03\_2}, F9_{03\_2}, F11_{03\_2}, F12_{03\_2}, F13_{03\_2}, F14_{03\_2}, F21_{03\_2}, F22_{03\_2}, F25_{03\_2}$	15
Subject-4/Se1 Subgroup-II	$F1_{04\_1}, F2_{04\_1}, F3_{04\_1}, F4_{04\_1}, F7_{04\_1}, F8_{04\_1}, F9_{04\_1}, F11_{04-1}, F12_{04\_1}, F13_{04\_1}, F17_{04\_1}, F21_{04\_1}, F22_{04\_1}, F25_{04\_1}$	14
Subject4/Se2 Subgroup-II	$F1_{04\_2}, F2_{04\_2}, F3_{04\_2}, F4_{04\_2}, F5_{04\_2}, F7_{04\_2}, F8_{04\_2}, F9_{04\_2}, F11_{04\_2}, F12_{04\_2}, F13_{04\_2}, F14_{04\_2}, F21_{04\_2}, F22_{04\_2}, F25_{04\_2}$	15
Subject5/Se1 Subgroup-II	$F1_{05\_1}, F5_{05\_1}, F6_{05\_1}, F8_{05\_1}, F13_{05\_1}, F16_{05\_1}, F17_{05\_1}, F18_{05\_1}, F28_{05\_1}$	9
Subject5/Se1 Subgroup-II	$F10_{5\_1}, F5_{05\_1}, F6_{05\_1}, F8_{05\_1}, F9_{05\_1}, F10_{05\_1}, F11_{05\_1}, F16_{05\_1}, F17_{05\_1}, F18_{05\_1}, F27_{05\_1}, F28_{05\_1}$	12
Subject6/Se1 Subgroup-II	$F1_{06\_1}, F4_{06\_1}, F5_{06\_1}, F6_{06\_1}, F7_{06\_1}, F8_{06\_1}, F17_{06\_1}, 8_{06\_1}, F22_{06\_1}, F24_{06\_1}$	10
Subject6/Se2 Subgroup-II	$F1_{06\_2}, F3_{06\_2}, F4_{06\_2}, F5_{06\_2}, F6_{06\_2}, F8_{06\_02}, F17_{06\_02}, F18_{06\_02}, F28_{06\_02}$	9
SubjectH1/Se1 Subgroup-III	$F1_{H\_01}, F5_{H\_01}, F6_{H\_01}, F8_{H\_01}, F10_{H\_01}, F16_{H\_01}, F17_{H\_01}, F18_{H\_01}, F22_{H\_01}$	9
SubjectH2/Se1 Subgroup-III	$F1_{02}, F2_{02}, F3_{02}, F4_{02}, F5_{02}, F7_{02}, F8_{02}, F9_{02}, F11_{02}, F12_{02}, F13_{02}, F14_{02}, F21_{02}, F22_{02}, F25_{02}$	15
SubjectH5/Se1 Subgroup-III	$F1_{05}, F2_{05}, F3_{05}, F4_{05}, F5_{05}, F7_{05}, F8_{05}, F9_{05}, F11_{05}, F12_{05}, F13_{05}, F14_{05}, F21_{05}, F22_{05}, F25_{05}$	15
SubjectH6/Se1 Subgroup-III	$F1_{H\_06}, F3_{H\_06}, F4_{H\_06}, F5_{H\_06}, F8_{H\_06}, F9_{H\_06}, F13_{H\_06}, F14_{H\_06}, F15_{H\_06}, F16_{H\_06}, F\ 17_{H\_06}, F18_{H\_06}, F22_{H\_06}$	13

(continued)

Table 3 (continued)

Subject	Selected features	Total
SubjectH7/Se1 Subgroup-III	$F1_{H\_07}$, $F5_{H\_07}$, $F6_{H\_07}$, $F7_{H\_07}$, $F8_{H\_07}$, $F10_{H\_07}$, $F16_{H\_07}$ $F18_{H\_07}$, $F28_{H\_07}$	9
SubjectH8/Se1 Subgroup-III	$F1_{H\_08}$, $F10_{H\_08}$, $F11_{H\_08}$, $F12_{H\_08}$, $F13_{H\_08}$, $F14_{H\_08}$, $F15_{H\_08}$, $F17_{H\_08}$, $F18_{H\_08}$, $F19_{H\_08}$	10
SubjectH9/Se1 Subgroup-III	$F1_{H\_09}$, $F4_{H\_09}$, $F5_{H\_09}$, $F6_{H\_09}$, $F7_{H\_09}$, $F8_{H\_09}$, $F10_{H\_09}$, $F12_{H\_09}$, $F13_{H\_09}$, $F16_{H\_09}$, $F17_{H\_09}$	11
SubjectH10/Se1 Subgroup-III	$F1_{H\_10}$, $F5_{H\_10}$, $F6_{H\_10}$, $F8_{H\_10}$, $F10_{H\_10}$, $F13_{H\_10}$, $F15_{H\_10}$, $F17_{H\_10}$, $F18_{H\_10}$, $F22_{H\_10}$	10

Table 4 Performance metrics result with ISRUC-Sleep subgroup-II data

	Pre (%)	Rec (%)	Spe (%)	F1-Sc (%)	Accu (%)	Kappa coefficient
Subject1_Se1	94.63	98.18	69.43	94.42	92.74	0.54
Subject1_Se2	96.74	97.82	97.18	96.44	95.75	0.67
Subject2_Se1	93.19	89.77	87.81	91.42	89.8	0.79
Subject2_Se2	95.72	97.71	87.88	96.72	94	0.89
Subject3_Se1	98.29	98.49	92.31	98.84	95.94	0.91
Subject3_Se2	97.51	100	85.46	98.68	98.74	0.90
Subject4_Se1	98.52	99.33	91.75	97.41	97.47	0.92
Subject4_Se2	98.98	99.16	93.58	98.57	98.73	0.82
Subject5_Se1	97.66	98.22	66.66	97.92	94.42	0.66
Subject5_Se2	97.74	96.64	72.78	98.19	94.95	0.72
Subject6_Se1	94.60	97.29	51.41	96.42	92.74	0.54
Subject6_Se2	97.83	99.18	84.86	97.47	95.74	0.85

Table 5 Performance result of the proposed Sleep study with healthy data using the SVM classifier

	Pre (%)	Rec (%)	Spe (%)	F1-Sc (%)	Accu (%)	Kappa coefficient
subjectH-1	93.32	97.67	71.81	95.44	92.53	0.74
SubjectH-2	96.82	98.70	84.84	97.75	92.26	0.86
SubjectH-5	93.61	97.28	49.42	95.41	91.73	0.53
SubjectH-6	92.82	98.56	1	95.60	91.6	0.007
SubjectH-7	97.26	97.97	92.71	97.61	96.53	0.91
SubjectH-8	70.73	64.12	68.51	67.26	66.13	0.32
SubjectH-9	96.08	98.30	74.25	97.18	95.06	0.77
SubjectH-10	95.85	96.47	79.36	96.16	93.6	0.76

Table 7 Overall accuracy performance values with ISRUC-Sleep Subgroup-II/III datasets

Subject Category	Subject Number	SVM (%)
ISRUC-Sleep/Subgroup-II Two Session Recordings	Subject1_Se1	91.73
	Subject_Se2	95.73
	Subject2_Se1	88.8
	Subject2_Se2	94
	Subject3_Se1	96.93
	Subject3_Se2	97.73
	Subject4_Se1	97.46
	Subject4_Se2	97.73
	Subject5_Se1	94.40
	Subject5_Se1	94.93
	Subject6_Se1	91.73
	Subject6_Se2	95.73
ISRUC-Sleep/Subgroup-III One Session Recordings	SubjectH-1	92.53
	SubjectH-2	92.26
	SubjectH-5	91.73
	SubjectH-6	91.6
	SubjectH-7	96.53
	SubjectH-8	66.13
	SubjectH-9	95.06
	SubjectH-10	93.6

6 Conclusion

The main reasons for the high accuracy of our proposed method are as follows. First, we adopted better feature screening techniques by combining an SVM classifier. The time and frequency domain information can be fully extracted to realize a more accurate feature extraction to study the behavior of the changes in sleep behavior. Finally, our proposed model was employed to classify two sleep states. Compared with that achieved on the two-classification task, our model's accuracy was somewhat improved. Moreover, our model required only a single-channel of EEG signals. On the one hand, our method has high accuracy and certain practicability and application value. In addition to the high precision, through the confusion matrix obtained from the two experimental results, the model proposed by the paper has high sleep staging accuracy in each sleep stage, which also shows that our method has vital for the clinical trial. In addition, it is worth mentioning that a notable feature of our proposed approach to be included in our proposed model is similar to other processing methods for other physiological signals (such as EOG, EMG, and ECG signals) shortly, which means that it may be suitable for application in other sleep-related disorders using such biological signal fields.

References

1. Panossian LA, Avidan AY (2009) Review of Sleep Disorders. Med Clin N Am 93:407–425
2. Smaldone A, Honig JC, Byrne MW (2007) Sleepless in America: inadequate sleep and relationships to health and well-being of our nation's children. Pediatrics 119:29–37
3. Hassan AR, Hassan Bhuiyan MI (2016) Automatic sleep scoring using statistical features in the EMD domain and ensemble methods. Biocybern. Biomed Eng 36(1):248–255
4. Satapathy SK, Loganathan D, Nanda R (2021) Automated analysis and classification of sleep stages based on machine learning techniques from a dual-channel EEG signal. In: Das B, Patgiri R, Bandyopadhyay S, Balas VE (eds) Modeling, simulation and optimization. smart innovation, systems and technologies, vol 206. Springer, Singapore
5. Obayya M, AbouChadi F (2014) Automatic classification of sleep stages using EEG records based on Fuzzy c-means (FCM) algorithm. In: Radio Science Conference (NRSC), pp 265–272
6. Alickovic E, Subasi A (2018) Ensemble SVM method for automatic sleep stage classification. IEEE Trans Instrum Mea 67(6):1258–1265
7. Satapathy SK, Bhoi AK, Loganathan D, Khandelwal B, Barsocchi P (2021) Machine learning with ensemble stacking model for automated sleep staging using dual-channel EEG signal. Biomed Signal Process Control 69:102898
8. Rechtschaffen A, Kales A (1968) A manual of standardized terminology, techniques and scoring systems for sleep stages of human subjects. U.G.P. Office, Public Health Service, Washington, DC, USA
9. Iber C, Ancoli-Israel S, Chesson AL, Quan SF (2007) The AASM manual for the scoring of sleep and associated events: rules, terminology and technical specification. Am Acad Sleep Med
10. Satapathy SK, Loganathan D, Machine Learning approaches with heterogeneous ensemble learning stacking model for automated sleep staging. Int J Comput Digit Syst Univ Bahrain J. https://doi.org/10.12785/ijcds/100109
11. Satapathy SK, Loganathan D (2021) Prognosis of automated sleep staging based on two-layer ensemble learning stacking model using single-channel EEG signal. Soft Comput 25:15445–15462
12. Yan R, Zhang C, Spruyt K, Wei L, Wang Z, Tian Lili, Li X, Ristaniemi T, Zhang J, Cong F (2019) Multi-modality of polysomnography signals' fusion for automatic sleep scoring. Biomed Signal Process Control 49:14–23
13. Shen H, Ran F, Xu M, Guez A, Li A, Guo A (2020) An automatic sleep stage classification algorithm using improved model based essence features. Sensors 20(17):4677
14. Huang W, Guo B, Shen Y, Tang X, Zhang T, Li D, Jiang Z (2020) Sleep staging algorithm based on multichannel data adding and multifeature screening. Comput Methods Programs Biomed 187:105253
15. Ghimatgar H, Kazemi K, Helfroush MS, Aarabi A (2019) An automatic single-channel EEG-based sleep stage scoring method based on hidden Markov model. J Neurosci Methods 108320
16. Satapathy SK, Loganathan D, Sharathkumar S, Narayanan P (2022) A machine learning model for automated classification of sleep stages using polysomnography signals. In: Khanna A, Gupta D, Bhattacharyya S, Hassanien AE, Anand S, Jaiswal A (eds) International conference on innovative computing and communications. Advances in intelligent systems and computing, vol 1387. Springer, Singapore
17. Diykh M, Li Y, Abdulla S (2020) EEG sleep stages identification based on weighted undirected complex networks. Comput Methods Programs Biomed 184:105116
18. Sriraam N, Padma Shri TK, Maheshwari U (2018) Recognition of wake-sleep stage 1 multichannel EEG patterns using spectral entropy features for drowsiness detection. Australas Phys Eng Sci Med 39(3):797–806
19. Memar P, Faradji F (2018) A novel multi-class EEG-based sleep stage classification system. IEEE Trans Neural Syst Rehabil Eng 26(1):84–95

20. Da Silveira TLT, Kozakevicius AJ, Rodrigues CR (2016) Single-channel EEG sleep stage classification based on a streamlined set of statistical features in wavelet domain. Med Biol Eng Compu 55(2):343–352
21. Wutzl B, Leibnitz K, Rattay F, Kronbichler M, Murata M (2019) Genetic algorithms for feature selection when classifying severe chronic disorders of consciousness. PLOS ONE 14(7)
22. Zhu G, Li Y, Wen PP (2014) Analysis and classification of sleep stages based on difference visibility graphs from a single-channel EEG signal. IEEE J Biomed Health Inform 18(6):1813–1821
23. Satapathy SK, Loganathan D, Automated Sleep stage classification based on multiple channels of electroencephalographic signals using machine learning algorithm. In: Dutta M, Krishna C, Kumar R, Kalra M (eds) Proceedings of international conference on IoT inclusive life (ICIIL 2019), NITTTR Chandigarh, India. Lecture notes in networks and systems, vol 116. Springer, Singapore
24. Satapathy SK, Loganathan D, Effect of linear features to determination of sleep stages classification from dual channel of EEG signal using machine learning techniques. In: Mandal J, Mukhopadhyay S, Roy A (eds) Applications of Internet of Things. Lecture notes in networks and systems, vol 137. Springer, Singapore
25. Khalighi S, Sousa T, Santos JM, Nunes U (2016) ISRUC-Sleep: a comprehensive public dataset for sleep researchers. Comput Methods Programs Biomed 124:180–192
26. Eskandari S, Javidi MM (2016) Online streaming feature selection using rough sets. Int J Approx Reason 69:35–57
27. Ilhan HO, Bilgin G (2017) Sleep stage classification via ensemble and conventional machine learning methods using single channel EEG signals. Int J Intell Syst Appl Eng 5(4):174–184
28. SandersTH, McCurry M, Clements MA (2014) Sleep stage classification with cross fre-quency coupling. In: 36th Annual International Conference IEEE Engineering in Medicine and Biology Society (EMBC), pp 4579–4582
29. Bajaj V, Pachori R (2013) Automatic classification of sleep stages based on the time-frequency image of EEG signals. Comput Methods Programs Biomed 112(3):320–328
30. Hsu Y-L, Yang Y-T, Wang J-S, Hsu C-Y (2013) Automatic sleep stage recurrent neural classifier using energy features of EEG signals. Neurocomputing 104:105–114
31. Zibrandtsen I, Kidmose P, Otto M, Ibsen J, Kjaer TW (2016) Case comparison of sleep features from ear-EEG and scalp-EEG. Sleep Sci 9(2):69–72
32. Berry RB, Brooks R, Gamaldo CE, Hardsim SM, Lloyd RM, Marcus CL, Vaughn BV (2014) The AASM manual for the scoring of sleep and associated events: rules, terminology and technical specifications. Am Acad Sleep Med
33. Sim J, Wright CC (2005) The kappa statistic in reliability studies: use, interpretation, and sample size requirements. Phys Ther 85(3):257–268
34. Liang S-F, Kuo C-E, Kuo Y, Hu, Cheng Y-S (2012) A rule-based automatic sleep stag-ing method. J Neurosci Methods 205(1):169–176
35. Satapathy SK, Loganathan D, Mitra A (2020) Effect of EEG dual-channel acquisition and gender specification subjects on the classification of sleep stages using machine learning techniques. In: Gupta S, Sarvaiya J (eds) Emerging technology trends in electronics, communication and networking. ET2ECN 2020. Communications in computer and information science, vol 1214. Springer, Singapore

Deep Learning-Based Prediction of Time-Series Single-Cell RNA-Seq Data

Dibyendu Bikash Seal, Sawan Aich, Vivek Das, and Rajat K. De

Abstract Single-cell transcriptomics or RNA-sequencing (scRNA-seq) technology enables unprecedented molecular profiling and provides novel insights into cellular heterogeneity and cell states, which bulk RNA-sequencing fails to capture. Transcriptomics data profiled at multiple time-points during developmental or disease stages, also known as time-series scRNA-seq, is now being increasingly available. Analysis of such data allows extraction of informative genes as compared to a static snapshot drawn at a single time-point. However, most of the present-day models ignore the time dimension within the data. This limits accurate inferencing of cellular trajectory and states encoded by the underlying genes. Utilizing the available time component could yield useful information regarding the underlying mechanisms at play during development and disease. Forecasting gene expressions for an advanced time-point is also necessary when data is degradable or missing. Hence, in this work, we have attempted to develop a deep neural network (DNN)-based prediction model for estimating gene expression values in time-series scRNA-seq data. The DNN regressor is capable of estimating data at advanced time-points of interest once trained on

D. B. Seal (✉)
A. K. Choudhury School of Information Technology, University of Calcutta, JD 2, Sector 3, Saltlake, Kolkata 700106, India
e-mail: dbsakc@caluniv.ac.in

S. Aich
Ramakrishna Mission Vivekananda Educational and Research Institute, PO Belur Math, Dist Howrah 711202, West Bengal, India

V. Das
Novo Nordisk A/S, Novo Nordisk Park 1, 2760 Maløv, Denmark

R. K. De
Machine Intelligence Unit, Indian Statistical Institute, 203 Barrackpore Trunk Road, Kolkata 700108, India
e-mail: rajat@isical.ac.in

R. Chaki et al. (eds.), *Applied Computing for Software and Smart Systems*,
Lecture Notes in Networks and Systems 555,
https://doi.org/10.1007/978-981-19-6791-7_13

multiple previous time-points. The performance of the proposed model compares favorably with some state-of-the-art regression models that exist at present.

Keywords Time-series · Single-cell RNA-seq · Gene expression · Deep learning

1 Introduction

In order to understand the biological mechanisms characterizing development and disease states, we need to model the cellular landscapes of an organism. Such level of precision and granularity is presently supported by single-cell omics technologies [5, 13]. They help us identify the overall cell populations, differential cell states, specific cell populations contributing to the development and disease states, and underlying gene signatures that drive either of the above [8]. However, most recent single-cell studies [11, 14] are based on a static snapshot of the developmental or disease stage(s) or at best, have a cross-sectional design of healthy and disease conditions, without proper longitudinal measurements. These studies though enable trajectory modeling or lineage tracing, they do not consider the inherent temporal behavior. Hence, they are often referred to as pseudotime analyses, which may capture progenitor progression into differentiated cells. Even though these studies have contributed immensely to the understanding of the transcriptomic continuum, these designs lack the time-to-event component.

Such limitations are presently addressed by new emerging single-cell time-series experimental designs and datasets. Time-series experiments [12] help us order cellular events as a function of time, thus enabling robust trajectory inferencing by identifying paths that the cells follow. The presence of multiple time-points also allows the investigation of stochastic changes in gene expression. Currently, multiple time-series scRNA-seq methods exist, which can determine specific developmental trajectories [12, 15] and regulatory mechanisms during cell differentiation [9]. However, when there are missing time components in time-series data, or the expression values profiled at a time-point of interest are not of good quality, analysis results can be compromised. Research is still limited when it comes to predicting an advanced time-point where an event of interest might occur during the course of cell development or decay. Currently, robust deep learning methods that can deal with the problem of estimating gene expression values for time-series single-cell RNA-seq datasets are missing. Drawing a novel inference is thus not always possible using existing regression models. This impacts the overall modeling of the gene expression landscape as a function of time.

Hence, in this work, a DNN-based regression model has been developed for predicting gene expression values for an advanced time-point in time-series scRNA-seq datasets. The proposed DNN regressor model has been evaluated on two publicly available datasets. Results compare favorably with other existing regression models.

2 Methodology

The methods used in this work for data acquisition and preprocessing, and gene expression estimation in time-series scRNA-seq data, have been explained in this section.

2.1 *Data Acquisition and Pre-processing*

The two publicly available datasets E-MTAB-7606 and E-GEOD-75140 used in this scope of work have been downloaded as normalized matrices from https://www.ebi.ac.uk/gxa/sc/home. These datasets have undergone further preprocessing as explained below.

- For each dataset, each cell has been subjected to min-max scaling which ensures that the gene expression values lie in [0, 1].
- For each assay, the metadata has been merged with the gene expression values.
- In order to remove genes lowly expressed across cells, we have discarded those falling under the first quantile.
- Assays have been replaced with time-points at which they were collected, in order to transform the dataset into ($gene \times time\text{-}points$) matrix.

For E-MTAB-7606 data, SAMEA5215593 assay has been dropped since its metadata is not available. The final number of genes obtained after pre-processing for E-MTAB-7606 and E-GEOD-75140 datasets are 10386 and 19037, respectively. Table 1 shows a summary of the time-series scRNA-seq datasets that have been used in this work.

2.2 *Deep Neural Network (DNN)-Based Methodology*

In this work, we have proposed a DNN-based regressor for predicting time-series scRNA-seq data. Once trained on data from multiple time-points, the deep neural network (DNN) regressor is capable of estimating gene expression values at advanced time-points of interest.

2.2.1 Multi-output Regression Using DNN

The DNN regressor used in this work is a multi-layer perceptron (MLP) model. An MLP is a feedforward artificial neural network with fully connected layers (input, hidden, and output). It is a supervised algorithm that is often used for the task of classification or regression (prediction). Usually, the task of prediction typically

Table 1 Detailed information regarding the datasets (time-series, scRNA-seq) used for method evaluation

Dataset	Description	#Cells	#Genes	#Time-points	Source
E-MTAB-7606	Single-cell RNA-seq of influenza tetramer specific Human CD8+ T cells from baseline, acute and long term memory time-points	210	14162	62 (Pre-infection), 66 (14th day), 15 (3rd month), 66 (1.5th year)	Koutsakos et al. [7]
E-GEOD-75140	Human cerebral organoids recapitulate gene expression programs of fetal neocortex development	734	25592	96 (53rd day), 80 (65th day), 79 (58th day), 74 (41st day), 71 (37th day), 68 (35th day), 40 (33rd day), 226 (Unknown)	Camp et al. [1]

employs a nonlinear activation function at the nodes (neurons) in the hidden layer(s) but a linear activation function at the nodes (neurons) in the output layer. Thus, the MLP accepts as input the input vector $\mathbf{x} \in \Re^n$ and maps it to an output $\mathbf{y} \in \Re^m$, where n denotes the number of time-points at the input and m denotes the number of time-points at the output.

In this work, the preprocessed ($gene \times time\text{-}points$) matrix obtained from each dataset has been fed as input to the DNN regressor. Let $\mathbf{D}$ be the training dataset dealing with expression values of k genes $\mathbf{x}_1, \mathbf{x}_2, \ldots, \mathbf{x}_k$, where each gene $\mathbf{x}_i$ is represented by a vector of n variables (time-points) $\mathbf{x}_i = [x_{i1}, x_{i2}, \ldots, x_{it}, \ldots, x_{in}]^T$. The output layer of the DNN regressor consists of the same number of neurons as the number of target variables corresponding to the advanced time-points to be predicted. Thus, for each gene $\mathbf{x}_i$, the output vector predicted by the DNN regressor consists of m target variables $\mathbf{y}_i = [y_{i1}, y_{i2}, \ldots, y_{it}, \ldots, y_{im}]^T$, which are the expression values at m time-points. For E-MTAB-7606 dataset, 66 time-points representing the data from 1.5 year timeframe have been considered as the target variables. These time-points represent an advanced stage of Influenza B as can be seen in Fig. 1. For the E-GEOD-75140 dataset, the target variables are 226 time-points that represent the final stage of a developed neuron as shown in Fig. 1.

The deep learning-based regression model has been trained using 'Adam' optimizer [6] for 10 epochs, with Mean-squared error (MSE) as the loss function. ReLU activation function has been used at the nodes in the hidden layers, while a linear activation function has been used at the nodes in the output layer. In order to prevent

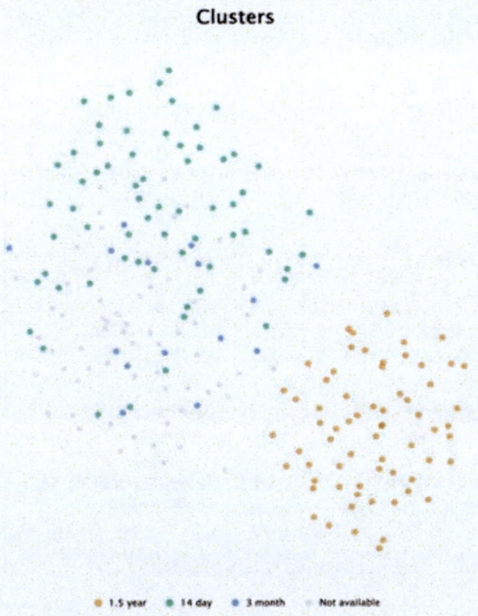

(a) Time

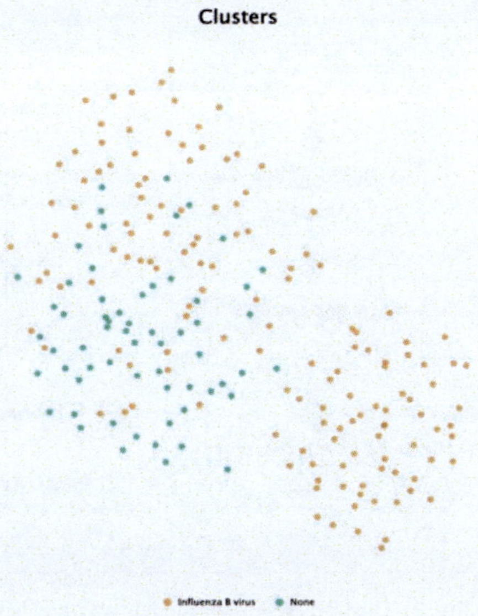

(b) Disease stage

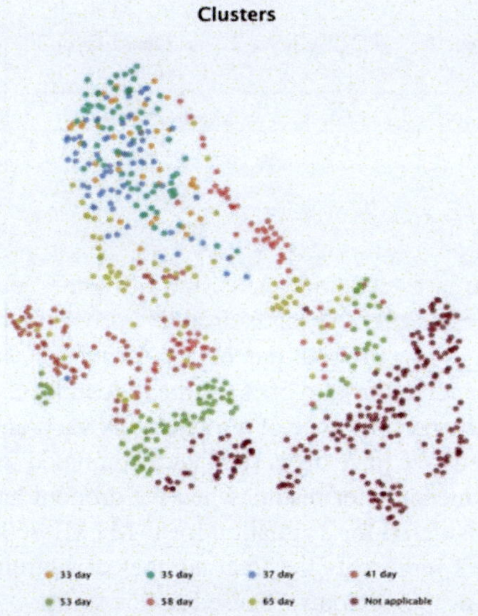

(c) Time

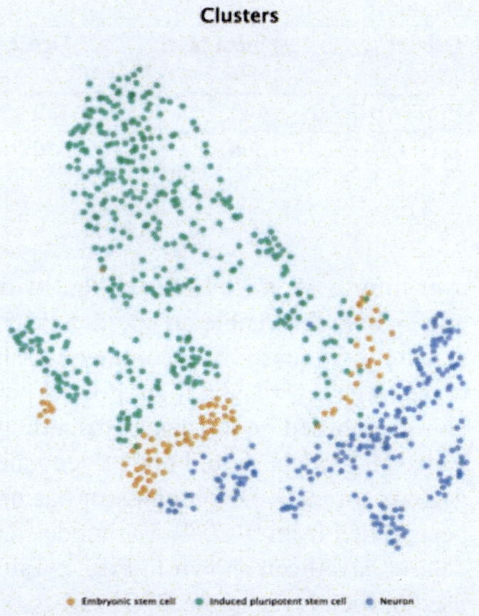

(d) Developmental stage

Fig. 1 **a** and **b** shows the UMAP representation of the cells in the E-MTAB-7606 dataset with respect to time points and the stages of disease Influenza B, whereas, **c** and **d** represents the UMAP representation of the cells in the E-GEOD-75140 dataset with respect to time and the stages of neuron development

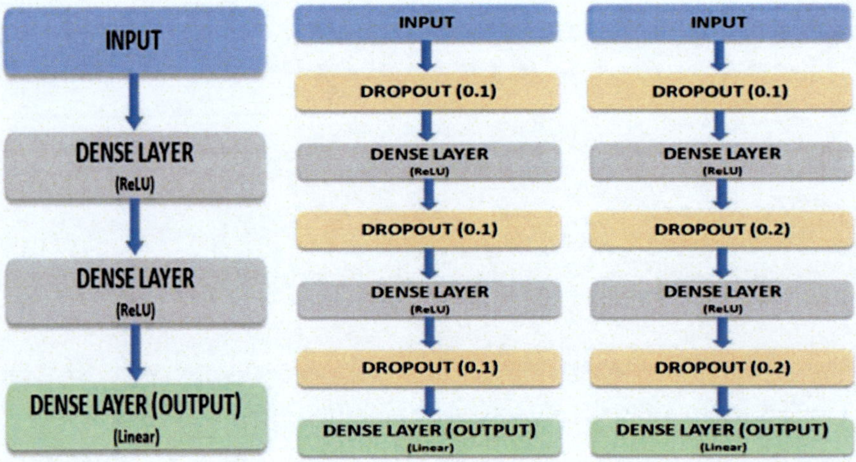

(a) DNN regressor with no dropout (b) DNN regressor with 10% dropout (c) DNN regressor with 10% dropout (input layer), 20% dropout (hidden layers)

Fig. 2 The DNN regressor models with and without dropout considered for evaluation in this work

Table 2 Total number of neurons in each layer of the DNN regressor for the two datasets used in this study

Dataset	Input layer	Dense layer 1	Dense layer 2	Dense layer 3 (Output)
E-MTAB-7606	143	130	75	66
E-GEOD-75140	508	350	275	226

over-fitting, we have used dropout which is a technique to switch off some neurons in both the visible and hidden layers randomly. As far as dropout is concerned, existing literature [10] recommends using a low dropout percentage (around 20%) at the input layer and a high dropout percentage (around 50%) at the hidden layers. However, based on the data distribution, the performance of a model may vary, and hence all models should not be generalized. We have observed that when tuned for performance, the DNN regressor has produced better results when the dropout has been varied from 0–20%. The models considered for evaluation for E-MTAB-7606 dataset have been shown in Fig. 2. Table 2 represents the total number of neurons used in each layer of the DNN regressor for the two datasets used in this study.

2.2.2 Performance Comparison of the proposed DNN-Based Regression Model with Other Existing State-of-the-Art Regression Models

The performance of the proposed DNN regressor has been compared with six other state-of-the-art regression models, viz., Linear regression, Ridge regression, Lasso regression, k-Nearest Neighbors (k-NN) regression, and Random Forest (RF) regression. Lasso is a linear regression model that adds sparsity to the coefficients by using an L1 penalty term, while Ridge linear model uses a regularization on the size of the coefficients by using an L2 penalty term [3]. Both Lasso and Ridge regression models are suited for making predictions closely comparable with true values. k-NN is another non-parametric method that works by finding the k-closest training examples to the test data [2]. Random forest [4], on the other hand, is an ensemble method that is used widely for the task of solving both classification and regression problems. All these above regression methods are often used as benchmark methods and represent the current state of the art.

3 Results

The DNN-based regression model developed in this work has been used to predict gene expression values for advanced time-points based on those at previous time-points. The problem of gene expression estimation from time-series scRNA-seq data has been modeled here as a multi-output/multi-target/multi-variate regression problem. In this work, for evaluation of the proposed regression model, we have considered three different models as shown in Fig. 2. The preprocessed datasets have been train-test split following an 80:20 ratio. Out of the training samples, 20% of them have been used for validation. The performance of the proposed regression model has been initially evaluated using Mean-Squared Error (MSE) and R^2 values. MSE in a regression problem is given by

$$MSE = \frac{1}{n}\sum_{i=1}^{n}(y_i - \tilde{y}_i)^2, \tag{1}$$

where y_i and $\tilde{y}_i$ are the actual and the estimated values, respectively, and n is the number of samples. R^2 statistic, on the other hand, provides a measure of the goodness of fit. It represents the ratio of variance explained and lies in [0, 1]. If the residual sum of squares (RSS) is given by $RSS = \sum_{i=1}^{n}(y_i - \tilde{y}_i)^2$, then R^2 can be evaluated as

$$R^2 = \frac{TSS - RSS}{TSS} = 1 - \frac{RSS}{TSS}, \tag{2}$$

where $TSS = \sum(y_i - \overline{y})^2$ is the total sum of squares and $\overline{y} = \frac{1}{n}\sum_{i=1}^{n} y_i$ is the mean of the observed values.

As shown in Fig. 3, for E-MTAB-7606 dataset, all the three DNN models have produced better R^2 values than all other regression methods used for comparison, with 'DNN (10% dropout–input, 20% dropout–hidden)' topping the chart. For E-GEOD-75140 dataset, the best R^2 value has been obtained using the 'DNN (no dropout)' model. When evaluated for the MSE scores, we have once again observed

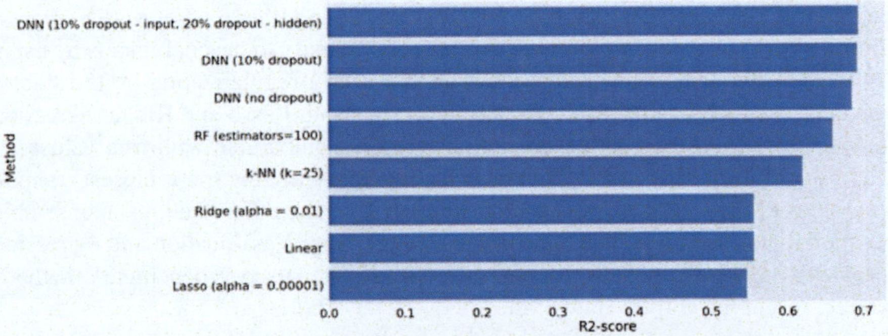

(a) R^2 values for E-MTAB-7606

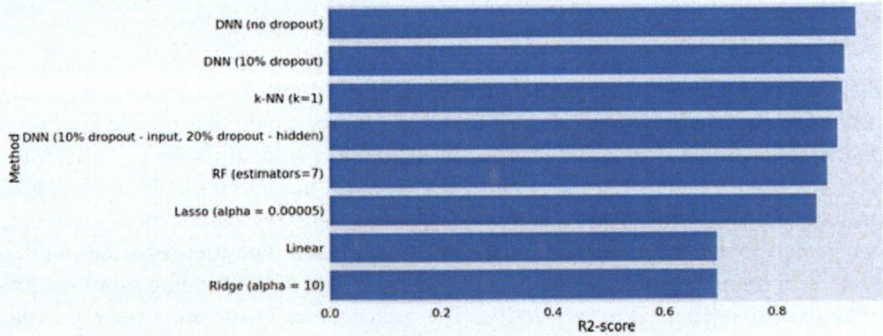

(b) R^2 values for E-GEOD-75140

Fig. 3 **a** shows R^2 values obtained using the proposed DNN-based regression model as compared to that obtained using other state-of-the-art regression methods for E-MTAB-7606 dataset; **b** shows similar results obtained for E-GEOD-75140 dataset

that all three DNN models have produced smaller MSE values as compared to other benchmark regression methods, with the 'DNN (10% dropout–input, 20% dropout–hidden)' model at the top once more for E-MTAB-7606 dataset. For E-GEOD-75140 dataset, the RF model (with 100 estimators) has, however, produced the best MSE value, while the results obtained using the proposed DNN models compare favorably with the others. This part of evaluation has been depicted in Fig. 4 which shows the MSE values produced by the proposed DNN regressor and other methods used for comparison, ranked by neglog(MSE).

Subsequently, we have tried to explore the amount of correlation between the observed and the predicted data for both the time-series scRNA-seq datasets. The Pearson correlation coefficient has been used for this purpose. As shown in Fig. 5, the test examples for both E-MTAB-7606 and E-GEOD-75140 datasets, when predicted using the proposed DNN-based models have shown better correlation than those predicted using the other methods.

Finally, we have ensured that the model is robust by repeating the train-test split multiple times and recording R^2, MSE values, and the Pearson correlation coefficients

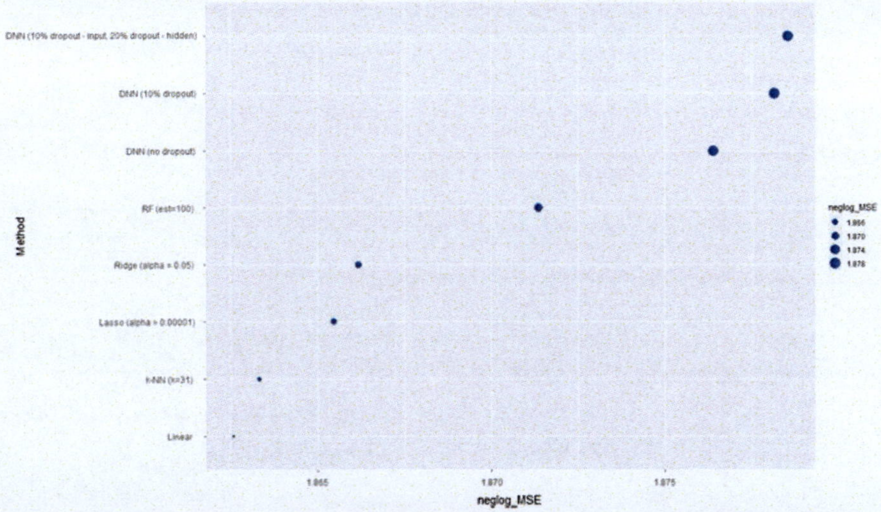

(a) MSE values for E-MTAB-7606 ordered by neglog(MSE)

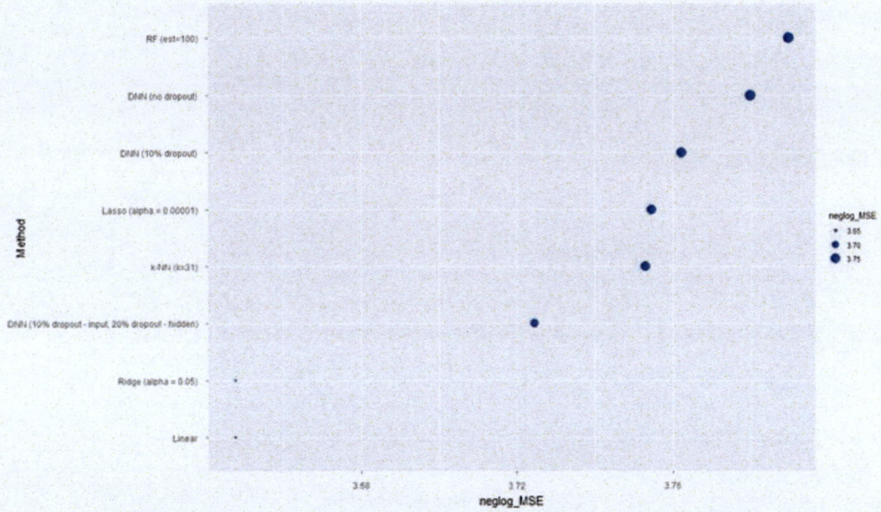

(b) MSE values for E-GEOD-75140 ordered by neglog(MSE)

Fig. 4 **a** shows MSE values obtained using the proposed DNN-based regression model as compared to that obtained using other state-of-the-art regression methods for E-MTAB-7606 dataset ordered by neglog(MSE); **b** shows similar results obtained for E-GEOD-75140 dataset ranked by neglog (MSE)

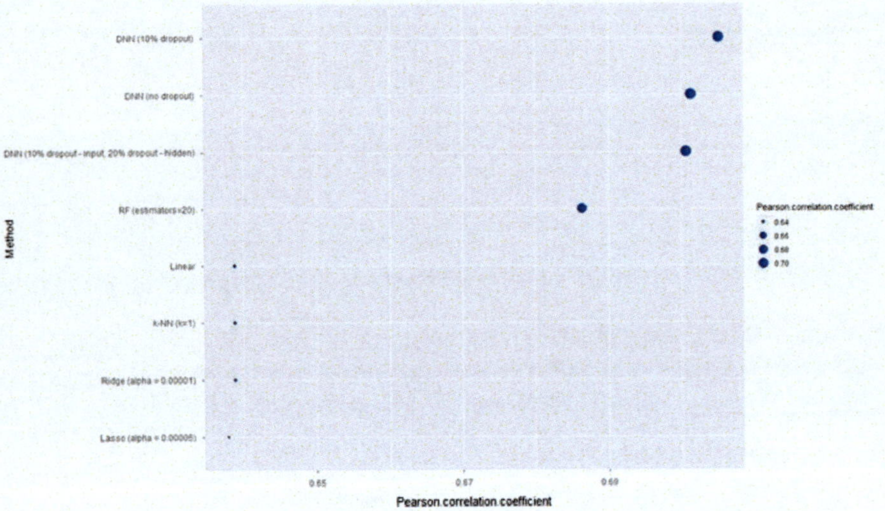

(a) Pearson correlation coefficients obtained for E-MTAB-7606

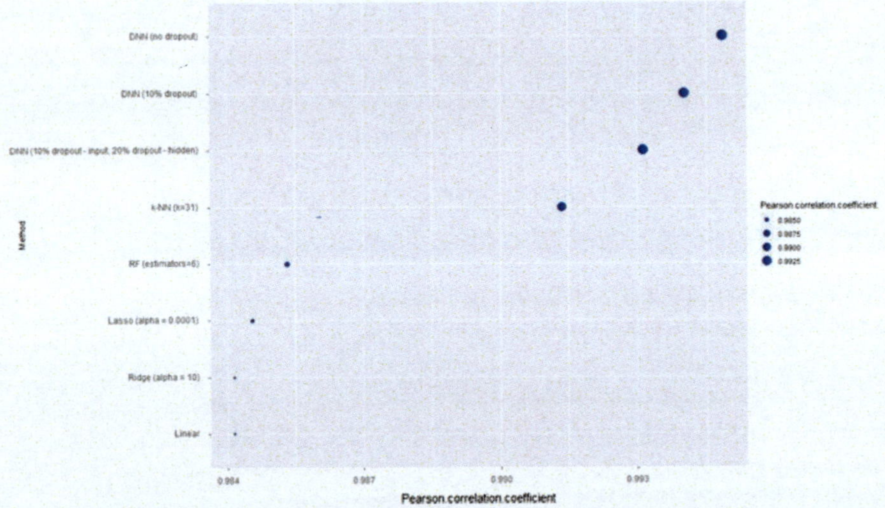

(b) Pearson correlation coefficients obtained for E-GEOD-75140

Fig. 5 **a** and **b** shows the amount of correlation between ground truth and predicted values as measured using Pearson correlation coefficient for the two datasets E-MTAB-7606 and E-GEOD-75140, respectively

measuring the amount of correlation between the observed and predicted values. Interestingly, we have observed that the proposed deep learning-based regression model has been robust enough against different train-test splits for both the datasets E-MTAB-7606 and E-GEOD-75140, which is proved by the fact that the MSE, R^2, and Pearson correlation coefficient values achieved by the proposed model have

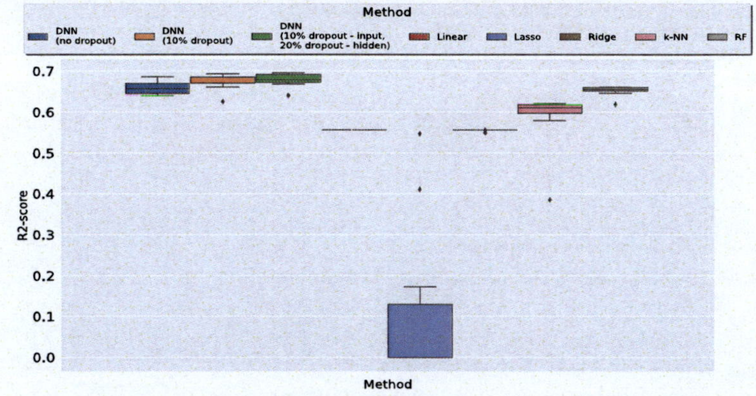

(a) R^2 values

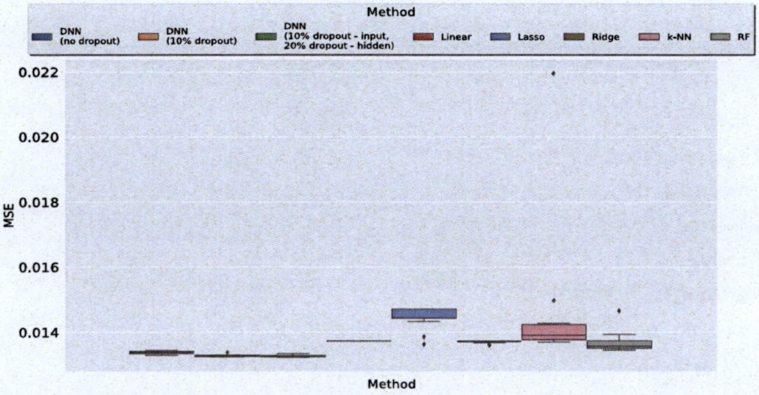

(b) MSE values

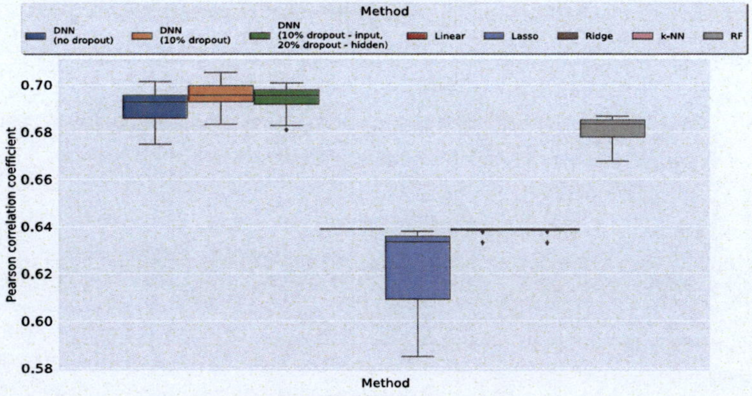

(c) Pearson correlation coefficients

Fig. 6 **a**, **b** and **c** show R^2 values, MSE values, and Pearson correlation coefficients obtained using the DNN regression model as compared with other state-of-the-art regression models on E-MTAB-7606 dataset

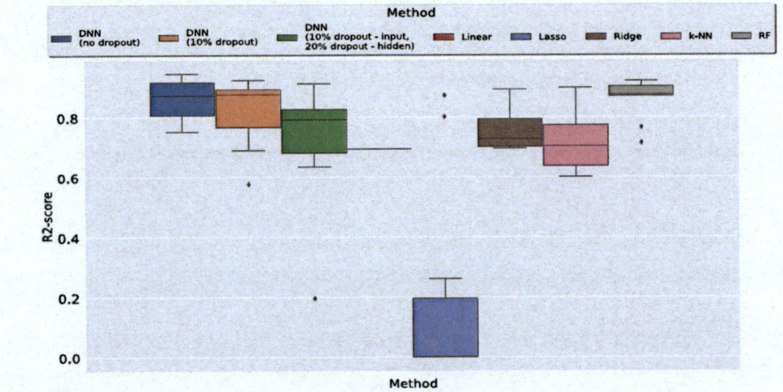

(a) R^2 values

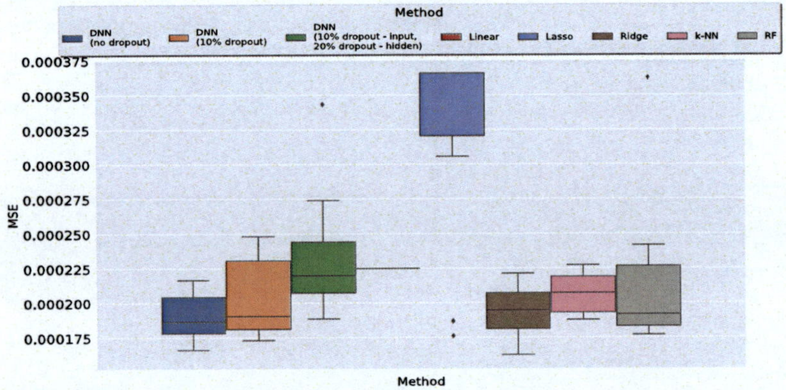

(b) MSE values

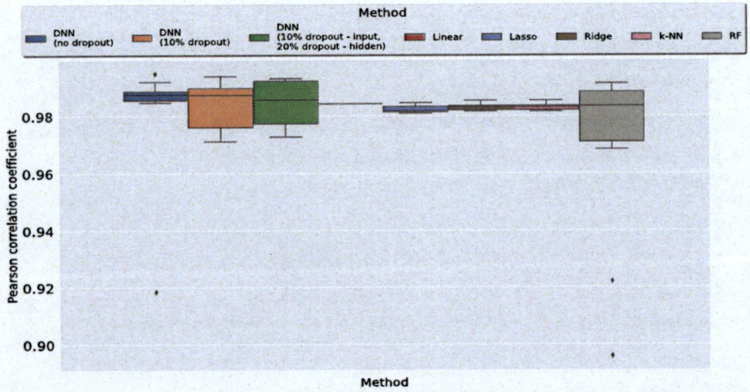

(c) Pearson correlation coefficients

Fig. 7 **a**, **b** and **c** show R^2 values, MSE values, and Pearson correlation coefficients obtained using the DNN regression model as compared with other state-of-the-art regression models on E-GEOD-75140 dataset been consistent with sufficiently low variance. The boxplots in Figs. 6 and 7 show the results from the multiple runs carried out on the E-MTAB-7606 and E-GEOD-75140 datasets, respectively.

4 Conclusion

In this study, we have developed a DNN regressor to predict gene expression estimates from scRNA-seq time-series data. The proposed regression model can be helpful when an advanced stage of development or disease is required to be estimated. It is also useful if expression values for an advanced timepoint are of low quality or not available at all. The present model has been evaluated on two publicly available datasets. In one dataset, we have predicted the gene expression for cells of an advanced diseased stage while in the other dataset, we have predicted the same for a developed adult neuron. Results have been competitive enough when compared with five existing state-of-the-art regression models. Currently, in this scope of work, we have confined our model to the prediction of gene expression values only, and not the developmental trajectory or stage. Predicting the stage of development, if not known, can be a future extension of this work. This work can also be extended to find out rare or novel clusters expressed in the later stages of cellular development.

Data and Code availability

This work has been implemented in Python 3. The codes to reproduce the results can be found at https://github.com/deeplearner87/timeSeries_scRNAseqPrediction. The datasets used in this work are available at https://www.ebi.ac.uk/gxa/sc/home.

Author's contribution

Conceptualization of methodology and framework: DBS, SA, VD, RKD. Data Curation, Data analysis, Formal analysis, Visualization, Implementation: DBS, SA. Investigation, Code review, Validation, Original draft preparation: DBS, SA, VD. Reviewing, Editing, Overall Supervision: RKD.

Conflict of Interest

VD works as a Senior Research Scientist at Novo Nordisk A/S, Måløv.

Acknowledgements DBS thanks the Data Science Laboratory, A. K. Choudhury School of Information Technology, University of Calcutta, Kolkata, India for providing the HPC infrastructure to execute and test the models. RKD acknowledges SyMeC Project grant [BT/Med-II/NIBMG/SyMeC/2014/Vol. II] given to the Indian Statistical Institute by the Department of Biotechnology (DBT), Government of India.

References

1. Camp JG, Badsha F, Florio M, Kanton S, Gerber T, Wilsch-Bräuninger M, Lewitus E, Sykes A, Hevers W, Lancaster M et al (2015) Human cerebral organoids recapitulate gene expression programs of fetal neocortex development. Proc Natl Acad Sci 112(51):15672–15677

2. Cover T, Hart P (2006) Nearest neighbor pattern classification. IEEE Trans Inf Theor 13(1):21–27 https://doi.org/10.1109/TIT.1967.1053964, https://doi.org/10.1109/TIT.1967.1053964
3. Hastie T, Tibshirani R, Friedman J (2001) The elements of statistical learning. Springer Series in Statistics, Springer, New York
4. Ho TK (1995) Random decision forests. In: Proceedings of the third international conference on document analysis and recognition (Volume 1), vol 1. ICDAR '95, IEEE Computer Society, Washington, DC, USA, p 278. http://dl.acm.org/citation.cfm?id=844379.844681
5. Hwang B, Lee JH, Bang D (2018) Single-cell rna sequencing technologies and bioinformatics pipelines. Exp Mol Med 50(8):1–14
6. Kingma DP, Ba J (2017) Adam: a method for stochastic optimization
7. Koutsakos M, Illing PT, Nguyen TH, Mifsud NA, Crawford JC, Rizzetto S, Eltahla AA, Clemens EB, Sant S, Chua BY et al (2019) Human cd8+ t cell cross-reactivity across influenza a, b and c viruses. Nat Immunol 20(5):613–625
8. Kumar P, Tan Y, Cahan P (2017) Understanding development and stem cells using single cell-based analyses of gene expression. Development 144(1):17–32
9. Pretschner A, Pabel S, Haas M, Heiner M, Marwan W (2021) Regulatory dynamics of cell differentiation revealed by true time series from multinucleate single cells. Front Genet 1730
10. Srivastava N, Hinton G, Krizhevsky A, Sutskever I, Salakhutdinov R (2014) Dropout: a simple way to prevent neural networks from overfitting. J Mach Learn Res 15(1):1929–1958. http://dl.acm.org/citation.cfm?id=2627435.2670313
11. Street K, Risso D, Fletcher RB, Das D, Ngai J, Yosef N, Purdom E, Dudoit S (2018) Slingshot: cell lineage and pseudotime inference for single-cell transcriptomics. BMC Genomics 19(1):1–16
12. Tran TN, Bader GD (2020) Tempora: cell trajectory inference using time-series single-cell rna sequencing data. PLoS Comput Biol 16(9):e1008205
13. Trapnell C (2015) Defining cell types and states with single-cell genomics. Genome research 25(10):1491–1498
14. Trapnell C, Cacchiarelli D, Grimsby J, Pokharel P, Li S, Morse M, Lennon NJ, Livak KJ, Mikkelsen TS, Rinn JL (2014) The dynamics and regulators of cell fate decisions are revealed by pseudotemporal ordering of single cells. Nat Biotechnol 32(4):381–386
15. Yeo GHT, Saksena SD, Gifford DK (2021) Generative modeling of single-cell time series with prescient enables prediction of cell trajectories with interventions. Nat Commun 12(1):1–12

Stress Analysis Using Machine Learning

B. Gowtham, H. Subramani, D. Sumathi,
and B. K. S. P. Kumar Raju Alluri

1 Introduction

Stress is the mental disturbance in an individual during adverse situations. It has many negative impacts on physical and mental health. There are many solutions for this problem but the actual problem is to self-diagnose and understand whether we are stressed or not and make necessary decisions. If we're confused about deciding, we consult a psychologist or doctor to understand ourselves better for which we have to pay the fee. As AI has begun to see solutions in different dimensions for problems, it is playing a significant role in the medical domain as well. Previous works have focused on using data from various sensors and surveys to analyze stress [1] but it is time-consuming and not computationally effective sometimes. As text is a cost-efficient and possible alternative approach instead of processing signals and research using text dataset i.e., on social media dataset fed to BERT base transformer model which has nearly 110 M parameters had achieved f1-score of 80.65% [2]. In this paper, a web-scrapped dataset was taken from Kaggle which is the composition of the text from Reddit posts and its binary labels like stress or not stress. Since being a social media dataset, it has lots of noise, those were preprocessed using the Spacy library. After preprocessing, the data was fed to machine learning and deep learning models and the models were evaluated based on the f1-score metrics. Here, we are proposing a hybrid model which is the combination of stacked transformer encoder layer and stacked Bi-directional LSTM layer on top of Fast-Text word embedding which has nearly 8.7 M parameters and resulted in an f1-score of 83.74%. In addition to improving the model, we've used the Layer Integrated Gradient technique to implement explainable AI for an embedding layer of the model and inspect the attribution of inputs in predicting the results.

B. Gowtham · H. Subramani · D. Sumathi (✉) · B. K. S. P. K. R. Alluri
VIT-AP University-Amaravati, Amaravati, AP, India
e-mail: sumathi.research28@gmail.com

R. Chaki et al. (eds.), *Applied Computing for Software and Smart Systems*,
Lecture Notes in Networks and Systems 555,
https://doi.org/10.1007/978-981-19-6791-7_14

The deep learning models we have today has evolved greatly. Starting with ANNs, researchers focused on building complex architectures to improve the effectiveness of the predictions. This process led to new models like RNNs, CNNs and transformers. Currently, transformers are being the best in the industry and more researchers are actively working on this. These have been resulting in advanced improvements in the performance and effectiveness of the models which have led to state-of-the-art models like BERT [3] and its varieties for text classification tasks and GPT-3 [4] for text generation tasks. These models have a significant impact on many domains like health, finances and others. Those have created a benchmark result in predicting. In addition to this, explainable AI has begun to be an important field of research that helps us to trust the models and their evaluations by inspecting the input attributions in predicting the results.

2 Literature Survey

Previous works mostly used models like CNNs, RNNs and state-of-the-art models like BERT on various kinds of signal data like ECG, EEG, survey-based data and social media-based text data. The authors of [5] used traditional Machine Learning algorithms on the WESAD dataset which is the collection of data from the sensors like ECG, EEG etc. and achieved an accuracy of 84%. The authors have collected physiological signals from sensors and fed those data into traditional machine learning algorithms. They achieved an accuracy ranging between 50% and 90%. So, to improve the performance, they have fed the data to CNN-1d and Multi-Layer Perceptron (MLP) to achieve the highest accuracy of 99.8%. Though the model has achieved high accuracy, this approach remains computationally costlier to train models using signals and also to serve those models in real-time [1]. Other research works used text data from Reddit posts and labelled the dataset. Later those data had been used for training both machine learning and deep learning models and achieved the highest f1-score of 80.65% using the BERT base transformer model which has nearly 110 M parameters. But in this paper, a more efficient model has been developed which perform better than the BERT base model and has a very smaller number of parameters i.e., nearly 8.7 M [2]. In the paper [6], the authors collected data from social media like Twitter and fed the preprocessed data to SVM and Naïve Bayes Model. Being SVM with WSD n-gram as the best model, precision of 65% and recall of 67%. As an alternative approach, the researchers used have used a publicly available dataset named "The Wearable Stress and Affect Detection Dataset" to build a CNN based model which takes raw data from the EDA sensor and processes it to predict the labels. In addition to that, they have also explored the personalization of the models which was developed. In this approach, they achieved an accuracy of 92.85% and an F1-score of 89%. Being a signal, it is computationally costlier again [7]. Survey-based research collected data from Jaypee Institute of Information Technology University which contains data of 206 students. Later, they have those data to train traditional Machine Learning models and achieved the highest accuracy of 85.71% being SVM as the best model. Since they have taken data from students, the questions asked to students differs from others like working professionals, kids etc., and it cannot be the generalized model [8].

3 Proposed Approach

3.1 Data Preprocessing and Preparation

The dataset contains 3.6 K samples. Since the text has been taken from social media, it has lots of ambiguity. So, the preprocessing techniques followed are removing URLs, handlers, emojis and special characters. I observed a few duplicates and a few empty sentences after preprocessing. So, all those samples have been removed. Later, a vocabulary of words was built. To make sentences of equal length i.e. 72 tokens, shorter sentences were padded with padding tokens and longer sentences were truncated to 72 tokens. After these, the data has been split into train and test datasets with a ratio of 80:20.

3.2 Traditional ML Models

Since the ML learning models take only two-dimensional data ie, (Samples, Features), the text data of the datasets have been encoded into TF-IDF vectors and passed those vectors to ML models like K-Nearest Neighbor (KNN), Logistic Regression (Log Reg), Decision Tree (DT), Random Forest (RF) and Support Vector Machines (SVM) algorithms. Being the best model as Random Forest, an f1-score of 71% had been achieved.

3.3 Bi-Directional LSTM (Bi-LSTM)

To train Bi-LSTM, the inputs should be in (batch, seq_len, Feature). So, the embedding layer has been used to create feature vectors for the words based on semantics and make the inputs of shape having the tokens (batch, seq_len) to (batch, seq_len, feature). Instead of initializing the embedding layer with random weights and training it, it may take a longer time to optimize its weights. So, this layer was initialized with Word2Vec and Fast-Text. Fast-Text has given better results compared with others. Bi-LSTM layers were used on top of the embedding layer. The Bi-LSTM takes additional parameters like the hidden size that is the number of features in the hidden state, the output of Bi-LSTM has been fed to Linear layers for predicting the labels. The highest f1-score of 74.44% occurred when the batch size is 32, the sequence length

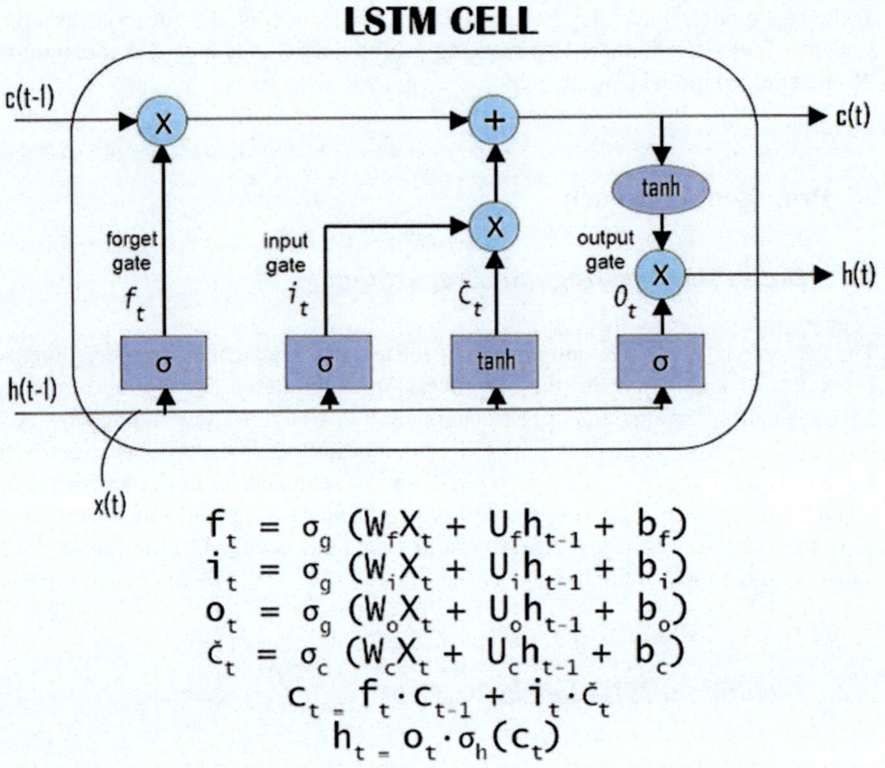

Fig. 1 Bi-LSTM architecture

is 72 and hidden size is 128. Also, the Adam optimizer is used with the learning rate being $3 * 10^{-4}$ to the loss function that is Cross-Entropy Loss and with learning rate decay of 10^{-1} for every 2 epochs to converge the model better and faster. Refer to Bi-LSTM in Fig. 1.

3.4 *Stacked Transformer Encoder Layer+Stacked Bi-LSTM*

Using the trainable Fast-Text embedding layer, the inputs were converted into (batch, seq_len, feature). The outputs of the embedding layer are passed to the Stacked Transformer Encoder Layer which has 2 layers of Transformer Encoder with 2 heads and the size of feed-forward network inside the encoder is 1024. Since the Transformer Encoder uses an attention mechanism, it helps the Bi-LSTM layer to understand the patterns better in the model. Later, the output of the transformer encoder layer has been sent to Stacked Bi-LSTM which had 2 layers of Bi-LSTM and a hidden size of 128 and its output has been fed into a Linear layer which takes of input size (batch,

256) generated by the Bi-LSTM layer and generates the output of shape (batch, 2). Again here, I used the Adam optimizer with the learning rate being $3 * 10^{-5}$ to the loss function that is Cross-Entropy Loss and the learning rate decayed by 10^{-1} for every 2 epochs to converge the model better. The highest f1-score of 83.74% on the test dataset.

3.5 Stacked Transformer Encoder Layer+CNN 1D

Using the trainable Fast-Text embedding layer, the inputs were converted into (batch, seq_len, feature). The outputs of the embedding layer are passed to the Stacked Transformer Encoder Layer which has 2 layers of Transformer Encoder with 2 heads and the size of feed-forward network inside the encoder is 1024. Since the Transformer Encoder uses an attention mechanism, it helps the CNN layer to understand the patterns better in the model. Later the output of the transformer encoder layer has been sent to CNN 1D with a kernel size equal to 3 and its output is sent to Max Pooling layer with a kernel size equal to sequence length-3. Later output has been fed into a Linear layer which takes input size (batch, sequence length) generated by the CNN 1D layer and generates the output of shape (batch, 2). Again here, I used the Adam optimizer with the learning rate being $3 * 10^{-5}$ to the loss function that is Cross-Entropy Loss and the learning rate decayed by 10^{-1} for every 2 epochs to converge the model better. The highest f1-score of 81.86% on the test dataset.

3.6 Explainable AI

It is important to understand what the model is looking for in the data when it comes to predicting the results. Because of this, the explainable AI has been implemented to the embedding layer of the best model using the Layer Integrated Gradient algorithm which follows the Integrated Gradient mechanism [9] to assign an attribution score to the inputs in predicting the outputs. Since it takes a single layer, an embedding layer is passed to check the attribution score of inputs. It evaluates the attribution score based on approximating the integral gradients of the model's prediction w.r.t inputs (Fig. 2).

4 Results

The performance of the models was evaluated based on F1-score as a primary metric and precision and recall as additional metrics, refer to Table 1. In this paper, we propose hybrid models which are the combination of stacked Transformer Encoder

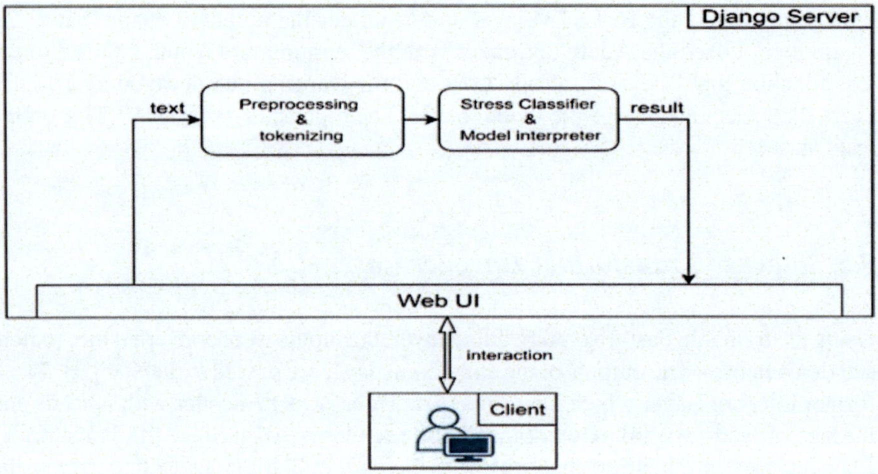

Fig. 2 Flow of proposed approach

Layer with Stacked Bi-directional LSTM and the combination of stacked Transformer Encoder layer with 1D CNNs on top of trainable Fast-Text embedding layer have given a better performance than the traditional models. Stacked transformer layer with stacked Bi-LSTM contains nearly 8.7 M parameters and Stacked transformer layer with CNN 1D contains nearly 7.2 M parameters which are comparatively very less than the BERT base model which has nearly 110 M parameters. Since its parameter is less than the BERT, it will have lesser computation and reduce the prediction time compared to BERT. To inspect input attribution in the model's prediction, the explainable AI has been implemented for the embedding layer. To test the model's explainability, the real-time data has been taken and fed to the model and its results and interpretation are shown in Figs. 3 and 4 respectively.

Explainable AI results on test data taken from the internet:

Explainable AI results on test data taken from the internet:

Table 1 Model results

Model	F1-score (%)	Precision (%)	Recall (%)
K Nearest neighbors	0.6817	0.6010	0.81
Logistic regression	0.6914	0.6520	0.7409
Decision tree	0.6651	0.6433	0.6902
Random forest	0.7143	0.6978	0.7322
SVM	0.6884	0.6298	0.7591
Bi-LSTM	0.7444	0.7309	0.7610
Stacked Trans Encoder + Stacked Bi-LSTM	**0.8377**	0.8221	**0.8560**
Stacked Trans Encoder + CNN 1D	0.8158	**0.8310**	0.8012

Legend: ■ Negative □ Neutral ■ Positive

True Label	Predicted Label	Attribution Label	Attribution Score	Word Importance
None	Stress (0.86)	Stress	3.59	today morning dark day #unk heavily overwhelmed self hatred #unk wanted end pain wrote doctor called parents tell end life fed said realised want rescue suicide choice left #unk mind #unk seek help googled numbers suicide helplines wanted talk time judge past called numbers picked failed attempts finally managed number girl line spoke #unk asked wanted life trying explain story line got disconnected realised life dark depressing maybe deserve die

Fig. 3 Interpretation report for input text with stress as original label

Legend: ■ Negative □ Neutral ■ Positive

True Label	Predicted Label	Attribution Label	Attribution Score	Word Importance
None	Non-Stress (0.85)	Non-Stress	-0.54	amazing person reasons incredible gift books latest excited add new #unk thanks steph #unk heart

Fig. 4 Interpretation report for input text with non-stressed as original label

(1) Suicide Note Article:

See Fig. 3.

(2) Tweet on felicitation

See Fig. 4.

5 Conclusion and Future Work

Stress is a predominant mental health issue. It needs to be identified and cured as early as possible to lead a healthy and happy life. Since the previous works were mostly based on signal analysis, the common people cannot make use of it and it requires investment for buying those sensors, also the evaluation of the models can be more computational which requires more time to predict the respective labels. In this project, We're proposing the model which is the combination of stacked transformer encoder layer with stacked Bi-LSTM layer has given a better result than the previous works carried out. It has given benchmark results of 83.74% f1-score and it contains 8.7 M parameters which makes it 12X computationally faster than the BERT base model [2] which contains 110 M parameters. Using the Layer Integrated Gradient algorithm, we can interpret the classifier and understand what features of the sentence contribute to predicting stress and not stress.

Since the dataset has 3.6 k samples and it is taken from social media. So, it had lots of ambiguity which could be one of the possibilities for resulting in less f1-score.

The results can be improved further by taking a good dataset with large data and with future models.

References

1. Li R, Liu Z (2020) Stress detection using deep neural networks. BMC Med Inf Decis Mak
2. Turcan E, McKeown K (2019) Dreaddit: a reddit dataset for stress analysis in social media. In: Proceedings of the tenth international workshop on health text mining and information analysis (LOUHI 2019)
3. Devlin J, Chang MW, Lee K, Toutanova K (2019) BERT: Pre-training of deep bidirectional transformers for language understanding. In: Proceedings of the 2019 conference of the north american chapter of the association for computational linguistics: human language technologies, vol 1 (Long and Short Papers). Association for Computational Linguistics, Stroudsburg, PA, USA, pp 4171–4186
4. Brown TB, Mann B, Ryder N, Subbiah M, Kaplan J, Dhariwal P, Neelakantan A, Shyam P, Sastry G, Askell A, Agarwal S, Herbert-Voss A, Krueger G, Henighan T, Child R, Ramesh A, Ziegler DM, Wu J, Winter C, Hesse C, Chen M, Sigler E, Litwin M, Gray S, Chess B, Clark J, Berner C, McCandlish S, Radford A, Sutskever I, Amodei D (2020) Advances in neural information processing systems 33 (NeurIPS 2020)
5. Faraz S, Ali SS, Adil SH (2018) Machine learning and stress assessment. In: IEEE-2018 3rd international conference on emerging trends in engineering, sciences and technology (ICEEST)
6. Reshma RB, Supriya K (2019) Detection and analysis of stress using machine learning techniques. Int J Eng Adv Technol (IJEAT) 9(1). ISSN: 2249–8958
7. Ramesh KS, Hassan G, Stress classification and personalization: getting the most out of the least. In: 17th IEEE-EMBS international conference on wearable and implantable body sensor networks (BSN'21).
8. Ravindra A, Alisha B (2019) Mental stress detection in university students using machine learning algorithms. In: International conference on pervasive computing advances and applications (PerCAA 2019)
9. Ahujaa R, Bangab A, Mental stress detection in university students using machine learning algorithms
10. Sundararajan M, Taly A, Yan Q (2017) Axiomatic attribution for deep networks. In: Proceedings of the 34th international conference on machine learning, vol 70 (ICML'17), pp 3319–3328

Deep Learning Towards Brain Tumor Detection Using MRI Images

Sanjib Roy and Ayan Kumar Das

1 Introduction

The uncontrolled division and growth of cells of living organisms may lead to tumor in the human body [1]. The unnatural growth can be developed in any part of the body and especially brain tumor may be occurred due to carcinogenesis part of the body. Depending on the location or tendency of malignancy tumors can be classified.

The manual detection of brain tumor by experts is a time-consuming process. Moreover, the segmentation applied on MRI images and the interpretation results by experts are varying under different circumstances. Therefore, the chance of survival from brain tumors can be possible by using some standard tools. Various pathological tests such as MRI images, Computed Tomography (CT) scan images and Biopsy reports are useful for detection of brain tumors [2]. MRI has certain advantages over CT images and nowadays these radiographic images have been popularly used for tumor detection to avoid the ionizing effect on human body.

An effective classification system is required for supporting the physicians to identify brain tumors. An appropriate segmentation process is required to choose before classification for better accuracy. Segmentation can effectively separate the tumor tissues from normal brain tissues. In this aspect, a new tissue segmentation algorithm [3] has been introduced with wavelets technique and neural networks which have produced effective segmentation of brain MRI images. There are several methods that have been applied for hand crafted feature extraction during the segmentation process like Random Forest, intensity based features, Markov Random Field,

S. Roy (✉) · A. K. Das
Birla Institute of Technology, Mesra, Patna Campus, Patna 800014, India
e-mail: phdcs10020.20@bitmesra.ac.in

A. K. Das
e-mail: das.ayan@bitmesra.ac.in

R. Chaki et al. (eds.), *Applied Computing for Software and Smart Systems*,
Lecture Notes in Networks and Systems 555,
https://doi.org/10.1007/978-981-19-6791-7_15

contextual and morphological features [4]. The hand crafted methods are not efficient as compared to deep learning methods [5]. Moreover, deep learning methods are more powerful as they have great ability to learn more discriminated features from images. A neural network based technique [6] have presented for brain tumor detection and classification. The method has achieved an accuracy of 83% with neural network based classifier. In [7], a capsule network has been introduced named as CapsNet that encapsulates MR images of the brain to characterize the brain tumor. The method obtained almost 91% precision. In another research [8], a framework has been suggested to identify glioma other than distinguishing high level and low level features with SVM and KNN. The model has achieved 85% precision along with 88% accuracy for multiple classifications. Though above described models have achieved very good performance in terms of different performance metrics, most of these approaches either suffer from the less number of samples present in the dataset or training data requires more memory or exhibits boundary inefficiencies on segmentation or increasing computational complexities.

The limitations of existing works have motivated the authors to propose a deep learning model to detect brain tumor. In this paper, a deep learning approach has been proposed to detect brain tumor by using feature extraction through CNN wavelet transformation along with classification through Capsule Network. Three ML techniques have been used separately with the above mentioned approach to get a useful quantitative performance analysis. The novel contributions of this paper are listed below.

i. An integrated approach has been proposed for early detection of brain tumor cases using MRI images.
ii. The integrated approach uses CNN wavelet transformation technique for feature extraction to achieve high accuracy.
iii. Capsule Network is used as a classifier in this approach.
iv. The performance of the proposed approach is tested on MRI images, which are collected from the standard dataset.
v. The quantitative analysis of the said approach has been compared with other three existing models by considering various performance metrics.

The succeeding part of the paper is organized as follows. Section 2 provides discussion on state-of-the-art research techniques that are used for brain tumor detection. Section 3 explains the proposed model. Section 4 describes the outcomes of the experiment on MRI brain tumor images and finally conclusion of the paper has been drawn in Sect. 5.

2 Background

Deep learning (DL), a field of AI technology, generates a prototype which is suitable for making multiple choices during disease detection. Convolution neural network

(CNN), under DL family has performed well to predict visual substances with negligible pre-processing. In [9], the authors have proposed a deep learning procedure for differentiating tumor and non-tumor cases from MR images. Bayesian Fuzzy C-Means algorithm has been used for segmentation purpose. Two processes namely WPTE and ST have been used exclusively for feature extraction and ultimately, deep auto-encoder is used for classification with an accuracy of 98.5%. An effective CNN method [10] is used for classification and segmentation of MR images for tumor detection. The method has achieved 97.5% classification accuracy along with 84% segmentation accuracy. An automatic deep learning classification method is used [11] to identify a tumor. On the other hand, in [12], a method has been proposed with different techniques using MRI images where feature extraction is accomplished by Daubechies Wavelet (DW) and feature reduction is done using PCA from MRI images. Finally, classification is done using SVM classifier. In another paper [13], different levels of wavelets have been used along with Support Vector Machine (SVM) as a classifier which in turn outperformed to classify benign and malignant tumors compared to Naïve Bayes classifier. A deep learning method [14] has been introduced to detect SAR image changes. In [15], deep learning structures have been compared with neural network using multiphase MRI images. The results show that deep learning structures perform well compared to neural network and thereby improved both sensitivity and specificity by 18%. A novel convolutional neural network named BrainMRNET [16] has been designed with Convolution Block Attention Module (CBAM) to get richer information about channel and spatial features of the images. The residual block along with hyper column technique has been applied used in this model for better generalization and efficient classification. The classification was achieved with 96.05% accuracy.

MRI data has a characteristic of high variability. Therefore, in brain tumor detection segmentation is one of the challenging tasks using MRI images. A segmentation method [17] has been proposed using genetic algorithm where the corners of the brain tumor regions have been extracted properly. In [18], segmentation process using MRI images not only detects the growth but also gives improved information about the centre as well as the unusual growth. Similarly, in [19] k-means clustering has been used for segmentation and the proposed algorithm shows good performance in edge detection. In [20] segmentation of brain tumors has been done using a multi-stage Fuzzy C-Means (FCM) framework. Similarly, in [21] another segmentation technique has been used with MRI images using Otsu's thresholding method.

Extraction of features is very important in image processing especially for disease detection through images. Various techniques have been applied such as transfer learning mechanism, super pixel method, principal component analysis, and wavelet transformation for feature extraction from MRI images. A deep learning model [23] has been used where modified ResNet50 is integrated with enhanced watershed segmentation algorithm for deep feature extraction and brain tumor classification. The brain tumor tissue images are classified effectively by this model with an accuracy of 90%. A CNN model [24] along with parametric optimization approaches like SFOA, FBIA and MGA has been proposed to analyze brain tumor from MRI images. The model has achieved high accuracy of 99% for effective classification of tumor in

Table 1 Comparative study of existing techniques

Method	Year	Technique used	Limitations
Multimodal MRI brain tumor segmentation with texture features using supervised learning [30]	2018	3D super voxel based learning method for segmentation, Random Forest (RF) classifier for classification	Detection of small volume of tumor with its boundary fails during RF method with Gabor filter
Automated lesion segmentation for chronic stroke using T1w anatomical MRI [31]	2019	Automated lesion identification, lesionGnb, LiNDA	Detection of small size tumor fails due to variable size data used
PGGAN based data augmentation for tumor detection [30]	2020	Progressive growing GANs used with conventional data augmentation	At least 25% cases fail to differentiate tumor and non-tumor
An encoder-decoder NN with 3D Squeeze for brain tumor segmentation [32]	2020	3D squeeze and excitation Vnet	Some of the cases are failed in tumor detection
Hybrid deep neural network for brain tumor detection [33]	2021	Adaptive fuzzy deep NN with frog leap optimization	Computational complexity is high. Lighter model is required for experiment

brain. A deep neural network [25] with a correlation learning mechanism has been applied for brain tumor detection. Finally, the model has achieved an accuracy of 96%. A comparative study of the state-of-the-art research is described in Table 1.

3 Methodology

This paper is designed to propose a useful detection of the presence of tumor in the brain using a deep learning approach where CNN wavelet transformation is used with capsule network for brain tumor detection. The proposed approach has shown very good performance in the detection of tumor. The flow diagram of the proposed approach is depicted in Fig. 1.

The entire working principle of the proposed method for brain tumor detection is tested on dataset [29] consisting of brain MRI images of different dimensions. The images are pre-processed to 224×224 so that it can work well with CNN wavelet transformation mechanism. After pre-processing, the final dimension of each image is $224 \times 224 \times 3$ and the images are rescaled and normalized to ensure that the model can capture the features. Now the pre-processed dataset is divided into training and testing datasets. The MRI data is split into 90:10 as training data and testing data respectively. The model is designed in such a way so that MRI images are working fine with architecture to extract features and classify the presence of tumor in brain effectively. The pre-processed training dataset is fed to CNN wavelet transformation

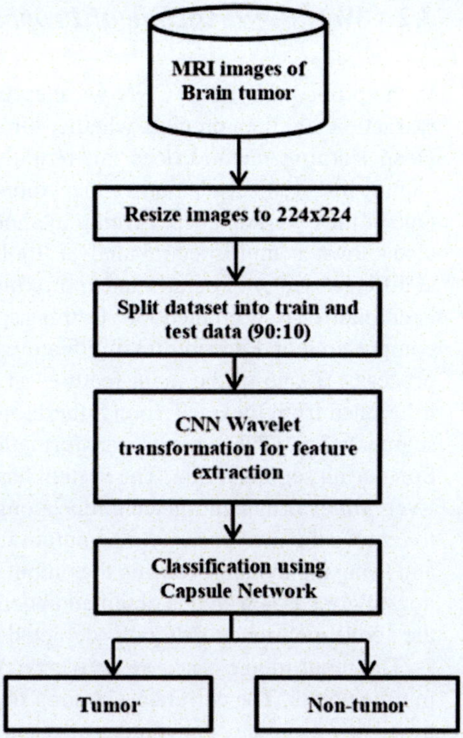

Fig. 1 Flow diagram for brain tumor detection using proposed approach

system to enhance the richer feature mapping. Finally, the model is able to correctly classify the presence of tumor in brain using capsule network. The model has achieved very good accuracy with other qualitative metrics using MRI images where the result of each metric is remarkable.

3.1 Dataset Description

The dataset [29] used for this approach contains 3060 number of MRI images of brain tumors. There are 1500 number of images of positive brain tumor and 1500 number of images has no symptoms of brain tumor that can be considered as normal patients. There are 60 numbers of images taken from the dataset as test data. The images are of varying sizes of 215×234 or 204×247 or 272×350 with a resolution of 96 dpi.

3.2 Working Principle of Proposed Approach

In the proposed system, CNN wavelet has been applied to MRI images for feature extraction and then predicts whether tumor is present or not using capsule network. Deep learning method does not require feature engineering. They automatically capture hierarchical attributes and features from raw data without feature engineering. One of the common deep learning methods is CNN wavelet transformation which is also known as multispectral analysis tool. It has the capability to parse input signals at different scales of details and approximations. Segmentation of brain MRI images is crucial for tumor detection. In this aspect, CNN wavelet has a unique feature of compressing and decompressing feature maps which in turn helps in segmentation process. As one of the main features of tumors in MRI images is their brightness difference from the background, therefore, edge detection is a useful clue for tumor segmentation. CNN wavelet transformation has performed effectively in edge detection during segmentation. The system has developed with nine convolutional layers. Every time normalization technique along with ReLU function, have been employed to remove the non-linearity. The output featured vector is fed to the primary capsule and using the dynamic routing the output image is correctly classified into tumor and non-tumor cases which is obtained under BT-CAP layer. The architectural layout of the above mentioned method is depicted in Fig. 2.

The input image $\mathrm{X_{m\times n}} = [\mathrm{x_1(t), x_2(t), \ldots x_n(t)}]$ consists of n pixels sampled at m time points. The correlation among row and column values of input matrix is not possible with square convolutional kernels as the information along row and column are sampled with different time constraints. Therefore, rectangular kernels are used to extract information from the input image as in Eq. 1.

$$Y_n^m = \sum_{v=1}^{2} k_{1v} \times X_{m,n+v-1} \quad (1)$$

where, X is the input matrix and Y is the output featured matrix respectively, k is a rectangular kernel, v is the number of columns of kernel k. The features extracted from the input layer l–1 are convoluted with kernel of l–1 to get the output feature map of lth layer using activation function and is defined in Eq. 2.

$$\mathrm{x_j^l} = \psi\left(\sum_{\mathrm{i=1\ldots M}} \mathrm{x_i^{l-1}} * \mathrm{k_{1j}^{l-1}} + \mathrm{b_j^l}\right) \quad (2)$$

where, $\mathrm{b_j^l}$ represents bias of the jth filter, M is the number of convolution kernels of jth convolution layer, $\mathrm{x_j^l}$ is the jth output feature map of layer l, $\mathrm{x_i^{l-1}}$ is the ith input feature map of layer l–1, ψ represents the scale transformation function.

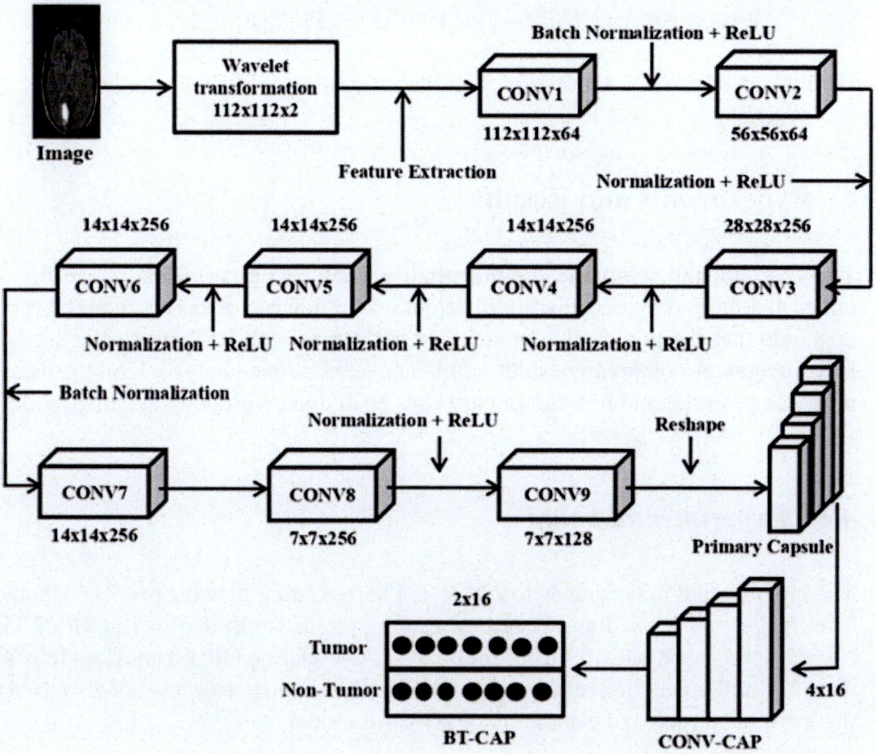

Fig. 2 Architectural construction using deep network for brain tumor detection

3.3 *Prediction of Presence of Tumor Using Capsule Network*

The reshaped features obtained from last convolutional layer of the proposed system used in this paper have been passed to the primary capsule layer and these features in terms of weight matrix can be used for prediction through Eq. 3.

$$\hat{U}_{j|i} = W_{ij} u_i \tag{3}$$

where, $\hat{U}_{j|i}$ is the predicted output feature vector that can be obtained through the multiplication of the features of the ith capsule with the weight matrix W_{ij}. The agreement between two different levels (i, j) capsules can be done on the basis of coupling coefficient c_{ij} which in turn depends on the log prior probabilities that can be updated using Eq. 4.

$$b_{ij} = b_{ij} + \hat{U}_{j|i} \times v_j \tag{4}$$

The loss function associated with the model is defined in Eq. 5.

$$\text{Loss} = \text{D}_\text{n}\max(0, \text{h}^+ - \|\text{v}_\text{n}\|)^2 + \lambda(1 - \text{D}_\text{n})\max(0, \|\text{v}_\text{n}\| - \text{h}^-)^2 \quad (5)$$

For nth capsule, if the tumor is present then the value of D_n is 1, otherwise 0.

4 Experiments and Results

The experimental setup and the quantitative analysis on performance are focused under this section by considering different evaluation metrics. The obtained results conclude that the proposed approach has performed well for tumor detection with MRI images. A comparative study of the above mentioned approach on the basis of presence of tumor and non-tumor cases has been done with other existing models.

4.1 Experimental Setup

The input images are reshaped to 224 × 224 before being fit to the proposed method. The images are trained and tested on Linux system composed of Intel(R) 2 GHz X86_64 processor, and 13 GB of on board RAM. Online GPU from Kaggle is also used for implementation purpose. In addition, Keras and tensorflow are also used for the implementation and evaluation of BT-CAP model.

4.2 Evaluation Metrics

The quantitative performance analysis of the suggested approach along with three different ML techniques has been assessed separately by calculating four significant evaluation metrics for proving the effectiveness of the model. The significant evaluation metrics are evaluated as:

(i) Accuracy: The metric by which a model can able to correctly predict the presence of tumor by considering all the cases and is defined in Eq. 6.

$$\text{Accuracy} = \frac{\text{TP} + \text{TN}}{\text{TP} + \text{TN} + \text{FP} + \text{FN}} \quad (6)$$

(ii) Precision: The metric by which a model can correctly identify positive tumor cases by considering all the cases which have the symptoms of brain tumor and is defined in Eq. 7.

$$\text{Precision} = \frac{\text{TP}}{\text{TP} + \text{FP}} \quad (7)$$

(iii) Recall: The metrics by which a model can correctly classify the presence of brain tumor by considering all the cases, which have the pre symptoms about brain tumor or not and is measured as in Eq. 8.

$$\text{Recall} = \frac{\text{TP}}{\text{TP} + \text{FN}} \tag{8}$$

(iv) F1-Score: It is used to quantify the model by classifying correctly the presence and absence of brain tumor from a test and is defined in Eq. 9.

$$\text{F1} - \text{score} = \frac{2}{\text{precision}^{-1} + \text{recall}^{-1}} \tag{9}$$

4.3 *Performance Comparison of Proposed Method with Different Machine Learning Techniques*

In the proposed system, CNN wavelet technique has been adopted for feature extraction from MRI images. Finally, the images with discriminant extracted features have been classified by capsule network. The comparative analysis with different performance metrics has been depicted with statistical representation in Fig. 3.

The confusion matrices are depicted in Fig. 4a, b, c, d, which revels that the three ML techniques have been used separately for detection of brain tumor and are quite good compared to the proposed model. Only MLP performs not so well compared to other methods. None of the techniques used above has shown high false negative cases thereby the prediction concludes by any of the techniques is not so dangerous

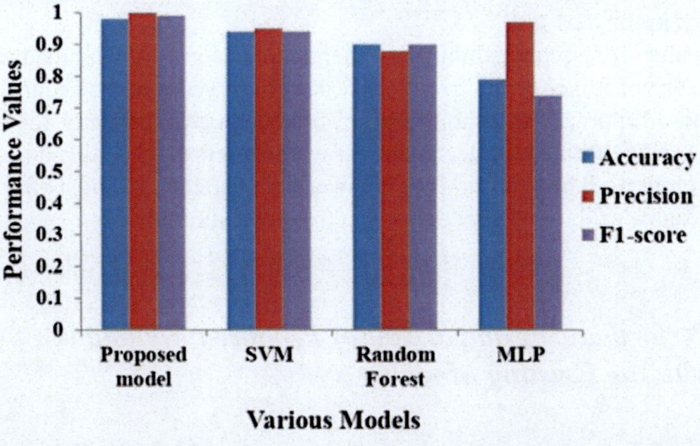

Fig. 3 Quantitative performance analysis of wavelet transformation and capsule network with ML techniques

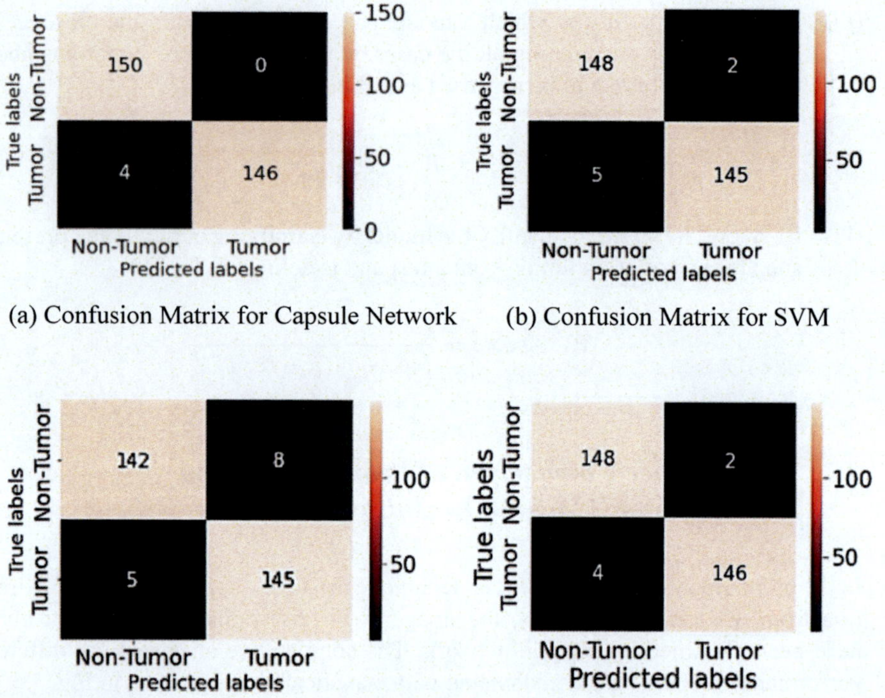

(a) Confusion Matrix for Capsule Network (b) Confusion Matrix for SVM

(c) Confusion Matrix for Random forest (d) Confusion Matrix for MLP

Fig. 4 Confusion matrices produced by proposed method and different ML techniques

for the infected people. False positive cases identified by different ML techniques and two models with capsule network are not as risky as those persons can be identified as normal in the next test.

The value of f1-score exhibits that the model has greatly classified the presence and absence of tumor in brain. In Fig. 5, it is observed that the accuracy curve for the proposed approach tends to be stable after 50 epochs. Similarly, the loss curve is almost starting to stabilize after 50 epochs which infers that learning through training is done properly. The loss curve goes lower than validation almost after 80 epochs. Thus it can be said that the model is free from over-fitting.

4.4 *Performance Comparison of Proposed Method with the Existing Models*

The researchers have already made great contributions by developing various models to detect brain tumor effectively. In [26], a deep learning model has been used for detection of tumor using MR images. Different transfer learning mechanisms

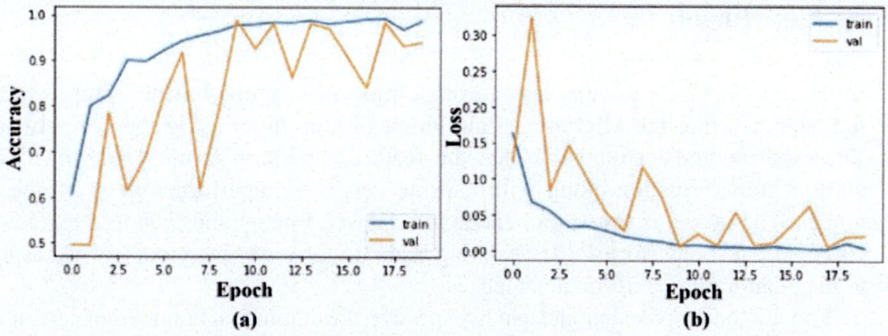

Fig. 5 **a** Accuracy curve and **b** Loss curve of proposed approach

like ResNet50, AlexNet, GoogleNet, DenseNet201 and InceptionV3 have been used separately to extract features inclusively and finally best suite is found by classifying the images properly. In [27], CNN is used as a base model for detection of brain tumor from the dataset comprised of 2D MR images. The model has applied fuzzy c-means clustering technique for segmentation along with six ML techniques separately for classification purpose. Similarly, another deep learning method [28] has been suggested for extraction of tumor regions along with local repetitive fields for classification purpose. The proposed model has performed well compared to the other discussed methods and classifies the brain tumor effectively. The comparative study with statistical representation is depicted in Fig. 6.

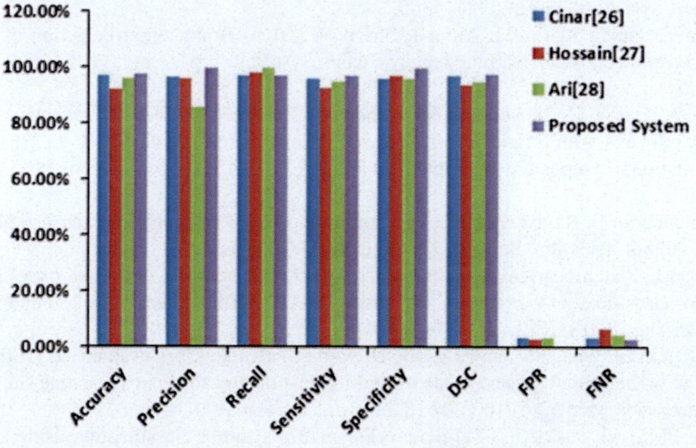

Fig. 6 Comparative analysis of two proposed models with existing models

5 Conclusion

In this article, CNN wavelet transformation has been proposed for feature extraction and classifies the MRI image into tumor or non-tumor using capsule network. The experimental results show that the feature extraction mechanism with CNN wavelet transformation along with capsule network outperforms other ML techniques. The model is tested on benchmark dataset. Finally, the proposed extractor-classifier pair outperforms three existing brain tumor screening models with respect to the quantitative performance metrics.

Though the proposed model usually speedup the automated brain tumor screening process, it cannot be said as a replacement of medical diagnosis as this approach has not been used for clinical study. The authors are aimed to make an extensive thorough investigation with large dataset towards more accurate prediction for future work.

References

1. Demirhan A, Törü M, Güler I (2014) Segmentation of tumor and edema along with healthy tissues of brain using wavelets and neural networks. IEEE J Biomed Health Inform 19(4):1451–1458
2. Charutha S, Jayashree MJ (2014) An efficient brain tumor detection by integrating modified texture based region growing and cellular automata edge detection. In: 2014 International conference on control, instrumentation, communication and computational technologies (ICCICCT). IEEE, pp 1193–1199
3. Lee H, Chen YPP (2015) Image based computer aided diagnosis system for cancer detection. Expert Syst Appl 42(12):5356–5365
4. Saba T, Sameh A, El- M, Amin J, Sharif M (2019) Brain tumor detection using fusion of hand crafted and deep learning features. Cogn Syst Res. https://doi.org/10.1016/j.cogsys.2019.09.007
5. Rao A, Ledig C, Newcombe V, Menon D, Rueckert D (2014) Contusion segmentation from subjects with traumatic brain injury: a random forest framework. In: 2014 IEEE 11th international symposium on biomedical imaging (ISBI). https://doi.org/10.1109/isbi.2014.6867876.
6. Damodharan S, Raghavan D (2015) Combining tissue segmentation and neural network for brain tumor detection. Int Arab J Inf Technol (IAJIT) 12(1)
7. Afshar P, Plataniotis KN, Mohammadi A (2019) Capsule networks' interpretability for brain tumor classification via radiomics analyses. In: 2019 IEEE international conference on image processing (ICIP). IEEE, pp 3816–3820
8. Zacharaki EI, Wang S, Chawla S, Soo D, Wolf R, Melhem ER, Davatzikos C (2009) Classification of brain tumor type and grade using MRI texture and shape in a machine learning scheme. Magn Reson Med 62(6):1609–1618. https://doi.org/10.1002/mrm.22147
9. Raja PMS, Viswasa A (2021) ScienceDirect Brain tumor classification using a hybrid deep autoencoder with Bayesian fuzzy clustering-based segmentation approach. Integr Med Res 1–14
10. Xu Y, Jia Z, Ai Y, Zhang F, Lai M, Eric I, Chang C (2015) Deep convolutional activation features for large scale brain tumor histopathology image classification and segmentation. In: 2015 IEEE international conference on acoustics, speech and signal processing (ICASSP). IEEE, pp 947–951

11. Pereira S, Pinto A, Alves V, Silva CA (2016) Brain tumor segmentation using convolutional neural networks in MRI images. IEEE Trans Med Imaging 35(5):1240–1251
12. Abd-Ellah MK, Awad AI, Khalaf AA, Hamed HF (2016) Classification of brain tumor MRIs using a kernel support vector machine. In: International conference on well-being in the information society . Springer, Cham, pp 151–160
13. Gurbina M, Lascu M, Lascu D (2019) Tumor detection and classification of MRI brain image using different wavelet transforms and support vector machines. In: 2019 42nd international conference on telecommunications and signal processing (TSP). https://doi.org/10.1109/tsp.2019.8769040.
14. Samadi F, Akbarizadeh G, Kaabi H (2019) Change detection in SAR images using deep belief network: a new training approach based on morphological images. IET Image Proc 13(12):2255–2264
15. Pan Y, Huang W, Lin Z, Zhu W, Zhou J, Wong J, Ding Z (2015) Brain tumor grading based on neural networks and convolutional neural networks. In: 2015 37th annual international conference of the ieee engineering in medicine and biology society (EMBC). IEEE, pp 699–702
16. Toğaçar M, Ergen B, Cömert Z (2020) BrainMRNet: brain tumor detection using magnetic resonance images with a novel convolutional neural network model. Med Hypotheses 134:109531
17. Kaushik D, Singh U, Singhal P, Singh V (2014) Brain tumor segmentation using genetic algorithm. Int J Comput Appl®(IJCA)(0975–8887); In: International conference on advances in computer engineering and applications (ICACEA2014) at IMSEC, GZB
18. Bathe K, Rana V, Singh S, Singh V (2021) Brain tumor detection using deep learning techniques. Available at SSRN 3867216
19. Dhanalakshmi P, Kanimozhi T (2013) Automatic segmentation of brain tumor using K-Means clustering and its area calculation. Int J Adv Electri Electron Eng 2(2):130–134
20. Szilagyi L, Lefkovits L, Benyo B (2015) Automatic brain tumor segmentation in multispectral MRI volumes using a fuzzy c-means cascade algorithm. In: 2015 12th international conference on fuzzy systems and knowledge discovery (FSKD). IEEE, pp 285–291
21. Rani R, Kamboj A (2019) Brain tumor classification for MR imaging using support vector machine. In: Progress in advanced computing and intelligent engineering. Springer, Singapore, pp 165–176
22. Ghassemi N, Shoeibi A, Rouhani M (2020) Deep neural network with generative adversarial networks pre-training for brain tumor classification based on MR images. Biomed Signal Process Control 57:101678
23. Sharma AK, Nandal A, Dhaka A, Koundal D, Bogatinoska DC, Alyami H (2022) Enhanced watershed segmentation algorithm-based modified ResNet50 model for brain tumor detection. BioMed Res Int
24. Nayak DR, Padhy N, Mallick PK, Bagal DK, Kumar S (2022) Brain tumour classification using noble deep learning approach with parametric optimization through metaheuristics approaches. Computers 11(1):10
25. Woźniak M, Siłka J, Wieczorek M (2021) Deep neural network correlation learning mechanism for CT brain tumor detection. Neural Comput Appl 1–16
26. Çinar A, Yıldırım M (2020) Detection of tumors on brain MRI images using the hybrid convolutional neural network architecture. Med Hypotheses 109684. https://doi.org/10.1016/j.mehy.2020.109684
27. Hossain T, Shishir FS, Ashraf M, Al Nasim MA, Shah FM (2019) Brain tumor detection using convolutional neural network. In: 2019 1st international conference on advances in science, engineering and robotics technology (ICASERT). IEEE, pp 1–6
28. Ari A, Hanbay D (2018) Deep learning based brain tumor classification and detection system. Turk J Electr Eng Comput Sci 26(5):2275–2286
29. https://www.kaggle.com/ahmedhamada0/brain-tumor-detection.
30. Han C, Rundo L, Araki R, Furukawa Y, Mauri G, Nakayama H et al (2020) Infinite brain MR images: PGGAN-based data augmentation for tumor detection. Neural approaches to dynamics of signal exchanges. Springer, New York, pp 291–303

31. Ito KL, Kim H, Liew SL (2019) A comparison of automated lesion segmentation approaches for chronic stroke T1-weighted MRI data. Hum Brain Mapp 40:4669–4685
32. Liu P, Dou Q, Wang Q, Heng PA (2020) An encoder-decoder neural network with 3D squeeze-and-excitation and deep supervision for brain tumor segmentation. IEEE Access 8:34029–34037
33. Deb D, Roy S (2021) Brain tumor detection based on hybrid deep neural network in MRI by adaptive squirrel search optimization. Multimed Tools Appl 80:2621–2645

Software and Systems Engineering

A Novel Unmanned Near Surface Aerial Vehicle Design Inspired by Owls for Noise-Free Flight

Rahma Boucetta, Paweł Romaniuk, and Khalid Saeed

1 Introduction

Researchers, even with assistance from computers, cannot usually find a solution for complex human problems, while nature, with specific mutations and gradual change of evolution, does things much better than humans. For these reasons, robotics remains the most significant field wherein biological characteristics in living beings are used as a knowledge base to develop original robotic designs and provide fruitful ground for bioinspired technologies [1].

Among these issues, bio-inspired movements are a recent subclass of design. It involves exploring concepts derived from nature and applying them to create engineering systems in the real world. Accordingly, most robots have some pattern of the natural locomotion system for a specific task of interest [2, 3].

In fact, designers are reproducing robots mimicking animals to create lifelike robotic counterparts: "The goal of bio-robotics is to design a machine that can interact with its environment and dynamic situations (…)"[4]. The Octobot, from Harvard University, is a soft autonomous robot that mimic octopus in its displacement using pneumatic controls [5]. At the Federal Polytechnic School of Lausanne, researchers had designed a robot that is moving like salamander by imitating its motions. Researchers at Carnegie Mellon University designed and developed a snake-shaped robot that kept crawling like a real snake.

R. Boucetta (✉)
Department of Physics, Faculty of Sciences, University of Sfax, Sfax, Tunisia
e-mail: rboucetta@yahoo.fr

P. Romaniuk · K. Saeed
Faculty of Computer Science, Bialystok University of Technology, Bialystok, Poland
e-mail: pawel.romaniuk@e-mcs.pl

K. Saeed
e-mail: k.saeed@pb.edu.pl

R. Chaki et al. (eds.), *Applied Computing for Software and Smart Systems*,
Lecture Notes in Networks and Systems 555,
https://doi.org/10.1007/978-981-19-6791-7_17

Considerable growth in the development of UAV allows to expand the application areas. Flexibility and falling costs have risen the use of such vehicles, not only in military missions but also in civilian tasks. Indeed, the use of UAVs covers many military and civilian fields, such as aerospace, demining, photography, detection, security, filming… that need accuracy, rapidity and immunity to ambient noises. In Gnanasekera's work [6], authors proposed an algorithm permitting UAV "to approach and follow a steady or moving herd of cattle using only range measurements. The algorithm is insensitive to the complexity of the herd's movement and the measurement noise" [6]. Bielecki and Smigielski developed "an algorithm designed for analysis and understanding a 3D urban-type environment by an autonomous flying agent, equipped only with a monocular vision" [7], the algorithm based on the structural representation of the analyzed scene showed in hierarchical form. OiV drone has fixed-wing and multirotor features. An autonomous drone is also informed when the event has been initially detected.

Our contribution in this paper is to develop a novel adequate mechanical concept of drones inspired of owls, capable to give the ability of flying in low height to follow moving objects and also the flexibility to wing extremely high with low noise, thanks to a well-studied choice of propellers ensuring low sound frequencies and a considerable power consumption reduction. Similar work was done by Lilley. To design an owl-inspired robot, a preliminary examination shows that owls have super-tuned senses helping them to hunt cleverly. Human cannot hear their flight at more than two meters [8].

2 Owl-Inspired Vehicle Design

2.1 Morphology

Owls are called as "silent predators of the night". Owls are able to fly just centimeters from their prey and have the tendency to being unnoticed. The secret of their silent flight is the specific design of their wings, especially the design of their trailing edge. To alter air turbulence accumulated over the wing, owls have few ways to reduce the created exuberant noise [5, 8].

Foremost, the feathers of the owl's wing in the leading edge own some petty structures, called serrations, which can break up the flowing air into small flows that when falling along the wing makes less turbulences. Furthermore, when the owl is very close to its prey, its wings are placed in a particular steep angle to provide clear noise reduction properly insured by the wing's serrated leading edge [8, 9]. A flexible fringe, presented on the trailing edge of the owl's wing, receives smaller airflows and breaks up the air flowing off the trailing edge to produce a large reduction in aerodynamic noise. Then velvety feathers on the owl's wings' trailing edge absorb high frequency sounds, sensitive for humans and prey [5, 10] (Figs. 1 and 2).

Fig. 1 **a–d** Owl wing postures in straight flight

Fig. 2 **a–d** Owl wing in L-shape postures in lowering flight or landing phase

2.2 Suggested Design

Analysis of the owl's shape both in flight and during the attack/landing phase determined the Owl Inspired Vehicle (OiV) shape. The body was constructed in a tube shape with a length of 40 cm and a diameter of 15 cm rounded at the front and from the point where the wings connect, tapering towards the back to form an owl-like tail.

The wings were placed in the center of gravity axis of the model. The point of attachment of the wings was selected experimentally depending on the modifications made to the model. The wings were raised relative to the upper edge of the body to lower the model's center of mass. The wingspan is 100 cm in length [11].

Location of the propellers was planned on the edges of the wings in holes cut in the wings to ensure air flow. With fixed angle of attack propellers, a small propeller was added at the front to control the direction of flight and OiV pitch.

2.3 Motor Set Analysis

To obtain a hovering position for the OiV, the weight of the different components needs to be approximated. A battery with 1800mAh weights about 200 g, the mass of a propeller is about 36 g and the frame model without motors can supposed be 350 g. Thus, the whole vehicle is weighed approximately 800 g with motors, battery and electronics. The weight is much important in view of the fact that 60% of thrust is necessary to stable the OiV and 100% of thrust to rise it up.

A standard problem associated with propeller rotation is the torque (rotational force) opposite to the direction of rotation of the propeller blades. This force causes the entire model to rotate in the direction opposite to the direction of the propeller blades. Single engine models require an additional rotor to counteract the torque of the main engine (Sikorsky system) or a controlled ejection of exhaust gases (NOTAR system). When two counter-rotating propellers are used (Piasecky and Flettner systems) the torques of individual propellers cancel each other out.

The torque is a pseudovector builded as a product of the distance vector and the force vector that depends on three quantities: applied force, the lever arm vector [12] and the angle between the force and lever arm vectors:

$$\boldsymbol{\tau} = \boldsymbol{r} \times \boldsymbol{F} \tag{1}$$

$$\tau = \|r\| * \|F\| * \sin\theta \tag{2}$$

where:

- τ torque vector,
- r position vector (a vector from the point about which the torque is being measured to the point where the force is applied),
- F is the force vector,
- $\times$ cross product, which produces a vector that is perpendicular to both r and F following the right-hand rule,
- θ is the angle between the force vector and the lever arm vector [13].

In the OiV model the antitorque motor cannot be used.

The other way for reduce torque forces is using the tandem rotor model. Frank Piasecki was the first to use it and is called as the Piasecki model. It "proposed a tandem configuration (rotors front and rear) to best meet the design conditions (…). The tandem design provided a significant increase in center of gravity travel, thus negating the need for shifting ballast, as was necessary in single rotor helicopters" [12].

Two single motors with propellers were placed at the end of each wing effects on the model with torque forces. The resultant force must be taken for general calculation. Piasecki modified the arrangement of the force calculations which will be more difficult and the resultant forces will affect the OiV model.

Coaxial propeller solves the problem of torque force by using two separate motors with propellers turning in opposite directions. The opposite torques generated by each propeller cancel each other out. The sum of the torque from the two propellers running on the same axis with the same angle speed but in opposite directions is zero. Also it is possible to control the torque size and direction by increasing the angle speed of one motor and decreasing it on the other.

The coaxial propeller was first constructed by Nikalai Kamov, a Russian constructor. This type is called the Kamov system and chosen for the OiV model.

2.4 Thrust Analysis

To find the right motors and propellers, the size proportion of the vehicle frame leads to use 5 to 6 inch propellers that look legit for the OiV. Most efficiency motors for flying drones are brushless direct-current electric motors (BLDC motors). To seek motors and propellers used for drones, any type of engine with different rotation speed and 36 g weight are considered to offer optimally 1000 g of thrust. An optimal motor for the 6 inch maximum propellers is the 2207 motor. Smaller motors, like 1807 will have no enough thrust (maximum thrust for 1807 motor with 5 inch 5045 propeller is 460 g). Bigger motors like 2409 are designed for 8 inch propellers and they have a big diameter for 6 inch propellers. Motor diameter reduces propeller efficiency by reducing the active surface and airflow area.

A summary of propeller characteristics with 2207 motor is gathered in the following shot Table 1.

By examining the collected features of the two Piasecki simple motors, a first selection can be directed towards the 2207/1700 kV propeller that ensures the lowest power consumption equal 246.96 W and the smallest current about 9.8A to give 1000 g of thrust.

Similarly, the two Kamov double motors are discussed because they give the best power usage and the highest efficiency, and they exist in three types of engine: 2207/2400 kV, 2207/1900 kV and 2207/1700 kV. About 300 g of thrust is chosen for each engine giving 1200 g of thrust. The lower motor in the Kamov system has

Table 1 Characteristics of two Piasecki simple motors

Motor	Propeller	Current (A)	Thrust (g)	Power (W)	Efficiency (g/W)	Rotation speed (RPM)	Motor mass (g)
2207/2400 kV	5.5*4*3	16.8	1000	282.24	3.54	15,900	76
2207/1900 kV	5046	11.0	1000	277.20	3.61	17,500	76
2207/1700 kV	5.5*4*3	9.8	1000	246.96	4.05	16,100	76

Table 2 Characteristics of two Kamov double motors

Motor	Propeller	Current (A)	Thrust (g)	Power (W)	Efficiency (g/W)	Rotation speed (RPM)	Motor mass (g)
2207/2400 kV	5.5*4*3	16.4	1200	275.52	4.36	12,600	152
2207/1900 kV	5045–3	11.2	1200	282.24	4.25	14,900	152
2207/1900 kV	5046	11.2	1200	282.24	4.25	13,800	152
2207/1700 kV	5.5*4*3	10.4	1200	262.08	4.58	12,500	152

30% less of its efficiency, thus it has only 1020 g of thrust, which is optimal for the one-kilogram OiV (Table 2).

For 1700 kV motor, the power consumption of the two motor systems is about 6% and for 2400 kV the better power consumption is 2.5% for the two double motors that are running with 12,500 RPM and generate a low frequency sound.

2.5 *Propeller Shape Analysis*

When building an OiV model, the noise generated by the propellers must be considered. In addition to minimize the number of revolutions to reduce noise at high frequencies, the shape of the propeller should be considered. In bigger scale the noise of the propeller was simulated and tested by Yuhang Wu et al. [16]. The smaller model is needed to take experiments with small propellers.

The overall shape of the propeller with its small size does not matter much with respect to the generated noise, but the shape of the end of the propeller is very important.

There are many different propeller tips in model flying. Examples are on Fig. 3.

For flying models of OiV size and a propeller size of 6 inches, there are four propeller tip shapes to consider:

- standard tip (2, 3)
- bull nose tip (1)
- hybrid bull nose tip (rounded 1)
- low-noise tip (4, 5).

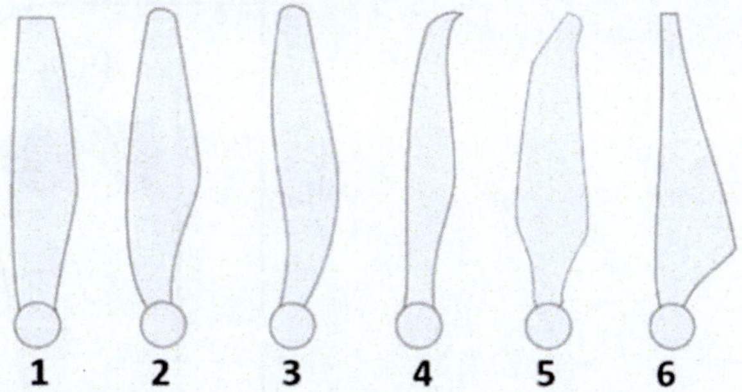

Fig. 3 Types of propeller tip

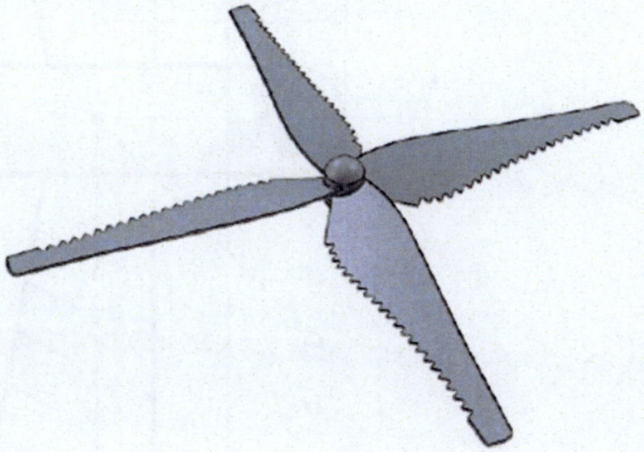

Fig. 4 Experimental propeller model with serrated tailing edge

- high stepping propeller shape (6)

For OiV, consider the bull nose tip and the low-noise tip.

There is also a new experimental propeller trailing edge shape [17]. The propeller has a serrated trailing edge, which matches the natural shape of the trailing edge of owl wings [17–19] and is shown in Fig. 4.

Practical testing has not been done to this date. This propeller should be tested after the model is built and the noise generated by the propeller should be measured.

3 Aerodynamics of the OiV

The proposed vehicle, illustrated by Figs. 5 and 6, is defined in a right-handed earth frame (e-frame) given by (x_e, y_e, z_e) and a right-handed vehicle frame (v-frame) given by (x_v, y_v, z_v) with positive x_v-axis towards the wings' rotors, positive y_v-axis to the left wing and positive z_v-axis upwards. Therefore, generalized coordinates for the OiV are designated by (x, y, z), corresponding to the position of the center of

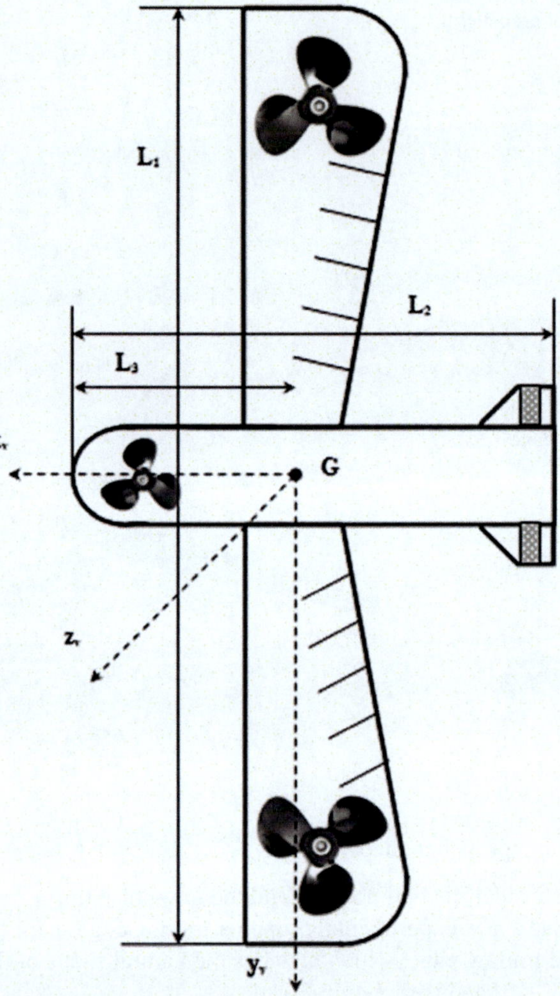

Fig. 5 The top view of the Owl-inspired Vehicle

mass G in the e-frame, and (ϕ, θ, ψ) relative to the orientation vector describing the rotation of the vehicle [20].

3.1 Kinematics

Any vector in the v-frame can be transformed into a vector in the e-frame using the following relation:

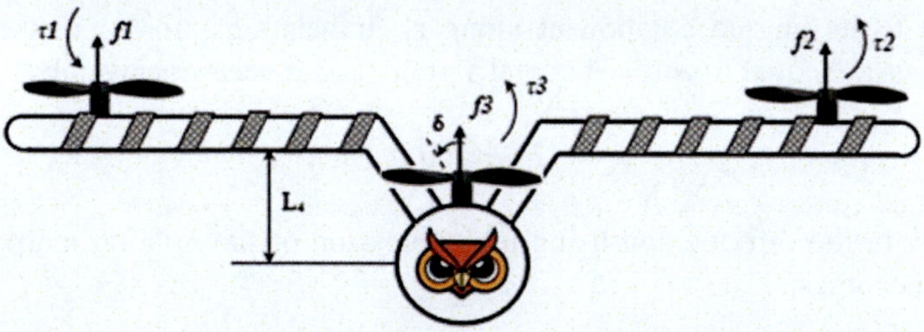

Fig. 6 The front view of the Owl-inspired Vehicle

$$\boldsymbol{q}_e = \boldsymbol{R}\boldsymbol{q}_v \tag{3}$$

where the rotation matrix $\boldsymbol{R}$ is considered using Euler angles and depending on time as:

$$\boldsymbol{R}(t) = \boldsymbol{R}_\psi \boldsymbol{R}_\theta \boldsymbol{R}_\phi \tag{4}$$

The derivative of $\boldsymbol{R}$ shows the linear dependence between $\boldsymbol{R}$ and (ϕ, θ, ψ) as:

$$\dot{\boldsymbol{R}} = \dot{\boldsymbol{\phi}} \boldsymbol{Ad}(i)\boldsymbol{R}_\phi \boldsymbol{R}_\theta \boldsymbol{R}_\psi + \dot{\boldsymbol{\theta}} \boldsymbol{R}_\phi \boldsymbol{Ad}(j)\boldsymbol{R}_\theta \boldsymbol{R}_\psi + \dot{\boldsymbol{\psi}} \boldsymbol{R}_\phi \boldsymbol{R}_\theta \boldsymbol{Ad}(k)\boldsymbol{R}_\psi \tag{5}$$

The vehicle angular velocity can be expressed as:

$$\boldsymbol{\Omega}_{/R_e} = \boldsymbol{Ad}^{-1}\left(\dot{\boldsymbol{R}}\boldsymbol{R}^T\right) = \begin{pmatrix} \boldsymbol{\Omega}_\phi \\ \boldsymbol{\Omega}_\theta \\ \boldsymbol{\Omega}_\psi \end{pmatrix} = \begin{pmatrix} \dot{\boldsymbol{\phi}} + \dot{\boldsymbol{\psi}} \boldsymbol{S}_\theta \\ \dot{\boldsymbol{\theta}} \boldsymbol{C}_\phi - \dot{\boldsymbol{\psi}} \boldsymbol{S}_\phi \boldsymbol{C}_\theta \\ \dot{\boldsymbol{\theta}} \boldsymbol{S}_\phi + \dot{\boldsymbol{\psi}} \boldsymbol{C}_\phi \boldsymbol{C}_\theta \end{pmatrix} \tag{6}$$

$\dot{\boldsymbol{R}}\boldsymbol{R}^T$ is a skew-symmetric matrix and $\boldsymbol{Ad}^{-1}$ designates the inverse transformation leading to extract the vector form, $\boldsymbol{C}$ and $\boldsymbol{S}$ are the respective abbreviations of *cos* and *sin.*

3.2 *Forces and Torques*

The total acting force on the vehicle remains usually vertical along the positive z_v axis and given by:

$$\boldsymbol{F}_{Z_v} = \boldsymbol{f}_1 + \boldsymbol{f}_2 + \boldsymbol{f}_3 \boldsymbol{C}_\delta \tag{7}$$

where $\boldsymbol{f}_1$, $\boldsymbol{f}_2$ and $\boldsymbol{f}_3$ are the thrust forces generated by the vehicle rotors and $\boldsymbol{\delta}$ is the tilt angle of the 3rd rotor about the y_v axis.

Due to its unique component along z_v, translational forces in the x_v and y_v directions are equal to zero. The total v-frame force vector is given by:

$$F_v = \begin{pmatrix} 0 & 0 & F_{z_v} \end{pmatrix} \tag{8}$$

Then in the e-frame and using the expression of the rotation matrix, the force vector becomes:

$$F_e = RF_v = \begin{pmatrix} (S_\psi S_\phi + C_\psi S_\theta C_\phi)F_{Z_v} \\ (-C_\psi S_\phi + S_\psi S_\theta C_\phi)F_{Z_v} \\ C_\theta C_\phi F_{Z_v} \end{pmatrix} \tag{9}$$

For a specific unsymmetrical trirotor vehicle, the torque vector can be expressed as:

$$\tau = \begin{pmatrix} \tau_\phi \\ \tau_\theta \\ \tau_\psi \end{pmatrix} = \begin{pmatrix} \frac{L_1}{2}(f_1 - f_2) \\ L_3 f_3 C_\delta \\ \tau_1 - \tau_2 - \tau_3 C_\delta + L_3 f_3 S_\delta \end{pmatrix} \tag{10}$$

where τ_1, τ_2 and τ_3 are the drag torques generated by the three rotors. The torque vector components will affect the roll, pitch and yaw movements of the vehicles.

3.3 Dynamics

The kinetic energy due to the translation of the vehicle with a mass m is given by:

$$E_t = \frac{1}{2}m(\dot{x}^2 + \dot{y}^2 + \dot{z}^2) \tag{11}$$

Furthermore, the kinetic energy due to the rotation of the vehicle in the e-frame can be written as:

$$E_r = \frac{1}{2}I_x(\dot{\phi} + \dot{\psi}S_\theta)^2 + \frac{1}{2}I_y(\dot{\theta}C_\phi + \dot{\psi}S_\phi C_\theta)^2 + \frac{1}{2}I_z(\dot{\theta}S_\phi + \dot{\psi}C_\phi C_\theta)^2 \tag{12}$$

where I_x, I_y and I_z are the moments of inertia about x_v, y_v and z_v respectively.

The potential energy of the vehicle, that is due to gravity, is defined by:

$$E_p = mgz \tag{13}$$

The Lagrangian, calculated from the difference between total kinetic and potential energies, can be written as follows:

$$L = E_t + E_r - E_p = \frac{1}{2}m\left(\dot{x}^2 + \dot{y}^2 + \dot{z}^2\right) + \frac{1}{2}I_x(\dot{\phi} + \dot{\psi}S_\theta)^2 + \frac{1}{2}I_y\left(\dot{\theta}C_\phi + \dot{\psi}S_\phi C_\theta\right)^2 + \frac{1}{2}I_z\left(\dot{\theta}S_\phi + \dot{\psi}C_\phi C_\theta\right)^2 - mgz \quad (14)$$

The dynamic model can be determined using the Euler–Lagrange formalism:

$$\frac{d}{dt}\left(\frac{\partial L}{\partial \dot{q}}\right) - \frac{\partial L}{\partial q} = \Gamma \quad (15)$$

where Γ is the effort vector including the force vector $F = \begin{bmatrix} F_x & F_y & F_z \end{bmatrix}^T$ and the torque vector as $\tau = \begin{bmatrix} \tau_\phi & \tau_\theta & \tau_\psi \end{bmatrix}^T$.

The dynamic model of the OiV is composed of six equations of movement: three equations for the translation movement written as follows:

$$\ddot{x} = \frac{1}{m}(S_\psi S_\phi + C_\psi S_\theta C_\phi)F_{z_v} \quad (16)$$

$$\ddot{y} = \frac{1}{m}(-C_\psi S_\phi + S_\psi S_\theta C_\phi)F_{z_v} \quad (17)$$

$$\ddot{z} = \frac{1}{m}C_\theta C_\phi F_{z_v} - g \quad (18)$$

Furthermore, the movement of rotations of the OiV are given by three equations such as:

$$I_x(\ddot{\phi} + \ddot{\psi}S_\theta + \dot{\psi}\dot{\theta}C_\theta) - (I_z - I_y)(\frac{1}{2}\left(\dot{\theta}^2 - \dot{\psi}^2 C_\theta^2\right)S_{2\phi} + \dot{\theta}\dot{\psi}C_\theta C_{2\phi} = \tau_\phi \quad (19)$$

$$\begin{aligned} &(I_y C_\phi^2 + I_z S_\phi^2)\ddot{\theta} - I_x C_\theta \dot{\psi}\dot{\phi} - \frac{1}{2}(I_x - I_y S_\phi^2 - I_z C_\phi^2)S_{2\theta}\dot{\psi}^2 \\ &+(I_z - I_y)\left(S_{2\phi}\dot{\phi}\dot{\theta} + \frac{1}{2}C_\theta S_{2\phi}\dot{\psi} + C_\theta C_{2\phi}\dot{\psi}\dot{\phi}\right) = \tau_\theta \end{aligned} \quad (20)$$

$$\begin{aligned} &I_x S_\theta \ddot{\phi} + \frac{1}{2}(I_z - I_y)S_{2\phi}C_\theta\ddot{\theta} + (I_x S_\theta^2 + (I_y S_\phi^2 + I_z C_\phi^2)C_\theta^2) + \frac{1}{2}(I_y - I_z)S_{2\phi}S_\theta\dot{\theta}^2 \\ &+(I_y - I_z)S_{2\phi}C_\theta^2\dot{\psi}\dot{\phi} + (I_x C_\theta + (I_y - I_z)C_{2\phi}C_\theta) \\ &+(I_x S_{2\theta} - (I_y S_\phi^2 + I_z C_\phi^2)S_{2\theta}) = \tau_\psi \end{aligned} \quad (21)$$

4 Conclusion

The above analysis of a new type of aircraft inspired by the shape and mechanical properties of a flying owl provides opportunities for future use along with preservation of flight characteristics. Analysis of the airflow on both, the leading and trailing edges of the wing, during the flight of the owl allowed the design of the appropriate wing shape and the trailing edge shape of the propellers used in the drone. The use of two, self-stabilizing Kamov propeller systems, allows for mechanical compensation of unwanted torque forces in the system. To control such a balanced system, it is sufficient to use a small motor located at the front of the drone with the ability to change the tilt angle of the motor from the Z-axis towards the positive X-axis turn. A new shape of the trailing edge of the carrier propellers has also been proposed allowing, in theory, to reduce the noise level generated, but testing under laboratory conditions as well as comparative testing under real conditions is required.

Acknowledgements This work was partially supported by the Ministry of the Higher Education and Scientific Research in Tunisia, and grant WI/WI-IIT/2/2021 from Bialystok University of Technology and funded with resources for research by the Ministry of Science and Higher Education in Poland.

References

1. Nathan S (2015) Forces of nature: Biomimicry in robotics. Engineer 41–42
2. Gopura R, Bandara A, Kiguchi R, Mann C (2016) Developments in hardware systems of active upper-limb exoskeleton robots: a review. Robot Auton Syst 203–220
3. Guizzo E (2012) Soft robotics [Turning Point]. Robot Autom Mag 19(1):128–125
4. www.machinedesign.com. https://www.machinedesign.com/mechanical-motion-systems/article/21835853/7-bioinspired-robots-that-mimic-nature. Last accessed 22 Nov 2021
5. Wagner H, Weger M, Klaas M, Schröder W (2017) Features of owl wings that promote silent flight. Interface Focus 7:20160078
6. Gnanasekera M, Katupitiya J, Savkin AV, De Silva AE (2021) A range-based algorithm for autonomous navigation of an aerial drone to approach and follow a herd of cattle. Sensors 21(21):7218
7. Bielecki A, Smigielski P (2021) Three-dimensional outdoor analysis of single synthetic building structures by an un-manned flying agent using monocular visio. Sensors 21(21):7270
8. Lilley G (2012) A study of the silent flight of the owl. In: 4th AIAA/CEAS aeroacoustics conference. Toulouse, France. Published online: 22 Aug 2012. Last accessed 23 Nov 2021
9. Bachmann T, Wagner H (2011) The three-dimensional shape of serrations at barn owl wings: towards a typical natural serration as a role model for biomimetic applications. J Anat 219(2):192–202
10. Ito S (2009) Aerodynamic influence of leading-edge serrations on an airfoil in a low reynolds number: a study of an owl wing with leading edge serrations. J Biomech Sci Eng 4(1):117–123
11. Bajd T, Mihelj M, Munih M (2013) Geometric robot model. In: Introduction to robotics. Springer Briefs in Applied Sciences and Technology, Springer, Dordrecht. ISBN: 978–94–007–6101–8
12. Tipler P, Mosca G (2004) Physics for scientists and engineers: mechanics, oscillations and waves, thermodynamics, 5th edn. W. H. Freeman. ISBN 978–07–167–0809–4
13. Torque https://en.wikipedia.org/wiki/Torque. Last accessed 22 Nov 2021
14. piasecki.com. https://piasecki.com/fnp-accomplishments/xhrp-x/2642/. Last accessed 26 Nov 2021
15. Kaadan A (2013) Multirotor Frame Configurations. Coptercraft. A Study Of Unmanned Aerial Systems Stability For Lasercom Applications. Master's Degree, University Of Oklahoma

16. Yuhang W et al (2019) A novel aerodynamic noise reduction method based on improving spanwise blade shape for electric propeller aircraft. Int J Aerosp Eng 3750451
17. Propeller shapes. Dockater. https://forum.dji.com/thread-115683-1-1.html. Last accessed 19 Aug 2021
18. Mysterious Facts About Owls. https://www.mentalfloss.com/article/68473/15-mysterious-facts-about-owls. Last accessed 19 Aug 2021
19. Liang GQ, Wang JC, Chen Y, Zhou CH, Liang J, Ren LQ (2010) The study of owl's silent flight and noise reduction on fan vane with bionic structure. Adv Nat Sci
20. Bouteraa Y, Boucetta R, Chabir A (2017) Trirotor mechatronic design and reduction of dynamic model inputs by aerodynamic forces identification. Int J Model Ident Control 27(1):14–21

Data Quality Driven Design Patterns for Internet of Things

Chouhan Kumar Rath, Amit Kr Mandal, and Anirban Sarkar

Abstract Many IoT applications are now using microservices design concepts and have developed as an emergent technology by leveraging containerization, modularity, autonomous deployment and loose coupling. The requirement of different software design patterns is essential to aid in the creation of scalable, interoperable and reusable solutions. In IoT systems and software development, several IoT patterns, such as IoT design patterns and IoT architectural patterns, have been studied. But, most of the studied design patterns are domain-specific, and they do not consider the impact of data quality in the design process. Also, in IoT environment data quality plays an important role while processing the data to produce accurate and timely decisions. Therefore, this paper presents a formal approach to incorporate the data quality dimensions in design patterns for the microservice based IoT applications. Here, data quality evaluation parameters are integrated with various microservice design patterns suitable to IoT applications such as event sourcing pattern, chained microservice pattern, API gateway pattern etc. to ensure the effective data communication and high-quality services provided by the IoT applications. Further, the proposed quality driven design patterns are systematically defined using Event-B language and validated through Rodin platform.

Keywords Design patterns · Data quality · Microservices · IoT · Event-B

C. K. Rath (✉) · A. Sarkar
Department of Computer Science and Engineering, National Institute of Technology Durgapur, Durgapur, India
e-mail: chouhanrath.nitdgp@gmail.com

A. Sarkar
e-mail: sarkar.anirban@gmail.com

A. K. Mandal
Department of Computer Science and Engineering, SRM University AP, Amaravati, India
e-mail: amitmandal.nitdgp@gmail.com

R. Chaki et al. (eds.), *Applied Computing for Software and Smart Systems*,
Lecture Notes in Networks and Systems 555,
https://doi.org/10.1007/978-981-19-6791-7_18

1 Introduction

Microservices transformed the IoT application design principles, it allowed the IoT applications to be defined as a collection of modular, customer-centric and self-contained services [1]. Microservice enables agile development and deployment of IoT applications by incorporating scalability, interoperability and maintainability [2]. It also enables creating an IoT application by combining heterogeneous services from different service providers. Therefore, support for various design patterns in microservice-based IoT applications became essential as it addresses a variety of issues, such as data management, communication, and integration [3]. Design patterns are considered as a way to develop an end-to-end solution in a well-defined manner and also to understand how different components works together in a system [4]. Majority of the IoT applications design pattern studied in literature are focused on collecting and organising services from various sources and converting it into an unified collection of microservices [5, 6]. A few of them also used design patterns at various abstract levels in domain-specific IoT applications [7, 8]. But, the studied design patters did not considered the data quality parameters while building the systems.

In an IoT environment, data quality provides the state of the generated dataset. The quality of the generated data can be measured by its objective elements such as completeness, accuracy, consistency etc [9]. However, its subjective measures can be considered as non-functional requirement of the IoT applications, such as dependability, availability, applicability etc. [10]. Therefore, the data quality dimension is a crucial factor in microservice for IoT environment as it influences the decision-making services offered by the IoT applications [11]. But, incorporating data quality dimension into the IoT design patterns is a challenging task as IoT network consists of a large number of interconnected devices, sensors, services, and applications that share data among themselves to form an intelligent decision-making systems [5]. The quality of data primarily depends on the devices where the data is generated and the communication protocol used to transmit it over the network [12]. Again, device profile contains the crucial information such as accuracy, precision, response time, I/O data formats, protocols etc. [13]. Besides, several other important factors such as data relevance, consistency, availability, privacy etc will depend heavily on the carrier microservice and underlying data distribution framework [5]. Most quality data-driven IoT frameworks are focused on data filtering techniques, data outlier detection, context extraction etc [14]. However, to accommodate data quality dimensions for IoT applications, relevant device profile parameters should be integrated into microservice design patterns.

This paper presents a formal approach which deal with multiple design patterns for microservices based IoT applications. First, a layered microservice based IoT model is developed, and then the system's behaviour is examined using several design patterns. Furthermore, it demonstrates the data quality evaluating parameters that are appropriate for making better decision making system and accurate data processing. A formal model for quality data evaluation at each level is created to ensure the quality of services provided by IoT systems. The main goal of this work is to contribute to a better understanding of design patterns enabled by microservice-based IoT systems, as well as the development of data quality driven microservices applications by construction. The methodology is formalised using the Event-B programming language, which is a modelling tool for formalising and designing discrete transition systems. It is based on the concept of events, and its primary goal is to assist in the creation of correct-by-construction systems.

2 Design Patterns for Microservice based IoT Applications

Design patterns are a way to build a robust and reusable solutions to understand how different system components interact in a system context. The following are the design patterns used to build a microservice based IoT application and various design patterns are presented in Fig. 1.

(a) *API Gateway Pattern*: In an IoT application, clients requires data in different data formats and in different protocols which may need to call multiple microservices associate with different device types and protocols. All clients should have a single gateway API that sends request to the relevant microservice or routes them to other microservices [4, 5]. It can also aggregate the findings before sending back to the consumer.

(b) *Aggregator Pattern*: The distributed and decentralized microservices are need to be identified and aggregated to meet a user demand since microservices are small in size and typically implements a single task [5, 6]. It provides a unified API to a client obtain data from various microservices.

(c) *Chained Microservice Pattern*: This pattern shows several dependencies for single or multiple microservices. Similar to aggregator pattern, it evaluates and triggers concurrent processes for the microservices that compose the response to the request. The chained pattern works by microservices calling other microservices and combining their previous responses to return a concatenated response [4].

(d) *Event Sourcing Pattern*: This pattern defines a method of handling data activities that are triggered by a series of events. Generally, a service typically needs to update the database and send messages/events automatically in order to avoid data inconsistencies [7]. Here, the IoT events are treated as the action performed by the IoT devices or the actions triggered when certain state changes occur.

(e) *Service Discovery Pattern*: Service discover patterns are beneficial since they assist us in finding the locations of services required to be invoked [4, 8]. A microservice can be invoked by other microservices to retrieve the information via service discovery mechanism. Microservices connect with the service registry to publish their locations, while clients use the registry to find the microservices that have been registered.

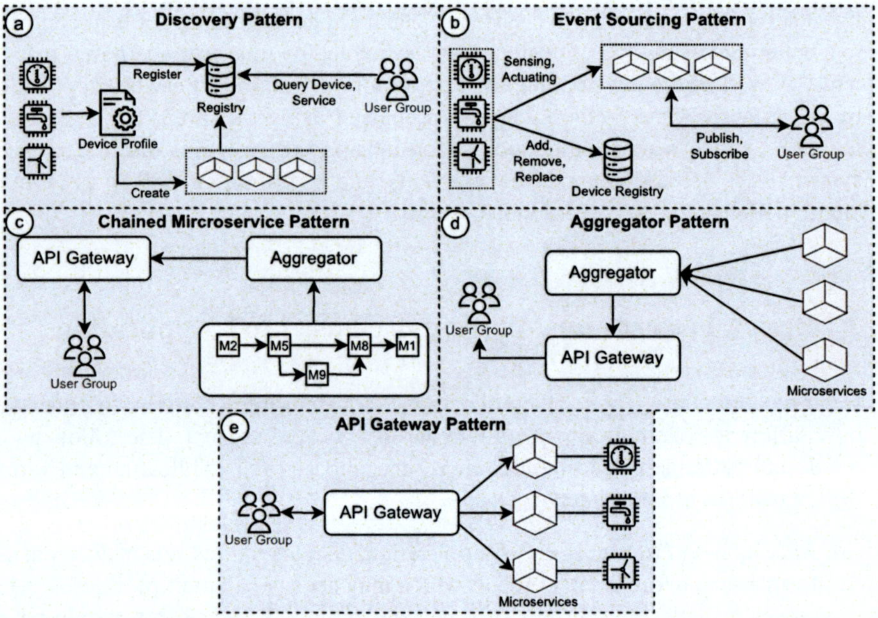

Fig. 1 Proposed design patterns for microservice based IoT applications

3 Related Work

A design pattern is a well-developed solution to a problem that arises repeatedly in a certain context [15]. Many design patterns exist in the IoT, in which application components deliver services to other components through a network using various communication protocols. Communication may involve simple data passing or include several coordinating services connecting to each other [1]. As the number of devices linked with different types of networks and communication patterns, IoT design patterns also increases. Qanbari et al. [16] have identified various design principles that serve as a seamless interface for heterogeneous things and are suited for implementation on resource-constrained devices. Bloom et al. [2] provided design patterns for improving functionality and automating tasks in an industrial IoT system. Lee et al. [17] have applied different design patterns to handle security issues and construct secure IoT systems. Similarly, Messina et al. [7] introduced a database pattern that integrates business logic with microservices to deliver database as a service. All of these patterns are used to create secure and reliable IoT systems, but they are restricted to a certain domain. Microservices are included in the IoT design patterns to make it easier to design and develop large and complex systems. Taibi et al. [6] and Márquez et al. [1] have conducted an exploratory literature review of microservice-based design patterns to assist developers in finding appropriate solution templates for microservice-based software architecture. Munonye et al. [18] examine several database patterns for microservices in order to create loose coupling between the service data while maintaining high data access performance. Authors in [8, 19], have implemented the API gateway patterns to take advantage of better customization and compatibility. Service discovery patterns have applied in [6, 20] to make communication, ease of maintenance, and development simpler. Service registry and API-gateway patterns are also used to expose the required services. In order to make a scalable, independent and secure mechanism database patterns were implemented in paper [7]. Similarly, circuit breaker patterns are applied in microservice architecture to detect failures and availability of external services [3]. Among the microservices architectural patterns the recent developments involve Service Registry pattern, Service Discovery, API gateway, Database patterns and Circuit breaker, but none of these patterns are implemented in IoT systems. This paper proposed a formal verification model for a microservice based IoT architecture with the support of various design patterns.

IoT is already being employed in a growing number of applications, and the importance of IoT data quality is widely recognized by practitioners and researchers. In different situations, the needs for data and its quality range from application to application or organisation to organisation. There are various techniques has been developed for defining, analysing and improving data quality. Zhang et al. [10] surveys various data quality frameworks and methodologies for IoT data and related international standards, comparing them in terms of data types, data quality definitions, dimensions and metrics. Mansouri et al. [14] analyse data quality challenges and summarize the existing data quality management approaches or standards. Liono et al. [9] proposed a quality data driven framework for storage and management of IoT applications. They employed a machine learning approach for quality data assessment but it is confined to the device and storage level. Fizza et al. [11] has demonstrated some of sensor data quality parameters through a smart agricultural application. Castillo et al. [21] investigated data quality management policies for smart connected devices in IoT. Similarly, Karkouch et al. [22] proposed a model driven approach to capture quality data streams effectively. Besides that, numerous domain specific data quality dimensions has been also defined for various specific applications, but there is no standard framework for evaluating quality data in each component of the IoT architecture. This model provides the formal verification of the data quality model, which adapts the underlying systems effectively based data quality standards.

4 Integrating Data Quality in Microservice Design Patterns

This section formally defines how to incorporate data quality in various design patterns for a microservice-based IoT system. The model is concerned with a number of essential functionalities, such as how clients access services in a microservices environment, how client requests are routed to an accessible service instance, and how each service interacts with the database. Moreover, this model integrates effective

Table 1 Design patterns presented in various events in the proposed formalisation model

Design patterns	Modellig events
Event Sourcing Pattern	Deployment, Sensing, Actuation, Result
Chained Microservice Pattern	Add_Device, Remove_Device, Replace_Device, Series_Composition, Parallel_Composition
Service Discovery Pattern	MS_Register, MS_Deregister, MSDiscovery
Aggregator Pattern	Publish_Data, Subscribe_Data
API Gateway Pattern	Subscribe_Topic, Unsubscribe_Topic, Send_Message, Receive_Message

data quality metrics using device profiles of the IoT devices to aid proper data utilisation and quality evaluation of the generated data. The requirement of this model is based on the notion of events i.e. transitions, and its primary goal is to assist in the development of correct-by-construction systems. Also, most of design patterns are described in natural language which lead to ambiguity in the development of correct IoT applications. Event-B models are based on first-order logic and set theory and are built by *Contexts* and *Machines*. *Contexts*, containing the static part of the system, whereas *Machines* containing the dynamic part of the system. The behavior of the system is described by machines or transitions which are dynamic in nature. They may include parameters, guards, and actions stating the conditions and events that occur. The system requirements are modeled using Event-B notations starting with abstract representations and refining to add precise requirements. The following are the prerequisites for this architectural model:

- Deployment of IoT devices.
- Create the microservices for different device types.
- Distribution of information among the network.
- Aggregation of data to fulfil user demands.
- Filtering out the quality data.
- Composition of microservices in case of dependency.

The Event-B language is used to create a systematic model based on above requirements and categorized into three important part such as device profile, corresponding microservices and data distribution mechanism. The *IoT Device Profile* defines the deployment, identification, and registration of devices through device template. The *Microservice* component explains how microservices are formed using device templates and the working principle of microservices in this model. Similarly, *Data Distribution* model shows how the data is distributed among the network. It also shows how the modified model takes into account the data quality evaluation criteria. Further, the data quality is evaluated considering the interaction among these components which is also defined using Event-B in a systematic and unambiguous manner. Table 1 shows the design patterns and supported events in the formalization model.

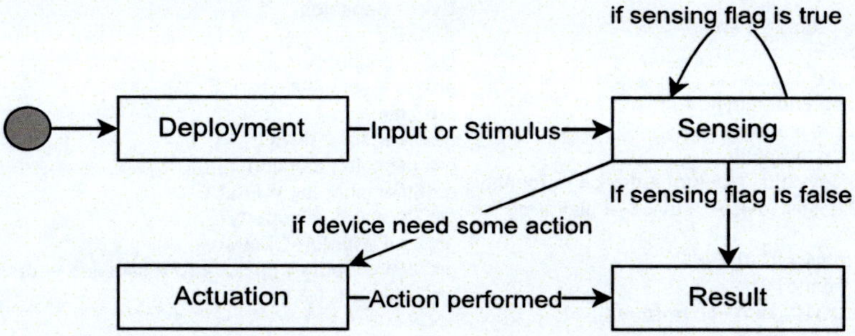

Fig. 2 State diagram of various events in IoT device profile model

4.1 *IoT Device Profile*

This section describes the properties of physical IoT devices, its features and capabilities by profiling the IoT devices. The device profile for IoT device can be decomposed into three main categories such as *Deployment, Functionalities*, and *Findings*. Figure 3 shows the event composition for IoT device profile and the state flow diagram of the machine is shown in Fig. 2.

1. **Deployment**: It specifies how an IoT device is deployed in a network. The device is initially registered using a device template that includes a unique resource identifier (uri) and is hosted by a platform.
2. **Functionalities**: It describes the functions of an IoT device, such as sensing, actuating, and triggering events. After being activated by some input or external events (stimulus), IoT devices begin to work, and generate events during run-time in order to perform various activities.
3. **Findings**: It depicts the data format and output data collected by the IoT device, where the context is derived from an observable property of a feature of interest.

4.2 *Microservices*

It describes the integration of IoT devices with microservices and the functionalities of microservices over a IoT network. The refinement strategies for microservices are *Device Integration* and *Registration*. Further, it is extended by taking the functionalities such as addition or removal of device, discovery of microservices, and its composition. Figure 4 shows the control flow of events for both concrete and refined model of microservices (as shown in Figs. 5 and 6).

1. **Device Integration**: This machine represents the relationship between microservices and IoT devices. Using a device template, different service instances are produced based on the types of devices, protocol, and data format. This machine

```
Event Deployment
WHERE
grd1:d ∈ device
grd2:uri ∈ Identity
grd3:p ∈ platform
grd4:t ∈ template
grd5:deviceId ◁ {d↦uri} ∈ device → Identity
grd6:host ◁ {d↦p} ∈ device → platform
THEN
act1:deviceId(d):=uri
act2:host(d):=p
act3:devRegistration ≔ {d ↦ t}
END

Event Result
WHERE
grd1:obs ∈ obsProperty
grd2:foi ∈ featureOfInterest
grd3:data ∈ sensingData
grd4:sensingFlag = TRUE
grd5:context = {obs↦foi}
grd6:time+1 ∈ ℕ
THEN
act1:output ≔ output ∪ {data↦{obs↦foi}}
act2:time ≔ time+1
END

Event Sensing
WHERE
grd1:i ∈ input
grd2:s ∈ stimulus
grd3:i≠s
grd4:sensor ∈ device
grd5:sensing ∈ action
grd6:sensingFlag = FALSE
grd7:obs ∈ obsProperty
grd8:foi ∈ featureOfInterest
grd9:(∅ ∈ input ↔ device)∨(∅ ∈ stimulus ↔ device)
grd10:time + 1 ∈ ℕ
THEN
act1:trigger ≔ {i↦sensor} ∩ {s↦sensor} ∩ ∅
act2:event ≔ {sensor ↦ sensing}
act3:sensingFlag ≔ TRUE
act4:context ≔ {obs↦foi}
act5:time ≔ time+1
END

Event Actuation
WHERE
grd1:actuator ∈ device
grd2:actuableProperty ∈ action
grd3:∅ ∈ device ↔ action
THEN
act1:event ≔ event ∩ {actuator ↦ actuableProperty} ∩ ∅
END
```

Fig. 3 Events for Device Profile model (m0_Device_Profile)

also explained how to create microservices as well as addition, removal and replacement of IoT devices with the help of template.

2. **Registration**: Microservices are registered in a registry, which is updated on a regular basis with the address and details about the linked device. The relevant services are discovered from the service registry when a user or another microservice requests an on-demand service. Further, discovered microservices are composed in various patterns to provide on-demand service.

4.3 Data Distribution

This section describes how the IoT data is available among various applications, i.e., exchange of information across various components in an IoT network. This machine decomposed into two main component such as *Data Storage* and *Data Communication* (Figs. 7, 8 and 9).

1. **Data Storage**: The data is distributed using a pub-sub mechanism where publishers and subscribers are treated as microservices. Publisher microservices collect data from physical devices and publish it to subscriber microservices, which subsequently disseminate it to other applications. Publishers and subscribers

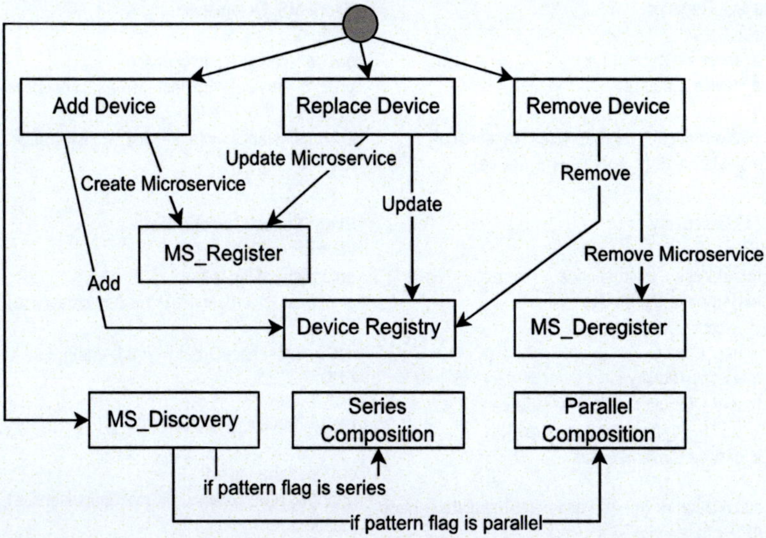

Fig. 4 State diagram of various events for microservice model

```
Event Add_Device
WHERE
grd1:a_service ∈ microservice
grd2:a_device ∈ device
grd3:t ∈ template
grd4:msInstance ◁
    {a_service ↦ a_device} ∈ microservice ↔ device
THEN
act1:devRegistration ≔ {a_device ↦ t}
act2:msInstance(a_service) ≔ a_device
END
```

```
Event Remove_Device
WHERE
grd1:a_device ∈ device
grd2:a_service ∈ microservice
grd3:t ∈ template
THEN
act1:devRegistration ≔ devRegistration \ {a_device ↦ t}
act2:msInstance ≔ msInstance \ {a_service ↦ a_device}
END
```

```
Event Relace_Device
WHERE
grd1:(device1 ∈ device)∧(device2 ∈ device)
grd2:t ∈ template
grd3:service ∈ microservice
grd4:devRegistration = devRegistration \ {device1↦t}
THEN
act1:devRegistration ≔ devRegistration ∪ {device2 ↦ t}
act2:msInstance ≔ {service ↦ device1} ◁ {service ↦ device2}
END
```

Fig. 5 Events for concrete microservice model (m0_Microservices)

```
Event MS_Register
WHERE
grd1:a_service ∈ microservice
grd2:a_device ∈ device
THEN
act1:microservice :∈ {microservice ∪ {a_service}}
act2:msInstance :∈ {a_service ↦ a_device}
END

Event MSDiscovery
WHERE
grd1:msAddress ∈ microservice
grd2:msTemplate ∈ template
grd3:a_microservice ∈ user
grd4:a_user ∈ user
grd5:msr ∈ msRegistry
grd6:(query = {a_user ↦ msr}) ∨ (query =
            {a_microservice ↦ msr})
grd7:a_service ∈ microservice
THEN
act1:msLocation ≔ {msAddress ↦ msTemplate ↦ msr}
act2:API ≔ {a_service ↦ a_user}
END

Event MS_Deregister
WHERE
grd1:a_service ∈ microservice
grd2:{a_service} ⩤ msInstance ∈ microservice ↔ device
THEN
act1:msInstance ≔ {a_service} ⩤ msInstance
END

Event Series_Composition
WHERE
grd1:patternFlag = 1
grd2:(ms1∈microservice)∧(ms2∈microservice)
THEN
act1:microserviceOrder ≔ {ms1↦ms2}
END

Event Parallel_Composition
WHERE
grd1:patternFlag = 2
grd2:(ms1∈microservice)∧(ms2∈microservice)
THEN
act1:microserviceOrder ≔ {ms1 || ms2}
END
```

Fig. 6 Events for refined Microservice model (m1_Microservices)

independently perform read/write actions on a topic in which records are stored and published.

2. **Data Communication**: The pub-sub mechanism implements a central messaging system for IoT called broker. Stream processing API enables complex aggregations or joins of input streams onto an output stream of processed data.

4.4 Data Quality Evaluation

For accessing quality data, this section discussed about different kinds of data and different quality measurement parameters. IoT data is collected from smart devices via networks in a variety of data formats. The appropriateness of the data need to deliver accurate services. Data quality dimensions are methods used for measurement and validation of quality data. The characteristics of data quality are primarily influenced by three factors of IoT: (a) Hardware integration, (b) Network Communication, and (c) Application Deployment. Different data quality dimensions for IoT are listed below:

- **Accuracy**: The numerical precision of data acquired from IoT devices is represented by data accuracy. It determines the closeness of captured values of data points with their original values.

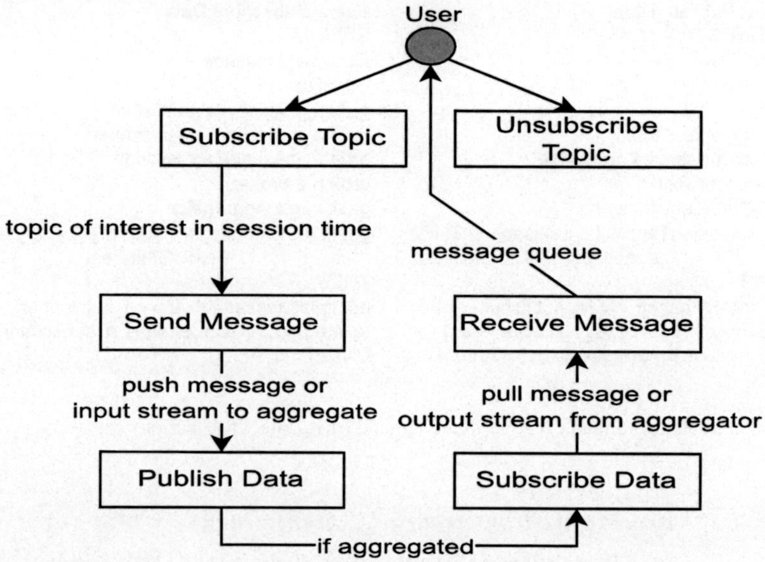

Fig. 7 Context and invariants for data distribution model

```
Event Subscribe_Topic
WHERE
grd1:a_subscriber ∈ subscriber
grd2:a_topic ∈ topic
THEN
act1:topicOfInterest :∈ {a_topic ↦ a_subscriber}
END

Event Send_Message
WHERE
grd1:a_publisher ∈ publisher
grd2:a_topic ∈ topic
grd3:a_message ∈ message
grd4:topicMessage ◁ {a_message ↦ a_topic}
          ∈ message → topic
grd5:sessionTime ◁ {a_message ↦ Time}
          ∈ message → 1..Time
THEN
act1:topicMessage(a_message) ≔ a_topic
act2:publisherMessage ≔ {a_message ↦
          a_publisher}
act3:sessionTime(a_message)≔Time
END
```

```
Event Unsubscribe_Topic
WHERE
grd1:a_subscriber ∈ subscriber
grd2:a_topic ∈ topic
THEN
act1:topicOfInterest ≔ topicOfInterest \ {a_topic ↦
          a_subscriber}
END

Event Receive_Message
WHERE
grd1:a_publisher ∈ publisher
grd2:a_subscriber ∈ subscriber
grd3:a_message ∈ message
grd4:topicMessage(a_message)↦a_subscriber
          ∈ topicOfInterest
grd5:{a_subscriber ↦ {a_message}}
          ∈ subscriber → ℙ(message)
THEN
act1:messageQueue :∈ {a_publisher ↦ a_subscriber}
act2:subscriberMessage ≔ {a_subscriber ↦
          {a_message}}
END
```

Fig. 8 Events for concrete data distribution model (m0_Data_Distribution)

```
Event Publish_Data
REFINES
Send_Message
WHERE
grd6:a_publisher ∈ publisher
grd7:a_topic ∈ topic
grd8:a_message ∈ message
grd9:b ∈ broker
grd10:ag ∈ aggregator
grd11:sessionTime ◁ {a_message ↦ Time}
        ∈ message→ 1..Time
THEN
act4:pushMessage ≔ {a_publisher ↦ b}
act5:inputStream :∈ {a_publisher ↦ ag}
act6:sessionTime(a_message):=Time
END
```

```
Event Subscribe_Data
REFINES
Receive_Message
WHERE
grd6:a_publisher ∈ publisher
grd7:a_subscriber ∈ subscriber
grd8:a_message ∈ message
grd9:b ∈ broker
grd10:ag ∈ aggregator
grd11:topicMessage(a_message) ↦ a_subscriber
        ∈ topicOfInterest
THEN
act3:pullMessage ≔ {b ↦ a_subscriber}
act4:outputStream :∈ {ag ↦ a_subscriber}
END
```

Fig. 9 Events for refined data distribution model (m1_Data_Distribution)

$$axm1 : partition(Accuracy, \{high\}, \{low\})$$
$$grd1 : (requiredData \in data) \vee (acquiredData \in data)$$
$$grd2 : requiredData \cong acquiredData$$
$$act1 : Accuracy := high$$

- **Applicability**: It refers to whether a particular device is suitable for a certain application or not. If the required data of the application lie within the device measurement range, the device is applicable for the application else, it become irrelevant.

$$inv1 : applicableData \in dataPoint..dataPoint$$
$$inv2 : applicability \in BOOL$$
$$grd1 : (D1 \in \mathbb{N}) \wedge (D2 \in \mathbb{N})$$
$$grd2 : datatype \in data$$
$$grd3 : (ApplicableData \in (D1..D2)) \vee (ApplicableData \in datatype)$$
$$act1 : applicability := TRUE$$

- **Relevance**: The context plays a major role in the relevance of data generated in the IoT. Context has a significant impact on what data a user of an application should perceive and given the current time and location. Application developers must be able to evaluate the data by utilizing the contextual relevance that IoT data provides.

$$
\begin{aligned}
&inv1 : context \in (obsProperty \rightarrow featureOfInterest) \leftrightarrow (time \times location)\\
&inv2 : relaventData \in BOOL\\
&grd1 : (requiredProperty \in obsProperty) \vee (foi \in featureOfInterest)\\
&grd2 : (requiredLocation \in location) \vee (timestamp \in time)\\
&act1 : context := \{\{requiredProperty \mapsto foi\} \mapsto\\
&\qquad (timestamp \mapsto requiredLocation)\}\\
&act2 : releventData := TRUE
\end{aligned}
$$

- **Timeliness**: It shows the freshness of the data acquired and their accurate timing in relation to the application context. It is mostly determined by the IoT device's own response time as well as network response time.

$$
\begin{aligned}
&inv1 : timeliness \in data \rightarrow ((obsProperty \rightarrow featureOfInterest) \leftrightarrow\\
&\qquad (time \times location))\\
&grd1 : acquiredData \in data\\
&grd2 : (requiredProperty \in obsProperty) \vee (foi \in featureOfInterest)\\
&grd3 : (l \in location) \vee (responseTime \in time)\\
&grd4 : context = \{\{requiredProperty \mapsto foi\} \mapsto (responseTime \mapsto l)\}\\
&act1 : timeliness := \{acquiredData \mapsto context\}
\end{aligned}
$$

- **Consistency**: Data consistency refers the data whose properties are coherent with other data without conflict. It might be a comparison of different data points from a single device or a comparison of data from multiple devices that produce similar data.

$$
\begin{aligned}
&inv1 : consistency \in data \rightarrow device\\
&inv2 : consistencyFalg \in BOOL\\
&grd1 : (device1 \in device) \vee (device2 \in device) \vee (device1 \neq device2)\\
&grd2 : properties \in data\\
&grd3 : consistency = \{properties \mapsto device1\} \Lleftarrow \{properties \mapsto device2\}\\
&act1 : consistencyFlag := TRUE
\end{aligned}
$$

- **Availability**: The availability of a device refers to the amount of time it is functioning and available to operate. Due to network difficulties or user authorization concerns, data accessibility is decreased or data from certain devices is inaccessible at a time.

$$inv1 : event \in device \rightarrow action$$
$$inv2 : accessible \in BOOL$$
$$grd1 : authenticate \in action$$
$$grd2 : dataTransmit \in action$$
$$grd3 : d \in device$$
$$act1 : (event(d) \notin authenticate) \vee (event(d) \notin dataTransmit) \Rightarrow accessible := FALSE$$

- **Privacy**: It specifies the types of data access, such as public, private, and protected. If the data is publicly available, it may be accessed by any other device, network, or microservice, whereas it is not accessible in private mode. The data is protected if it is only accessible with some authentication.

$$axm1 : partition(Privacy, \{public\}, \{private\}, \{protected\})$$
$$inv1 : privacyFlag \in 1..3$$
$$act1 : (privacyFlag := 1 \Rightarrow Privacy := \{public\}) \vee (privacyFlag := 2 \Rightarrow Privacy := \{private\}) \vee (privacyFlag := 3 \Rightarrow Privacy := \{protected\})$$

- **Security**: It determines the data that has been accessed is secure or not. IoT security includes both physical device and network security, and has an influence on the procedures, technology, and safeguards required to protect IoT devices and networks.

$$inv1 : secureData \in data \rightarrow secureStrategy$$
$$inv2 : security \in BOOL$$
$$grd1 : (encryption \in secureStrategy) \wedge (authenticate \in secureStrategy)$$
$$grd2 : trustDevice \in device$$
$$grd3 : (data \mapsto encryption) \vee (data \mapsto authenticate) \vee (data :\in trustDevice)$$
$$act1 : security := TRUE$$

4.5 Proof of Validation of the Model

The proposed formalization model is written in Event-B language and validated through Rodin platform. In Rodin, the correctness of the model is determined by discharging the proof obligation rules (*INV, WD, FIS*). The invariant preservation rule (*INV*) states that anytime the values of variables change, each invariant in a machine is preserved. The well-definedness (*WD*) ensures the guards are properly defined. Similarly, action feasibility (*FIS*) ensures that a non-deterministic action is feasible. Each proof obligation is identified by its semantics. If a statement is $H \mapsto G$, it indicates that the G is provable goal under the conditions H. Three proof obligation rules specific to this model are shown below.

(a) *Invariant Preservation Rule(INV):* This proof obligation rule ensures that each invariant in a machine is preserved whenever values of variables change by each event. Rodin theorem prover automatically discharges this proof using different rules of inference, i.e., "*time* ∈ N ↦ time+1 ∈ *N*" get a sequence of form which further applied another inference rule in axioms applied for natural numbers.
(b) *Well-definedness of guard (WD):* This proof obligation rule ensures that a guard is well-defined. For example,

$$inv1 : deviceId \in device \rightarrow Identity$$
$$grd1 : d \in device$$
$$grd2 : deviceId \mathbin{\lhd\!\!\!-} d \mapsto uri \in device \rightarrow Identity$$
$$act1 : deviceId(d) := uri$$

The action "*deviceId(d) := uri*" is defined only when *grd1* and *grd2* are well defined.
(c) *Feasibility proof obligation rule (FIS):* The purpose of this proof obligation is to ensure that a non-deterministic action is feasible. For example, let the initial statement of a model is,

$$inv1 : applicableData \in N..N$$
$$act1 : applicableData :\in 1..1$$
$$axm1 : d \in N$$
$$|-applicableData(d) \in N$$

the action of the initialization is feasible only when "$N \neq \phi$".

A total of 65 proof obligations are created, including 41 *INV*s, 18 *WD*s, and 6 *FIS*s among them. All proof obligations are automatically discharged, indicating that the model is correctly defined and constructed. Due to various consequences, data quality concerns may arise at each stage of the data transmission process. The data quality in *Hardware Integration* may be impacted by a variety of hardware issues, such as device upgrades, power issues, mechanical failures, and so on, resulting in missing and erroneous values. The response time of an IoT device can be determined by verifying the *Timeliness* of a data. Taking data points from different devices, the *Consistency* of data could be measured. By verifying the *Availability* of data, the aforementioned errors might be detected and corrected. The impacting elements in *Network Communication* could be an unreliable network, bad weather, security, and so on, all of which pose a threat to the IoT data quality. The inaccuracy of the data might be seen by checking the *Accuracy*. IoT device interfaces and connections must be secure in order for IoT objects to operate effectively. The *Security* and *Privacy* data quality parameters are utilised in various interfacing and connectivity states to determine the efficiency of an IoT system. Similarly, data quality in the *Application Deployment* may be affected by stream processing, missing records, and incorrect data format, among other things. By analysing the *Applicability, Relevance,* and *Availability* of IoT data, these issues can be found and corrected. All of the aforementioned concerns might be identified and addressed by combining quality evaluation parameters with various design patterns in a microservice-based IoT application.

5 Illustration of the Model Through a Case Study

This section presents a systematic study of the proposed model using a Smart-City based application. Various applications such as traffic control, weather monitoring, pollution control, etc., are implemented in a smart city. Some cases are considered for the illustration of the case study to demonstrate the design flow of distinct design patterns in this model (shown in Fig. 10).

IoT devices are deployed to monitor data such as temperature, humidity, traffic volume, crowd, and so on, and these devices are registered in the network via device templates. Microservices are now created for each device type, e.g., separate microservices for temperature, humidity, traffic volume, and crowd monitoring. In machine *m0_Device_Profile*, the deployment strategies of IoT devices through device templates shows the scalable property since the devices could automatically configured and created microservices according to different data formats, protocols and device types. Machine *m0_Microservices* shows the flexibility and interoperability while adding, removing and replacing new devices in a huge network, using data from the device profile. Device profiles are used to specify both *privacy* and *security*. For example, A speed detection sensor in a public transportation vehicle may be available to others, but in a personal motor vehicle, it may only have private access. As shown in Fig. 10, a registry is used to store the location information of both devices and microservices so that they may be discovered when an user or service client queries for them. Data *relevance* and *applicability* shows the device should be identified depending on the needed context, for example, if a user queries for temperature, the system will determine whether the required temperature data is for a location, a class room, or the user's body temperature. Various events might be created in the network by IoT devices or microservices to make the flexibility in real-time processing as shown in Fig. 10. For example, when an user requests information on a vehicle's speed, the system creates an event that triggers the speed detection sensor of that particular location. Here, system's *availability* and *timeliness* are observed, i.e., the sensor must be available to detect the vehicle at a certain time and exact timing is required in order to detect the speed of all vehicles. Similarly, other events such as sensing, actuating, publishing and subscribing the services, as

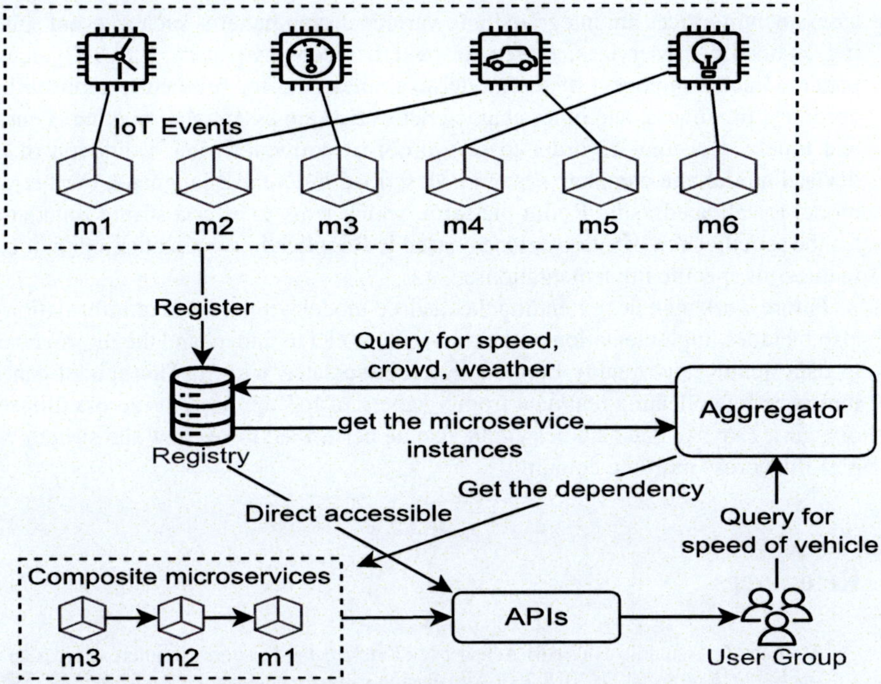

Fig. 10 Illustration of case study with different design patterns in a microservice based IoT applications

well as addition, deletion and replacement of IoT devices are generated. There may be some dependency occurs while creating the sensing events for monitoring speed detection, i.e., the speed of a vehicle may dependent on current traffic volume of that location. Machine *m1_Microservices* used chained microservice pattern to execute the dependencies (referred in Fig. 10). Further, the discovered microservices are aggregated to provide on-demand services, i.e., according to the speed of the vehicle and traffic volume the system able to provide the decision that the speed should be reduced (shown in Fig. 10). Data *consistency* is observed in both the aggregator and chained microservice patterns because the threat of data conflict is relatively high. All the aggregated services are accessible to service client through suitable APIs as shown in Fig. 10.

6 Conclusion

This paper presents a formalisation model for incorporating data quality parameters into various microservices-based IoT design patterns. To achieve effective data communication and the quality in services provided by IoT systems, data quality assessment matrices are integrated into various design patterns such as event sourcing, chained microservices, aggregator pattern, discovery pattern, and API gateway pattern. Each component of the model examines data for relevance, applicability, accuracy, timeliness, and other characteristics that are essential to making accurate and timely decisions. In order to give proof of correctness and validation of the devised model, the complete system is described in Event-B language. Further, the model is validated using Rodin platform, which aims to offer a strong foundation for the specification of microservices-based IoT applications by evaluating their correctness of specific implementations.

Future work includes extending the devised model with contextual information. It also includes implementation of the proposed model to understand the improvement in data quality and quality of the services associated with it. This model can be improved by collecting real data from a variety of IoT applications across different domains. Experimental studies could also be performed to see how the system will perform across multiple domain.

References

1. Márquez G, Astudillo H (2018) Actual use of architectural patterns in microservices-based open source projects. In: 2018 25th Asia-Pacific software engineering conference (APSEC). IEEE, pp 31–40
2. Bloom G, Alsulami B, Nwafor E, Bertolotti IC (2018) Design patterns for the industrial internet of things. In: 2018 14th IEEE international workshop on factory communication systems (WFCS). IEEE, pp 1–10
3. Vergara S, González L, Ruggia R (2020) Towards formalizing microservices architectural patterns with event-b. In: 2020 IEEE international conference on software architecture companion (ICSA-C). IEEE Computer Society, pp 71–74
4. Javeri P (2018) Architectural patterns for iot—micro services design patterns. Accessed 09 March 2022
5. Washizaki H, Ogata S, Hazeyama A, Okubo T, Fernandez EB, Yoshioka N (2020) Landscape of architecture and design patterns for iot systems. IEEE Internet Things J 7(10):10091–10101
6. Taibi D, Lenarduzzi V, Pahl C (2018) Architectural patterns for microservices: a systematic mapping study. In: CLOSER 2018: Proceedings of the 8th international conference on cloud computing and services science. Funchal, Madeira, Portugal <error l="305" c="Invalid <error l="303" c="Invalid
command: paragraph not started." /> command: paragraph not started." />
7. Messina A, Rizzo R, Storniolo P, Urso A (2016) A simplified database pattern for the microservice architecture. In: The eighth international conference on advances in databases, knowledge, and data applications (DBKDA), pp 35–40
8. Balalaie A, Heydarnoori A, Jamshidi P, Tamburri DA, Lynn T (2018) Microservices migration patterns. Softw Pract Exp 48(11):2019–2042
9. Liono J, Jayaraman PP, Kai Qin A, Nguyen T, Salim FD (2019) Qdas: quality driven data summarisation for effective storage management in internet of things. J Parallel Distrib Comput 127:196–208
10. Zhang Lina, Jeong Dongwon, Lee Sukhoon (2021) Data quality management in the internet of things. Sensors 21(17):5834
11. Kaneez F, Jayaraman PP, Banerjee A, Georgakopoulos D, Ranjan R (2021) Evaluating sensor data quality in internet of things smart agriculture applications. IEEE Micro

12. Byabazaire J, O'Hare G, Delaney D (2020) Data quality and trust: review of challenges and opportunities for data sharing in iot. Electronics 9(12)
13. Rath CK, Mandal AK, Sarkar A (2022) An Event-B based device description model in IoT with the support of multimodal system. Springer, Singapore, pp 3–19
14. Mansouri T, Moghadam MRS, Monshizadeh F, Zareravasan A (2021) IoT data quality issues and potential solutions: a literature review. Comput J 11
15. Yussupov V, Breitenbücher U, Krieger C, Leymann F, Soldani J, Wurster M (2020) Pattern-based modelling, integration, and deployment of microservice architectures. In: EDOC, pp 40–50
16. Qanbari S, Pezeshki S, Raisi R, Mahdizadeh S, Rahimzadeh R, Behinaein N, Mahmoudi F, Ayoubzadeh S, Fazlali P, Roshani K et aL (2016) Iot design patterns: computational constructs to design, build and engineer edge applications. In: 2016 IEEE first international conference on Internet-of-Things design and implementation (IoTDI). IEEE, pp 277–282
17. Lee W-T, Law P-J (2017) A case study in applying security design patterns for iot software system. In: 2017 international conference on applied system innovation (ICASI). IEEE, pp 1162–1165
18. Munonye K, Martinek P (2020) Evaluation of data storage patterns in microservices archicture. In: 2020 IEEE 15th international conference of system of systems engineering (SoSE). IEEE, pp. 373–380
19. Lin J, Lin LC, Huang S (2016) Migrating web applications to clouds with microservice architectures. In: 2016 international conference on applied system innovation (ICASI), pp 1–4
20. Brown K, Woolf B (2016) Implementation patterns for microservices architectures. In: Proceedings of the 23rd conference on pattern languages of programs, pp 1–35
21. Perez-Castillo R, Carretero AG, Rodriguez M, Caballero I, Piattini M, Mate A, Kim S, Lee D (2018) Data quality best practices in iot environments. In: 2018 11th international conference on the quality of information and communications technology (QUATIC). IEEE, pp 272–275
22. Karkouch A, Mousannif H, Al Moatassime H, Noel T (2016) A model-driven architecture-based data quality management framework for the internet of things. In: 2016 2nd international conference on cloud computing technologies and applications (CloudTech). IEEE, pp 252–259